VOLUME FIFTY

ADVANCES IN
ECOLOGICAL RESEARCH
Eco-Evolutionary Dynamics

ADVANCES IN ECOLOGICAL RESEARCH

Series Editor

GUY WOODWARD
Imperial College London
Silwood Park Campus
Ascot, Berkshire, United Kingdom

VOLUME FIFTY

Advances in
ECOLOGICAL RESEARCH
Eco-Evolutionary Dynamics

Edited by

JORDI MOYA-LARAÑO
Functional and Evolutionary Ecology
Estación Experimental de Zonas Áridas - CSIC
Carretera de Sacramento s/n
La Cañada de San Urbano
Almería, Spain

JENNIFER ROWNTREE
Faculty of Life Sciences,
The University of Manchester,
Manchester, United Kingdom

GUY WOODWARD
Imperial College London
Silwood Park Campus
Ascot, Berkshire, United Kingdom

AMSTERDAM • BOSTON • HEIDELBERG • LONDON
NEW YORK • OXFORD • PARIS • SAN DIEGO
SAN FRANCISCO • SINGAPORE • SYDNEY • TOKYO
Academic Press is an imprint of Elsevier

Academic Press is an imprint of Elsevier
32 Jamestown Road, London NW1 7BY, UK
The Boulevard, Langford Lane, Kidlington, Oxford OX5 1GB, UK
225 Wyman Street, Waltham, MA 02451, USA
525 B Street, Suite 1800, San Diego, CA 92101-4495, USA

First edition 2014

ISBN: 978-0-12-801374-8
ISSN: 0065-2504

For information on all Academic Press publications
visit our website at store.elsevier.com

Working together
to grow libraries in
developing countries

www.elsevier.com • www.bookaid.org

CONTENTS

CONTRIBUTORS

Joseph K. Bailey
Department of Ecology and Evolutionary Biology, University of Tennessee, Knoxville, Tennessee, USA

Francisco Baldó
Instituto Español de Oceanografía, Centro Oceanográfico de Cádiz, and Instituto de Ciencias Marinas de Andalucía (CSIR), Cádiz, Spain

Gabriel Barrionuevo
Department of Informatics, University of Almeria, Cañada de San Urbano S/N, Almería, Spain

Ronald D. Bassar
Department of Environmental Conservation, University of Massachusetts, Amherst, Massachusetts, USA

Tim G. Benton
School of Biology, Faculty of Biological Sciences, University of Leeds, Leeds, United Kingdom

José Román Bilbao-Castro
Department of Informatics, University of Almeria, Cañada de San Urbano S/N, Almería, Spain

Tom C. Cameron
Environmental & Plant Sciences, School of Biological Sciences, University of Essex, Colchester, United Kingdom

Leocadio G. Casado
Department of Informatics, University of Almeria, Cañada de San Urbano S/N, Almería, Spain

Samantha K. Chapman
Department of Biology, Villanova University, Villanova, Pennsylvania, USA

Tim Coulson
Department of Zoology, University of Oxford, Oxford, United Kingdom

Jacques A. Deere
Department of Zoology, University of Oxford, Oxford, United Kingdom

Pilar Drake
Instituto de Ciencias Marinas de Andalucía (CSIR), Cádiz, Spain

Stephen P. Ellner
Department of Ecology and Evolutionary Biology, Cornell University, Ithaca, New York, USA

Regis Ferriere
Department of Ecology and Evolutionary Biology, University of Arizona, Tucson, Arizona, USA

Mark A. Genung
Department of Ecology and Evolutionary Biology, University of Tennessee, Knoxville, Tennessee, USA

Enrique González-Ortegón
Instituto de Ciencias Marinas de Andalucía (CSIR), Cádiz, Spain

Benjamin Gosney
School of Biological Sciences, and National Centre for Future Forest Industries, University of Tasmania, Hobart, Tasmania, Australia

Nelson G. Hairston Jr.
Department of Ecology and Evolutionary Biology, Cornell University, Ithaca, New York, USA, and Swiss Federal Institute of Aquatic Science and Technology, Eawag, Dübendorf, Switzerland

Matthew Hamilton
School of Biological Sciences, and National Centre for Future Forest Industries, University of Tasmania, Hobart, Tasmania, Australia

Teppo Hiltunen*
Department of Ecology and Evolutionary Biology, Cornell University, Ithaca, New York, USA

Giles Hooker
Department of Biological Statistics and Computational Biology, Cornell University, Ithaca, New York, USA

Laura E. Jones
Department of Ecology and Evolutionary Biology, Cornell University, Ithaca, New York, USA

Mouhammad Shadi Khudr
Faculty of Life Sciences, The University of Manchester, Manchester, United Kingdom

J. Adam Langley
Department of Biology, Villanova University, Villanova, Pennsylvania, USA

Andrés López-Sepulcre
Laboratoire Ecologie et Evolution, CNRS Unité Mixte de Recherche, École Normale Supérieure, Paris, France

Sara Magalhães
Centro de Biologia Ambiental, Faculdade de Ciências da Universidade de Lisboa, Lisbon, Portugal

*Present address: Department of Food and Environmental Sciences/Microbiology, University of Helsinki, Helsinki, Finland.

Blake Matthews
Department of Aquatic Ecology, Swiss Federal Institute of Aquatic Science and Technology, Switzerland

Carlos J. Melián
Department of Fish Ecology and Evolution, Center for Ecology, Evolution and Biogeochemistry, Swiss Federal Institute of Aquatic Science and Technology, Switzerland, and National Center for Ecological Analysis and Synthesis, University of California, Santa Barbara, California, USA

Marta Montserrat
Instituto de Hortofruticultura Subtropical y Mediterránea "La Mayora" (IHSM-UMA-CSIC), Consejo Superior de Investigaciones Científicas, Algarrobo-Costa, Málaga, Spain

Jordi Moya-Laraño
Department of Functional and Evolutionary Ecology, Estación Experimental de Zonas Áridas, EEZA-CSIC, Carretera de Sacramento s/n., Almería, Spain

Marianne Mugabo
School of Biology, Faculty of Biological Sciences, University of Leeds, Leeds, United Kingdom

Julianne M. O'Reilly-Wapstra
School of Biological Sciences, and National Centre for Future Forest Industries, University of Tasmania, Hobart, Tasmania, Australia

Stuart B. Piertney
Institute of Biological and Environmental Sciences, University of Aberdeen, Aberdeen, United Kingdom

Stewart Plaistow
Institute of Integrative Biology, University of Liverpool, Liverpool, United Kingdom

Tomos Potter
Faculty of Life Sciences, The University of Manchester, Manchester, United Kingdom

Brad M. Potts
School of Biological Sciences, and National Centre for Future Forest Industries, University of Tasmania, Hobart, Tasmania, Australia

Richard F. Preziosi
Faculty of Life Sciences, The University of Manchester, Manchester, United Kingdom

David Reznick
Department of Biological Science, Florida State University, Tallahassee, Florida, and Department of Biology, University of California, Riverside, California, USA

Jennifer Rowntree
Faculty of Life Sciences, The University of Manchester, Manchester, United Kingdom

Dolores Ruiz-Lupión
Department of Functional and Evolutionary Ecology, Estación Experimental de Zonas Áridas, EEZA-CSIC, Carretera de Sacramento s/n., Almería, Spain

Thomas W. Schoener
Department of Evolution and Ecology, University of California, Davis, California, USA

Jennifer A. Schweitzer
Department of Ecology and Evolutionary Biology, University of Tennessee, Knoxville, Tennessee, USA

John K. Senior
School of Biological Sciences, University of Tasmania, Hobart, Tasmania, Australia

Isabel M. Smallegange
Institute for Biodiversity and Ecosystem Dynamics (IBED), University of Amsterdam, Amsterdam, The Netherlands

Joseph Travis
Department of Biological Science, Florida State University, Tallahassee, Florida, USA

René E. Vaillancourt
School of Biological Sciences, University of Tasmania, Hobart, Tasmania, Australia

César Vilas
IFAPA, Centro El Toruño, El Puerto de Santa María, Cádiz, Spain, and Marine Science Institute, University of California, Santa Barbara, California, USA

Tim Wardlaw
Forestry Tasmania, Hobart, Tasmania, Australia

Carmen Whiteley
School of Biological Sciences, University of Tasmania, Hobart, Tasmania, Australia

Dean Williams
Forestry Tasmania, Hobart, Tasmania, Australia

Richard J. Williams
Microsoft Research Ltd., Cambridge, United Kingdom

PREFACE

Thomas W. Schoener*, Jordi Moya-Laraño†, Jennifer Rowntree‡, Guy Woodward§

*Department of Evolution and Ecology, University of California, Davis, California, USA
†Department of Functional and Evolutionary Ecology, Estación Experimental de Zonas Áridas, EEZA-CSIC, Almería, Spain
‡Faculty of Life Sciences, University of Manchester, Manchester, United Kingdom
§Imperial College London, Silwood Park Campus, Ascot, Berkshire, United Kingdom

The growing realisation that ecology and evolution can often operate over similar timescales (Carroll et al., 2007; Hairston et al., 2005; Schoener, 2011) has led to exciting new avenues of investigation into the links between them, including the burgeoning field of eco-evolutionary dynamics, the dynamic interplay between evolution and ecology in real time. This volume deals with the latest advances in this multidisciplinary research endeavour, including examples of major topics of current interest to both empiricists and theoreticians, as well as providing clear signposts towards the future development of the field.

The first five contributions (Chapters 1–5) consider the feedback loops between ecology and evolution, i.e., how ecological dynamics determine the selective pressures that shape trait dynamics, as well as vice versa, how trait evolution influences ecological dynamics. Of the two directions, ecology to evolution is by far the more studied to date, focusing primarily on selective pressures that affect trait evolution (Endler, 1986; MacColl, 2011). The opposite direction, which addresses how trait evolution affects ecological patterns and processes, is only now a research area that is coming into its own. At the micro-evolutionary scale, this is the field of community genetics (Antonovics, 1992; Hersch-Green et al., 2011); at the macro-evolutionary scale, this could include community phylogenetics, or how phylogenetic distances affect ecological outcomes (Webb et al., 2002). The last four papers in the volume (Chapters 6–9) deal with this part of the reciprocal eco-evolutionary loop.

Below we provide a brief overview of the current volume, highlighting for each chapter its main themes and how each contributes to the advance of eco-evolutionary research. We conclude with a discussion of the current state of the field, especially in light of the contents of this volume, and we give some thoughts about how we envision the field developing in the foreseeable future.

In the first paper, Travis et al. (2014) synthesise a large and growing body of work on the interactions that drive eco-evolutionary dynamics in the Trinidadian guppy model system, which has provided some intriguing and often counterintuitive insights into this new field. They used a multipronged empirical and theoretical approach that encompasses mathematical modelling, laboratory common-garden experiments, replicated introductions in the wild, experiments in manipulated artificial streams *in situ*, as well as altering stream productivity through canopy manipulation. They review the widespread differences in top-down ecological effects of replicated lines of guppies evolved in high- versus low-predation environments and grown at two different densities, and how these effects feed back to determine fitness among experimental groups. They also demonstrate how unexpected indirect ecological effects can arise from the eco-evolutionary feedback loop, as seen in the dampening of a trophic cascade by cycling of limiting nutrients. Finally, the authors provide an overview of ongoing research devised to understand the evolution of the low-predation phenotypes by documenting both the effects of evolutionary changes in the guppies and the strong concomitant changes they induce in the ecosystem, including the evolutionary response of competing species.

The second paper, by Hiltunen et al. (2014), focuses on another model system in which the authors have combined predictions from differential equation models of population dynamics, either with or without evolution, with sophisticated laboratory chemostat experiments on predator prey interactions (e.g. Becks et al., 2012; Yoshida et al., 2003, 2007). Here, the authors build on recent work (Hiltunen et al., 2013) by incorporating evolution within a simple food web that includes intraguild predation. They found that the dynamics of intraguild predators and prey were sensitive to genetic variation, and thus evolution in the shared prey. These dynamics are "intriguingly complex" and their outcome depends on the trade-off in prey defence between the predators. The authors suggest that with prey evolution, the dominance of each predator essentially "takes turns" as in a type of dynamics known as a "canard cycle", having a very fast transition of dynamical behaviour within a small range of a "control" parameter. This concept is apparently described here in the context of population ecology for the first time (see May and Leonard, 1975, however, for a similar kind of behaviour with Lotka–Volterra competition equations).

The third paper (Moya-Laraño et al., 2014) deals with simulations in more diverse food webs that include up to 20 species and their spatial context. The authors have extended a recent Individual-Based Model platform (Moya-Laraño et al., 2012) to enable eco-evolutionary dynamics to be simulated for multiple genetic quantitative traits in complex soil food webs. Their results suggest that highly connected webs in islands at intermediate distances, when allowing for evolution in the 20 constituent species (i.e., high genetic variation), were generally more persistent despite otherwise widespread species extinctions. They also report diverse and fast trait-evolutionary dynamics (i.e., oscillating or monotonic changes in the means of several traits across a few generations) occurring during ecological dynamics. The authors acknowledge that these initial forays into simulations of evolving food webs can, at present, only capture a small portion of the complexity of nature, but they also suggest ways in which their framework could be developed in a Feedback Research Program to help understand real systems, including the engineering of eco-evo webs for biological pest control in the future.

In the fourth paper, Smallengange and Deere (2014) show the results of a harvesting experiment on male morphs of bulb mites. Each morph has a different sexual strategy (fighters vs. scramblers), and the authors found that, contrary to traditional theoretical predictions (Tomkins and Hazel, 2007), regardless of which morph was harvested, the frequency of scramblers always increased. The authors show how this apparently unexpected result is due to an eco-evolutionary feedback and conclude that the ecological background against which evolution is studied must always be taken into account to fully understand the drivers of trait evolution. A key aspect of the novelty of their results is that to fully understand the evolution of sexual selection (not just natural selection), one has to also consider ecological feedbacks of trait evolution: Smallegange et al.'s results provide a new benchmark in studies of sexual selection, as they show how the evolution of such traits can also be tightly coupled to ecological dynamics.

In the fifth paper, Cameron et al. (2014) introduce and review an extensive body of research on soil-mite population dynamics, returning us to an experimental approach that neatly complements Travis et al. Cameron et al. demonstrate explicitly how intimately ecological and evolutionary dynamics of living systems are linked, how complex even seemingly simple systems are in reality and how difficult it is to distinguish "ecological" from "evolutionary" effects without carefully designed experiments. The body of research covered by Cameron et al. highlights how current and historic parental states and environments interact to generate and maintain pheno-typic variation of life history traits (e.g. growth rate and the trade-off between age and size to maturity) and how this can influence population dynamics at intergenerational scales. The novel experimental data presented introduce a harvesting (or management) regime that is akin to that of com-mercial fisheries. These experiments aim to "close the eco-evolutionary loop" by determining whether the observed changes in life history traits (a delay in developmental time linked to an increase in fecundity) are caused by selection leading to increased population growth rates. They use molec-ular techniques to show that, in addition to drift, selection causes the reported differences, leading to a clear feedback loop. They also ask if these selective forces can change the evolutionary trajectory of a population over time, for example, by rescuing it from extinction. Crucially, Cameron et al. show that both selection and drift play a role in determining future evolu-tionary trajectories of populations under different environmental and harvesting conditions, and that evolution can rescue populations even if they are facing classic extinction vortices, as is the case for many global fisheries.

In the sixth paper, Melián et al. (2014) present an individual-based modelling approach in which they link phenotypic variability in predator selectivity to species diversity in food webs. Predator selectivity is modelled using two concepts from foraging theory (Schoener, 1987; Stephens and Krebs, 1986): (1) per–prey-item profitability (net energy/handling time) and (2) a type of learning, in which previous consumptions act as an agent of reinforcement predilecting future selectivity. The resulting individual-trait-based models contain two basic parameters, the speed of learning and the strength of prey selection. Approximate Bayesian computational methods are then used to uncover which model is the best predictor of tro-phic links in a massive food-web data set consisting of the diets of >5500 individual fishes. They conclude that the maintenance of diversity in food webs may depend on patterns of predatory behaviour and relative abun-dance of prey. Strongly connected predators (having a high number of prey items, as might result from high feeding rates) preferred common prey and vice versa for weakly connected predators, and the balance between them drove patterns of species diversity. The chapter offers a glimpse into eco–evolutionary dynamics: in the authors' words, "we were able to infer different types of density-dependent prey selection from individual preda-tors for a given ecological time scale, but we were unable to estimate how such variation in prey selectivity might drive frequency dependent selection pressures . . ."

In the seventh paper, Khudr et al. (2014) bring us squarely into the field of community genetics, presenting novel data from a genetically explicit plant–herbivore system. They deconstruct levels of diversity and examine the strength of competition among pea aphid species and genotypes in a genetically variable plant–host environment. Although competition effects among herbivores sharing a host are implicit in many of the classic community genetics systems, few, if any, of these studies have examined the interactions with the degree of detail presented here. Khudr et al. measured both inter- and intra-specific competition, alone and in combination, using a model system of aphid herbivores and faba bean host plants. In addition, they dissected the intra-specific competition to distinguish competition effects within and among different clones of the same species. They found that the strength of competition increased as relatedness among aphids increased and that the effects of the aphids on each other were stronger than the influence of the plant host. However, in all scenarios tested, plant genotype influenced or mediated competitive interactions among the aphids. These findings complement those of Cameron et al., once again clearly demonstrating the complexity that can exist within even seemingly simple ecological systems. This work also highlights the intractability of understanding the ecological and evolutionary responses of species interactions if the underlying genetics of a system are not taken into account. The use of aphids, particularly clonal ones, has proved very profitable for studies of eco-evolutionary dynamics in other contexts: in a multigenerational field experiment, Turcotte et al. (2011) studied the existence and consequences of eco-evolutionary dynamics by in essence controlling for evolution: single-clone treatments (no evolution) were contrasted with multi-clone treatments (evolution, defined as a change in gene frequency, occurs when clones change frequencies).

In the eighth paper, O'Reilly-Wapstra et al. (2014) build on previous work using a semi-natural plant genetic resource: multiple common gardens of *Eucalytus* species, in this case *E. globulus*, containing replicated families and populations, originally collected from across the Australian landscape. As with Khudr et al., they examine the influence of within-species genetic diversity at multiple scales (within and among populations), across abiotic environmental gradients, on plant–herbivore and pathogen interactions. The deconstruction of genetic diversity in this case is, however, within the plants rather than the plant pests. Again, the importance of genetic diversity within plants in determining the outcome of interactions with plant pests is highlighted, as is the difficulty in predicting under what circumstances, and at what level, this is important. More specifically, O'Reilly-Wapstra et al. determine genetic correlations among the plants and the various pest species and investigate the consistency of the patterns observed. This approach enables them to explore the stability of plant–pest interactions across different environments (the common gardens) and determine if the genetic response of the plants to any of the enemies influences the response to others (i.e., could the occurrence of multiple plant pests facilitate or hinder an evolutionary response to attack?). It is particularly notable that the team found that most of the interactions between the plant genetic variation and the plant pests were independent of each other and that there was a consistency of plant genetic response across abiotic environments. However, a family-level (but not population-level) genetic correlation was observed between the plant genetic response to a fungal leaf pathogen and the response to sawfly larvae. This suggests that an evolutionary response of the plants to one of these enemies would influence and potentially drive the response to the other, but that different evolutionary forces were acting on the interactions within and among populations.

The ninth paper stays with *Eucalyptus* as a model system but looks at the impact of global change in atmospheric carbon dioxide and soil nitrogen levels on multiple congeners of *Eucalyptus*. Genung et al. (2014) take us from the micro- to the macro-evolutionary scale and ask how evolutionary history and phylogenetic relationships among species influence the contemporary responses of the plants to these abiotic factors in multi-species communities. They show that plant productivity is positively correlated with contemporary range size, but that range size is a product of evolutionary history. That is, past evolution seems to have driven current distributional ranges through increased plant fitness. Similar to the earlier contribution by Khudr et al., the authors also explore the impact of individual relatedness among competitors on the productivity of the plants under competitive conditions. They show that plant productivity of monocultures (where conspecifics compete) is reduced compared to mixtures, but only for those mixtures in which the competitors were of intermediate relatedness. This demonstrates the contingency in responses among different organisms (cf. the findings of Khudr et al.) to competition. Overall, Genung et al. (2014) demonstrate that there are complex relationships between phylogenetic similarity, productivity and the influence of environmental stressors on the focal plants.

In summary, this volume offers a timely coverage of an important emerging field at the nexus of ecology and evolution. Its chapters represent the current strong focus on certain systems, i.e., aquatic systems and plant–herbivore systems; indeed, the latter have provided the first two published examples (Agrawal et al., 2013; Turcotte et al., 2011) of multigenerational experimental studies of eco-evolutionary dynamics in nature (Fussmann et al., 2007; Schoener, 2011). Concentration on these systems has clear ramifications for the ecosystem services (fisheries and crop production) they provide to human societies. Very different empirical approaches, model systems and species are represented in this volume, but nonetheless some common themes are evident. Most importantly, these chapters add compelling evidence to the growing but still unconfirmed view (Schoener, 2011; Thompson, 1998) that eco-evolutionary dynamics are not simply idiosyncratic curiosities but rather are ubiquitous and potentially hugely powerful forces in nature. Travis et al. (2014) show that for all four approaches used in their studies of the Trinidadian guppy—comparative demography, mesocosm experiments, common-garden laboratory experiments and introduction experiments—eco-evolutionary dynamics play major explanatory roles. Hiltunen et al. (2014) show that observed patterns of population cycling of two predators (flagellates and rotifers, which latter also consume the former) and a prey (algae) depend upon whether evolution of prey defences is present or absent; the very unusual patterns produced when evolution is present require an eco-evolutionary framework for their explanation. Moya-Laraño et al. (2014) model food webs across space and find for simulations incorporating genetic variation that "widespread trait-evolutionary changes [were] indicative of eco-evolutionary dynamics". Smallengange and Deere (2014) find in bulb mites that the tendency of a particular male mating morph (scramblers) to increase with harvesting of either that morph or the other morph (fighters) is only explainable by an eco-evolutionary approach—prediction from the traditional purely evolutionary theory is incorrect. Cameron et al. (2014) using a model system of soil mites provide evidence for a full eco-evolutionary loop, showing that under poor and unpredictable environments, harvesting leads to a decrease in developmental time, which in turn translates into higher fitness and an increase in population growth. The tight interdependency of ecology and evolution in their system cause the authors to remark that "evolutionary and ecological effects on dynamics can . . . be almost impossible to partition . . ." Khudr et al. (2014) conclude that genetic variation in herbivores, in addition to the genetic variation in their host plants that has been the usual object of study for plant/herbivore systems, is necessary for a full understanding of eco-evolutionary dynamics in agricultural and natural settings.

Finally, O'Reilly-Wapstra et al. (2014) show for two pest species (a fungal pathogen and a sawfly larva) that damage inflicted by one of the species is genetically correlated with damage inflicted by the other species; hence, a response to selection by the plant against one of the pest species will affect the population dynamics of the other, in what they call an indirect eco-evolutionary feedback loop.

While the studies presented here are still embryonic in terms of their ability to capture the entire complexity of natural systems having eco-evolutionary dynamics, they certainly represent a major leap forward with their novel ways of linking ecology and evolution more closely than before; as such, they represent some of this new research direction's first steps towards an ultimate goal of untangling and understanding Darwin's "entangled bank" of interacting species in space and time. The multi-disciplinary approaches seen in this volume do, however, fairly reflect the scale of the challenge, foreshadowing the degree to which the next generation of eco-evolutionary research will need to employ a huge range of expertise—both empirical and theoretical—to predict how natural systems are likely to respond to future change.

REFERENCES

Agrawal, A.A., Johnson, M.T.J., Hastings, A.P., Maron, J.L., 2013. A field experiment demonstrating plant life-history evolution and its eco-evolutionary feedback to seed predator populations. Am. Nat. 181, S35–S45.

Antonovics, J., 1992. Toward community genetics. In: Fritz, R.S., Simms, E.L. (Eds.), Plant Resistance to Herbivores and Pathogens: Ecology, Evolution, and Genetics. University of Chicago Press, Chicago, IL, pp. 426–449.

Becks, L., Ellner, S.P., Jones, L.E., Hairston Jr., N.G., 2012. The functional genomics of an eco-evolutionary feedback loop: linking gene expression, trait evolution, and community dynamics. Ecol. Lett. 15, 492–501.

Cameron, T.C., Plaistow, S., Mugabo, M., Piertney, S.B., Benton, T.G., 2014. Eco-evolutionary dynamics: experiments in model system. In: Moya-Laraño, J., Rowntree, J., Woodward, G. (Eds.), Advances in Ecological Research, Vol. 50: Eco-evolutionary Dynamics, pp. 171–206.

Carroll, S.P., Hendry, A.P., Reznick, D.N., Fox, C.W., 2007. Evolution on ecological timescales. Funct. Ecol. 21, 387–393.

Endler, J.A., 1986. Natural Selection in the Wild. Princeton University Press, Princeton, NJ.

Fussmann, G.F., Loreau, M., Abrams, P.A., 2007. Eco-evolutionary dynamics of communities and ecosystems. Funct. Ecol. 21, 465–477.

Genung, M.A., Schweitzer, J.A., Senior, J.K., O'Reilly-Wapstra, J.M., Chapman, S.K., Langley, J.A., Bailey, J.K., 2014. When Ranges Collide: Evolutionary History, Phylogenetic Community Interactions, Global Change Factors, and Range Size Differentially Affect Plant Productivity. In: Moya-Laraño, J., Rowntree, J., Woodward, G. (Eds.), Advances in Ecological Research, Vol. 50: Eco-evolutionary Dynamics, pp. 297–350.

Hairston, N.G., Ellner, S.P., Geber, M.A., Yoshida, T., Fox, J.A., 2005. Rapid evolution and the convergence of ecological and evolutionary time. Ecol. Lett. 8, 1114–1127.

Hersch-Green, E.I., Turley, N.E., Johnson, M.T.J., 2011. Community genetics: what have we accomplished and where should we be going? Philos. Trans. R. Soc. Lond. B 366 (1569), 1453–1460.

Hiltunen, T., Jones, L.E., Ellner, S.P., Hairston Jr., N.G., 2013. Temporal dynamics of a simple community with intraguild predation: an experimental test. Ecology 94, 773–779.

Hiltunen, T., Ellner, S.T., Hooker, G., Jones, L.E., Hairston Jr., N.G., 2014. Eco-evolutionary dynamics in a three-species food web with intraguild predation: intriguingly complex. In: Moya-Laraño, J., Rowntree, J., Woodward, G. (Eds.), Advances in Ecological Research, Vol. 50: Eco-evolutionary Dynamics, pp. 41–74.

Khudr, M.S., Potter, T., Rowntree, J., Preziosi, R.F., 2014. Community genetic and competition effects in a model pea aphid system. In: Moya-Laraño, J., Rowntree, J., Woodward, G. (Eds.), Advances in Ecological Research, Vol. 50: Eco-evolutionary Dynamics, pp. 243–266.

MacColl, A.D.C., 2011. The ecological causes of evolution. Trends Ecol. Evol. 26, 514–522.

May, R.M., Leonard, W.J., 1975. Nonlinear aspects of competition between three species. SIAM J. Appl. Math. 29, 243–253.

Melián, C., Baldó, F., Matthews, B., Vilas, C., González-Ortegón, E., Drake, P., Williams, R.J., 2014. Individual trait variation and diversity in food webs. In: Moya-Laraño, J., Rowntree, J., Woodward, G. (Eds.), Advances in Ecological Research, Vol. 50: Eco-evolutionary Dynamics, pp. 207–242.

Moya-Laraño, J., Verdeny-Vilalta, O., Rowntree, J., Melguizo-Ruiz, N., Montserrat, M., Laiolo, P., 2012. Climate change and eco-evolutionary dynamics in food webs. In: Woodward, G., Jacob, U., Ogorman, E.J. (Eds.), Advances in Ecological Research, Vol. 47: Global Change in Multispecies Systems, Pt 2. 1–80.

Moya-Laraño, J., Bilbao-Castro, J.R., Barrionuevo, G., Ruiz-Lupión, D., Casado, L.G., Montserrat, M., Melián, C., Magalhães, S., 2014. Eco-evolutionary spatial dynamics: rapid evolution and isolation explain food web persistence. In: Moya-Laraño, J., Rowntree, J., Woodward, G. (Eds.), Advances in Ecological Research, Vol. 50: Eco-evolutionary Dynamics, pp. 75–144.

O'Reilly-Wapstra, J.M., Hamilton, M., Gosney, B., Whiteley, C., Bailey, J.K., Williams, D., Wardlaw, T., Vaillancourt, R.E., Potts, B.M., 2014. Genetic correlations in multi-species plant/herbivore interactions at multiple genetic scales; implications for eco-evolutionary dynamics. In: Moya-Laraño, J., Rowntree, J., Woodward, G. (Eds.), Advances in Ecological Research, Vol. 50: Eco-evolutionary Dynamics, pp. 267–296.

Schoener, T.W., 1987. A brief history of optimal foraging theory. In: Kamil, A.C., Krebs, J., Pulliam, H.R. (Eds.), Foraging Behavior. Plenum Publishing Corporation, New York, NY, pp. 5–67.

Schoener, T.W., 2011. The newest synthesis: understanding the interplay of evolutionary and ecological dynamics. Science 331, 426–429.

Smallengange, I., Deere, J.A., 2014. Eco-evolutionary interactions as a consequence of selection on a secondary sexual trait. In: Moya-Laraño, J., Rowntree, J., Woodward, G. (Eds.), Advances in Ecological Research, Vol. 50: Eco-evolutionary Dynamics, pp. 145–170.

Stephens, D.W., Krebs, J.R., 1986. Foraging Theory. Princeton University Press, Princeton, NJ.

Thompson, J.N., 1998. Rapid evolution as an ecological process. Trends Ecol. Evol. 13, 329–332.

Tomkins, J.L., Hazel, W., 2007. The status of the conditional evolutionarily stable strategy. Trends Ecol. Evol. 22, 522–528.

Travis, J., Reznick, D., Bassar, R.D., López-Sepulcre, A., Ferriere, R., Coulson, T., 2014. Do eco-evolutionary feedbacks help us understand nature? Answers from studies of the Trinidadian guppy. In: Moya-Laraño, J., Rowntree, J., Woodward, G. (Eds.), Advances in Ecological Research, Vol. 50: Eco-evolutionary Dynamics, pp. 1–40.

Turcotte, M.M., Reznick, D., Hare, J.D., 2011. The impact of rapid evolution on population dynamics in the wild: experimental test of eco-evolutionary dynamics. Ecol. Lett. 14, 1084–1092.

Webb, C.O., Ackerly, D.D., McPeek, M.A., Donoghue, M.J., 2002. Phylogenies and community ecology. Annu. Rev. Ecol. Syst. 33, 475–505.

Yoshida, T., Jones, L.E., Ellner, S.P., Fussmann, G.F., Hairston Jr., N.G., 2003. Rapid evolution drives ecological dynamics in a predator-prey system. Nature 424, 303–306.

Yoshida, T., Ellner, S.P., Jones, L.E., Bohannan, B.J.M., Lenski, R.E., Hairston Jr., N.G., 2007. Cryptic population dynamics: rapid evolution masks trophic interactions. PLoS Biol. 5, 1868–1879.

CHAPTER ONE

Do Eco-Evo Feedbacks Help Us Understand Nature? Answers From Studies of the Trinidadian Guppy

Joseph Travis[*,1,2], David Reznick[*,†,2], Ronald D. Bassar[‡], Andrés López-Sepulcre[§], Regis Ferriere[¶], Tim Coulson[‖]

[*]Department of Biological Science, Florida State University, Tallahassee, Florida, USA
[†]Department of Biology, University of California, Riverside, California, USA
[‡]Department of Environmental Conservation, University of Massachusetts, Amherst, Massachusetts, USA
[§]Laboratoire Ecologie et Evolution, CNRS Unité Mixte de Recherche, École Normale Supérieure, Paris, France
[¶]Department of Ecology and Evolutionary Biology, University of Arizona, Tucson, Arizona, USA
[‖]Department of Zoology, University of Oxford, Oxford, United Kingdom
[1]Corresponding author: e-mail address: travis@bio.fsu.edu
[2]Sharing credit as joint first authors.

Contents

Abstract

The bulk of evolutionary ecology implicitly assumes that ecology shapes evolution, rather than vice versa, but there is increasing interest in the possibility of a two-way interaction. Dynamic feedbacks between ecological and evolutionary processes (eco-evo feedbacks) have long been recognized in the theoretical literature, and the observation of rapid evolution has since inspired empiricists to explore the consequences of these feedbacks. Laboratory studies prove that short-term evolutionary change can significantly alter ecological dynamics, particularly in pair-wise interactions. We know

Advances in Ecological Research, Volume 50
ISSN 0065-2504
http://dx.doi.org/10.1016/B978-0-12-801374-8.00001-3

1

far less about whether these reciprocal dynamics are important in more complex natural systems. Here, we outline our approach to that question, focusing on the Trinidadian guppy and the stream ecosystems it inhabits. We summarize results from several types of studies: comparative demography in two types of communities, experiments in mesocosms, common garden laboratory experiments and replicated introduction experiments. The latter were designed as perturbations to the natural steady state that allow us to follow the joint ecological and evolutionary dynamics of guppies and their ecosystem. In each approach, we replicated experiments across multiple independent origins of guppy population types and found that eco–evo feedbacks play major roles in guppy evolution. There are three possible sources for these feedbacks, all of which have some support in our data, which will form the focus of future research efforts.

1. INTRODUCTION

Feedbacks between ecology and evolution occur when an organism modifies some feature of its environment and, by extension, changes the nature of selection that it experiences (Cameron et al., 2014, chapter 5 of this volume; Ferriere et al., 2004; Kokko and López-Sepulcre, 2007). This change in the nature of selection may elicit a genetic response that alters the impact the organism has on its environment, which can in turn change the nature of selection again, creating a feedback loop that links the dynamics of phenotypes and those of their constituent alleles and genotypes with the dynamics of ecological variables. Understanding the prevalence and importance of these so-called eco–evo feedbacks is important for two reasons. First, they can generate outcomes of simple ecological interactions that differ from those that prevail in the absence of the evolutionary feedback (Hiltunen et al., 2014, chapter 2 of this volume). For example, a selective feedback from predator to prey can cause prey to evolve resistance to the predator, which may in turn stabilize an otherwise unstable system or destabilize an otherwise stable system (Abrams and Matsuda, 1997). Second, these feedbacks may determine which traits evolve and how they do so. The optimal phenotype when there are eco-evo feedbacks can be quite different from that in their absence (MacArthur and Wilson, 1967).

In this chapter, we refer to 'feedbacks between ecology and evolution' as 'eco–evo feedbacks'. This convenient shortcut distils a potentially complex process into two phases: an ecological impact of adaptive genetic change and an evolutionary feedback loop that propels further change. To appreciate this, recall that the parameters used to characterize ecological and evolutionary feedbacks in mathematical models of population and genetic dynamics are emergent properties of the interactions among individuals within and among species and between individuals and their abiotic environment.

These interactions change as adaptive evolution modifies how individuals respond to the dynamics of ecological variables like per capita food levels or encounter rates with predators.

The first phase, the ecological impact, occurs as adaptive evolution modifies the traits of one species. As these traits change, the demography of its population will change (Cameron et al., 2014, chapter 5 of this volume). The altered demography of our focal species can, in turn, affect variables like resource replenishment rates or the demography of a competitor or predator. The result is a change in the ecological dynamics of a system and perhaps its emergent properties (Moya-Laraño et al., 2014, chapter 3 this volume). The effects of adaptive change on stability in predator–prey systems (e.g. Abrams and Matsuda, 1997) illustrate this result.

The second phase, the evolutionary feedback loop, may or may not follow. The evolutionary feedback loop will occur if the effects of the focal species on the demography of a competitor or predator provoke an evolutionary response in that second species. The feedback loop can also act directly on the focal species if the effects of the focal species on ecological variables alter the nature of selection on the focal species (Cameron et al., 2014, chapter 5 of this volume). Whether the second phase of the eco-evo feedback propels further evolutionary change in the focal species or another species depends on whether different genotypes respond differently to the effects of these ecological impacts. Thus, when we refer to eco-evo feedbacks, we are focusing on an evolutionary process driven by how genotypes *respond* to the dynamics of one or more ecological variables.

The current explosion of interest in the empirical study of eco-evo feedbacks is relatively new, but the underlying concepts are well established. The principles of population genetics that underlie such interactions, such as frequency- or density-dependent selection (Pimentel, 1961, 1968), are well defined in theory (e.g. Charlesworth, 1971; Clarke, 1972; Cockerham et al., 1972; MacArthur, 1962; Roughgarden, 1971; Smouse, 1976; Wallace, 1975). The connections between these forms of selection and the dynamics of ecological variables, such as population density or the abundance of competitors, predators or pathogens, have also been long recognized in population genetic theory (e.g. Jayakar, 1970; Leon, 1974; Levin and Udovic, 1977; Roughgarden, 1976). These and many subsequent papers since share the common theme of modelling joint ecological and evolutionary dynamics driven by the reciprocal influences of ecological variables and genetic variation.

The burgeoning interest in eco-evo feedbacks, however, has been inspired by more recent experiments that have shown feedbacks that are strong enough to make the joint dynamics of ecological and genetic variation visible (Ellner, 2013; Schoener, 2011). Joint dynamics occur when (a) different genetically based phenotypes have different effects on ecological variables and (b) the selection coefficients generated by the feedback loop from the ecological variables to the fitness of those phenotypes are large (Otto and Day, 2007). These feedbacks are best known in a few model ecosystems, such as in the integrated theoretical and empirical work on predator–prey oscillations performed by Hairston, Ellner and colleagues (Ellner, 2013; Hiltunen et al., 2014, chapter 2 of this volume).

These laboratory studies provide the proof of concept for the potential importance of eco-evo feedbacks for the outcome of ecological dynamics. There are some well-known case studies from nature that illustrate how such feedbacks affect the trait distributions we observe in natural populations. For example, interactions between hosts and pathogens and the interlocking roles of evolving immunity and cycling population densities are among the earliest and most striking examples of eco-evo feedbacks (Duffy and Sivars-Becker, 2007; Duffy et al., 2009). The question now facing us is whether eco-evo feedbacks play similar prominent roles when we move from strong pair-wise interactions to the more general case of complex ecosystems with many interconnected species (Schoener, 2011).

We distil the challenge of understanding the importance of eco-evo feedbacks into two questions. First, how pervasive are reciprocal feedbacks between ecology and evolution in natural ecosystems? Second, are these interactions necessary to explain the nature of adaptation and ecological interactions in nature. Here, we describe how answering these questions can be done and illustrate by reviewing our work to date with the Trinidadian guppy, *Poecilia reticulata* (Poeciliidae).

2. OPERATIONAL FRAMEWORK

First, we consider a general template for understanding the role of eco-evo feedbacks in a natural community. We begin by imagining the study of some focal organism, how it is evolving in its natural environment and how its evolution might influence and interact with ecological processes. We must describe our focal organism in terms of individual attributes (e.g. age, gender, developmental stage), ecologically important phenotypic traits (e.g. body size, morphology, coloration), individual life history parameters (age and size at maturity, reproductive output), population attributes (population size, population structure) and demographic processes (age- or stage-specific survival and reproductive rates).

Having characterized the population, we must be able to describe the evolutionary response of our focal species to the varied agents of selection. To do so, we must fulfil three requirements. First, we must be able to follow individuals through a generation to enable us to estimate individual reproductive success, the distribution of fitness, the selection gradients on key traits. Second, we must be able to construct pedigrees for individuals and use those pedigrees to estimate the genetic parameters governing the response to selection. We can meet these first two requirements through building an individual-based mark–recapture study that includes data on individual genotypes. Third, we must follow the focal population over many generations to document which traits are changing directionally and thus potentially evolving. This means extending our mark–recapture data into a genuinely long-term study. With all of these data, we can distinguish evolution from other factors that can cause phenotypic change. By combining the results of individual-based mark–recapture with the observations of population, phenotypic and a genetic dynamics, we can elucidate the drivers of evolutionary change and characterize the role of eco-evo feedbacks (Coulson et al., 2010).

At the same time, we must characterize the community attributes (e.g. the abundance and population structure of key interacting species, such as predators and competitors and their rate of consumption) and ecosystem attributes (e.g. standing crops and renewal rates of resources, distribution and dynamics of limiting nutrients). Finally, we need to address the potential influence of abiotic factors, such as rainfall, light availability, temperature and temporal/seasonal variation in these factors. These factors can be agents of classical 'hard selection' (sensu Wallace, 1975) on the focal species, but more interestingly, variation in abiotic conditions can have more subtle effects by modifying the target species' population dynamics, carrying capacity and its interactions with other species (Dunson and Travis, 1991; Fowler and Pease, 2010). Through these modifications, variation in abiotic factors could influence the entire structure of the eco-evo feedback loop. In this light, it becomes clear that we cannot rely on observation alone: experiments that can disentangle these effects are needed (Fowler et al., 2006).

Mathematical modelling plays an essential role in this process because it can help refine the empirical description of the ecosystem. A key to the successful characterization of the ecosystem is to determine which variables are crucial and which can be ignored. We must refine what might be measured into which factors *must* be measured to understand the mechanisms through which feedbacks occur. While strong ecological knowledge is essential for making these decisions, mathematical models can assess the sensitivity of dynamic ecological and evolutionary processes to variations in specific parameters and thereby define what must be measured.

Mathematical models contribute in other critical ways. First, models illustrate the consequences of our assumptions about how a given process might unfold. For example, models describing 'scramble' and 'contest' competition reveal how those different forms of density dependence produce different effects on population dynamics and stability (e.g. Kot, 2001). Second, models allow us to generate predictions that follow from combining several component processes into an overall model of system dynamics (Bassar et al., 2012). Third, models can assess effects that might be playing an important role in eco–evo interactions yet cannot be detected from feasible experiments (Bassar et al., 2012): this is especially important in studying complex, multi-species interactions in which indirect effects can play prominent roles (Travis and Lotterhos, 2013; see also Smallegange and Deere, 2014, chapter 4 of this volume).

Even with all of these elements in place, eco–evo feedbacks may not be visible in communities at their steady state. The feedbacks may have played out in the path to this condition, as will be the case if feedbacks have driven adaptive, directional evolution. Seeing and characterizing eco–evo feedbacks may thus demand a perturbation of the system to cause an ecological and evolutionary disequilibrium that ignite their dynamics (Smallegange and Coulson, 2013).

An ideal perturbation is one that represents a facsimile of some known, natural phenomenon. The resulting return to equilibrium can enhance our understanding of how the organism has adapted to change and how adaptation was integrated with its impact on the local ecosystem. One such perturbation could be the invasion of a new habitat, which would enable quantification of the trajectories of both ecological and evolutionary responses. The resulting time-series responses will not necessarily reveal the mechanisms, however, since they represent a complex of interconnected causes and effects. To understand the mechanism of eco–evo interactions, we must use additional experiments, observations and modelling to diagnose those mechanisms.

We endorse a combination of observations, experiments and modelling in diagnosing eco–evo interactions, reflecting our conviction that science is most powerful when it combines data from diverse sources (Coulson, 2012). Here, we describe how we have done so to investigate the role eco–evo interactions may have played in driving the evolution of life histories in the Trinidadian guppy and the structure of the ecosystems in which they are found. We will describe what we have done, what progress we have made and what remains to be done.

3. POPULATION BIOLOGY OF GUPPIES

3.1. Natural history and evolution

The Northern Range Mountains of Trinidad offers a natural laboratory for studying interactions between ecology and evolution. The rivers draining these mountains flow over steep gradients punctuated by waterfalls that create distinct fish communities above and below waterfall barriers. Species diversity decreases upstream as waterfalls block the upstream dispersal of some fish species. The succession of communities is repeated in many parallel drainages, providing us with natural replicates.

These streams also offer the opportunity of performing experimental studies of evolution because rivers can be treated like giant test tubes, as fish can be introduced into portions of stream bracketed by waterfalls to create *in situ* experiments (Endler, 1978, 1980). Downstream guppies co-occur with a diversity of predators, which prey on the adult fish (high predation, or HP). Waterfalls often exclude predators but not guppies, so when guppies are found above waterfalls they have greatly reduced predation risk and increased life expectancy (low predation, or LP). Hart's killifish, *Rivulus hartii* (Rivulidae), the only other fish found in many of these localities, rarely preys on guppies and tends to focus on the small, immature size classes (Endler, 1978; Haskins et al., 1961). In some headwater streams, *Rivulus* is the only fish species present because they are capable of overland travel on rainy nights (*Rivulus* only, or RO localities). Population genetic analyses reveal that at least some of these rivers represent independent replicates of the evolution of guppies adapted to HP and LP environments (Alexander, et al., 2006) and that LP and HP populations are more genetically distinct than expected under migration-drift equilibrium (Barson et al., 2009).

Guppies adapted to HP environments mature at an earlier age, devote more resources to reproduction, produce more offspring per brood and produce significantly smaller offspring than LP guppies (Reznick and Endler, 1982; Reznick et al., 1996b). All of these differences are consistent across replicate HP–LP comparisons in multiple watersheds, and are also consistent with predictions derived from theory that models how life histories should evolve in response to selective predation on juveniles (LP environments) versus adults (HP environments) (Charlesworth, 1994; Gadgil and Bossert, 1970; Law, 1979; Michod, 1979; see Chapter 5). HP and LP guppies also differ in male colouration (Endler, 1978), courtship behaviour (Houde, 1997), schooling behaviour (Seghers, 1974; Seghers and Magurran, 1995), morphology (Langerhans and DeWitt, 2004), swimming performance (Ghalambor et al., 2004) and diet (Zandonà et al., 2011). Male colouration evolves in response to the combined, conflicting influences of natural and sexual selection (Endler, 1978, 1980). Laboratory studies confirm that the differences in life histories, colouration, behaviour and body shape have a genetic basis (Endler, 1980; O'Steen et al., 2002; Reznick, 1982; Reznick and Bryga, 1996). Genetic diversity is consistently greater in the higher order streams than in the headwaters. This pattern, combined with the observation that guppies periodically invade or are extirpated from headwaters, implies a dynamic process of invasion and adaptation to LP environments.

We exploit the presence of barrier waterfalls to perform experimental studies of evolution in natural populations of guppies. We can manipulate the mortality risks of guppies by transplanting them from HP localities below barrier waterfalls into previously guppy-free portions of streams above barrier waterfalls. In this way, we can simulate a natural invasion. Introduced guppies evolve delayed maturation and reduced reproductive allocation, as seen in natural LP communities. In previous experimental introductions, male traits evolved in 4 years or less (Endler, 1980; Reznick and Bryga, 1987; Reznick et al., 1990, 1997). Our inferences that guppies had evolved were derived from laboratory experiments performed on the laboratory-reared grandchildren of wild-caught parents collected from the introduction sites and the ancestral HP sites (Reznick and Bryga, 1987; Reznick et al., 1990, 1997). Other attributes of guppies, including behaviour (O'Steen et al., 2002), also evolved rapidly. These results argue that the presence or absence of predators imposes intense selection on life histories and other features of guppy phenotypes.

Mark–recapture studies on natural populations support the role of predators in shaping guppy evolution, but at the same time, it suggest that resource availability is important. HP guppies experience substantially higher mortality rates than LP guppies (Reznick and Bryant, 2007; Reznick et al., 1996a), which suggests that predator-induced mortality is a candidate cause for the evolution of the HP phenotype. However, guppy populations in LP sites tend to have higher population densities, slower individual growth rates and size distributions shifted towards larger fish. These differences in population structure are attributable in part to demography: HP guppies have higher birth and death rates (Reznick et al., 2001; Rodd and Reznick, 1997). They are also attributable to evolved differences in life histories: LP guppies mature at a later age and have lower birth rates. We hypothesize that these ecological differences are an indirect effect of the absence of predation in LP environments, as guppy populations increase and per capita resource availability declines. If eco-evo interactions prevail, then the evolution of the LP phenotype must be driven in part by the way burgeoning populations of guppies deplete resources and modify the ecosystem, and then adapt to these changes. These differences in population biology are confounded with other ecological differences between LP and HP locations. The LP environments tend to be smaller streams with lower light availability and productivity. They are not associated with differences in other features of the physical environment, such as water chemistry. This confounding provides alternative explanations for higher growth rates in HP environments (Reznick et al., 2001).

Additional results are consistent with a hypothesis that guppies in LP environments are resource limited and LP guppies are adapted to life at high population densities. Natural populations of LP guppies have asymptotic body sizes that are about 30% smaller than HP guppies (Reznick and Bryant, 2007; Reznick et al., 2001) and produce offspring that are 40–50% larger (Reznick, 1982; Reznick et al., 1996b). Experimental and theoretical studies show that larger offspring size is an adaptation to a food-limited environment (Bashey, 2008; Jorgensen et al., 2011). In nature, LP guppies produce only about one-third as many offspring per brood as HP guppies (Reznick, 1982; Reznick et al., 1996b). The magnitude of this difference in fecundity is much smaller in the lab, when LP and HP guppies are reared on the same food rations (LP brood sizes are about 75–85% of those in HP: Reznick and Bryga, 1996). This pattern suggests that the field results are the combined product of a genetic predisposition of LP guppies to produce fewer offspring and also lower food availability in their natural habitats.

Our empirical estimation of mortality rates in natural populations provides a second line of evidence that the direct effects of predation alone cannot explain guppy life history evolution. The original life history theory that successfully predicted the way by which guppies will adapt to life with and without predators also assumed to be density-independent population growth (Charlesworth, 1994; Gadgil and Bossert, 1970; Law, 1979;

Michod, 1979). This is a different way of saying that predation alone, without any eco-evo feedback, can explain how guppies adapt to HP and LP environments. However, this can only happen if there is also an asymmetry in age-specific mortality rates between HP and LP populations: adult mortality rates must be higher in HP localities but juvenile mortality rates must be the same or lower in HP localities. Laboratory studies were consistent with this assumption, as predators from HP localities preyed preferentially on larger guppies (Haskins et al., 1961; Mattingly and Butler, 1994). Stomach content analyses of wild-caught R. hartii, the only fish that co-occurs with guppies in LP localities, indicated that they prey preferentially on smaller guppies (Seghers, 1973).

We sought to confirm this difference between HP and LP localities in age-/size-specific mortality risk with mark–recapture studies of natural populations. Guppies from HP environments do indeed have higher mortality risk, but this increased risk is equal across all size classes (Reznick et al., 1996a). The same body of theory that predicts the evolution of the observed life history patterns under asymmetric mortality risk also predicts that if a change in mortality risk is equal across all age classes, then life histories will not evolve, but we had already seen these life histories evolve in our introduction experiments.

The fact that the LP life history evolves in spite of an absence of asymmetry in mortality risk suggests that the change in predator-induced mortality alone is insufficient to explain guppy life history evolution and a different agent of natural selection must be responsible. One candidate agent, in theory, would be density-dependent selection. This hypothesis is a natural candidate, given the higher densities in LP populations and the evidence that they are resource limited. For density-dependent selection to be playing this role, LP populations would have to be more tightly regulated than HP populations and the mortality imposed by that regulation would have to act differently across age classes (Charlesworth, 1994; Michod, 1979). While the mortality rates exposed by our mark–recapture results might seem, at first glance, to belie the requirement for differential mortality effects of regulation, we did not manipulate density in the LP populations in those studies; so, we cannot draw conclusions from those results about how increases and decreases in density would affect age-specific mortality rates.

A third line of evidence indicating that risk of predation alone is not sufficient to explain the evolved differences between HP and LP life histories emerged from our comparative study of senescence in HP and LP guppies (Reznick et al., 2004). Medawar's (1952), Williams' (1957) and Kirkwood's

(1993) theories for the evolution of senescence all predict that increased mortality rates and decreased life expectancy should drive the evolution of a shorter intrinsic lifespan. Our common garden comparison of HP and LP guppies showed the opposite: HP guppies mature at an earlier age and have a higher rate of investment in reproduction early in life, but they also have lower intrinsic mortality rates throughout their lives and longer mean lifespans. They continue to reproduce at a higher rate throughout their lives and do so into later ages than LP guppies. These results, which were derived from fish kept at densities of one per aquarium on controlled food availability, suggest that HP guppies are unconditionally superior to LP guppies. If true, then the LP life history should never evolve, yet it has repeatedly evolved in LP environments. We call this result the 'super guppy paradox'. It begs the question 'what is it about the LP environment that causes the LP phenotype to evolve?' One possibility is that the ecological setting of the LP environment, which includes high population densities, limited resources and co-occurrence with *R. hartii*, have in some way combined to favour the evolution of the LP phenotype.

3.2. The importance of density regulation

All of the previous lines of evidence suggest that predation–driven mortality differences alone cannot explain life history differences among HP and LP populations. Theory suggests that density dependence may play an important role in shaping how guppy life histories evolve. This leads naturally to asking whether there is evidence that LP populations are more tightly regulated than HP populations and then, if so, how might the effects of increased and decreased densities in these populations be expressed.

Population biomasses at LP sites are nearly six times higher, and individual growth rates are lower, than HP populations (Reznick and Bryant, 2007; Reznick et al., 2001). Primary production in LP streams tends to be light-limited and lower than HP localities (Reznick et al., 2001). The combination of higher population density with lower productivity in LP sites suggests lower per capita resource availability. The presence of density regulation can be tested experimentally by manipulating density and measuring the demographic responses of the population. We have conducted a series of such experiments in 10 natural LP populations (Bassar et al., 2013; Reznick et al., 2012), whereby guppies were removed from three pools, marked, then reintroduced at either the ambient density (control) or at an increased (1.5× or 2×) or a decreased density (0.5×). We recollected the guppies approximately 25 days later and compare the size-specific somatic growth, fecundity, reproductive status, offspring size, survival and fat-content of the remaining fish. Guppies from increased density treatments had lower somatic growth rates, offspring that were smaller at birth and lower recapture rates of larger fish, but there were no differences in fecundity. Surviving fish also had lower fat reserves compared with the control populations. Guppies from decreased density pools had increased somatic growth rates, an increased number of offspring and a decreased mortality in the smallest size classes. These asymmetrical responses to density across age classes are the *sine qua non* for density-dependent selection. Although the responses of increased and decreased density were asymmetrical, they imply an increase in population growth rate when we reduce density and vice versa. Density manipulations in HP localities indicate that there is no pattern of association between density and the expression of the life history in these sites (David Reznick and Ronald Bassar, unpublished data).

We then used the demographic responses to these density manipulations to develop integral projection models (IPMs) to estimate the population growth rates in the three density treatments (Bassar et al., 2013). IPMs are analogues of age- or stage-structured demographic models. Age- or stage-structured models represent vital rates like survival and fecundity as step functions of age or size (constant within an age- or size-class). IPMs represent vital rates as functions of an underlying continuous variable like body size (Easterling et al., 2000). Guppy populations lend themselves to IPMs because vital rates appear as functions of body size and body size is continuously distributed without obvious breaks in most populations. We were able to parameterize IPMs from what we learned about recruitment, size-specific survival, size-specific growth and size-specific fecundity in each experimental pool. Our IPMs revealed that guppy populations at ambient densities tended to be stable with population growth rates about equal to replacement values. Reduced densities would cause increases in population size, and increased densities would cause decreases in population size. The changes in population growth rate in response to the density manipulation would thus tend to return the population to its initial size. This combination of perturbation and modelling illustrates how the two approaches can be integrated to fully understand what unfolds in nature. These results indicate that LP populations are tightly regulated and that elevated densities are an important feature of the LP environment.

Another line of evidence for an increased role of density in LP sites is the comparative diets of each phenotype. We sampled resource availability and stomach contents of guppies from paired HP and LP communities in two rivers (Zandonà et al., 2011). Guppies from HP sites ate more invertebrates and preyed selectively on higher quality invertebrates. We inferred quality from the ratio of carbon to nitrogen because lower ratios of C:N imply a higher protein content in the prey. Guppies from LP sites had consumed more algae and detritus, which are lower quality food resources. The invertebrates consumed by LP guppies represented a non-selective sample of what was available in the natural environment, rather than a selective subset of higher quality prey. The LP guppies were thus much more generalist consumers, a pattern expected when the availability of high-quality food is limited (Lauridsen et al., 2014; Pulliam, 1974). Limited food availability is likely caused by some combination of higher population densities of guppies, a consequence of the release from predation or reduced light availability and primary productivity, a consequence of the heavier canopy cover in LP locations.

There is also a hint of other differences in feeding behaviour in guppies from HP and LP environments. LP guppies feed at a higher rate than HP guppies (de Villemereuil and Lopez-Sepulcre, 2010; Palkovacs et al., 2011). Palkovacs et al (2011) also found that the LP guppies were less discriminating about where they forage for prey. Both studies were done in the laboratory and made use of only one type of prey, so it is difficult to know how these results would play out in a natural environment. LP guppies are confronted with lower food availability and a lower risk of predation, so one possibility is that they have evolved to feed more quickly and to be less discriminating about where they feed.

4. EXPERIMENTAL STUDIES OF ECO-EVO DYNAMICS

4.1. Hypotheses for eco-evo feedbacks in the evolution of LP guppies

While the release from predation alone may be necessary for the LP phenotype to evolve, it is certainly not sufficient and one or more additional agents of selection must be operating. The comparative ecology of the stream system, along with our knowledge that the direction of evolution is most likely *from* a HP-like ancestor *to* the LP descendant, suggests three non-exclusive hypotheses for those agents.

First, HP guppies may affect the ecosystem in a way that facilitates a selective advantage for individuals with a more LP-like phenotype. This could occur if, for example, the selective foraging of HP guppies on higher quality invertebrates created a trophic cascade that favoured a more LP-like feeding habit. If HP guppies were to substantially reduce the abundance of invertebrates, many of which are algal grazers, algal abundance might increase if those grazers had been controlling algal abundance. The net effect could provide ample resources for a more LP-like feeding habit. This could initiate a directional evolution towards the LP phenotype.

Second, as HP guppies colonize a habitat and their population density increases, density-dependent selection via intra-specific competition should favour individuals with a more LP-like phenotype. This could occur if the depletion of resources that would follow an invasion by a HP-like guppy was to create conditions favouring either higher feeding rates (as observed in LP guppies) or greater efficiency of resource use.

Third, individuals with a more LP-like phenotype may have the advantage in interactions with *Rivulus* in the absence of other fish. We describe some of the possible ways this could occur below in our presentation of experimental studies of the *Rivulus*–guppy interaction.

All of these hypotheses include implicit eco-evo feedbacks because, in each case, an effect of HP guppies on the ecosystem would create conditions that alter the selective milieu towards favouring a more LP-like phenotype. While we have described some mechanisms for these feedbacks, these are by no means the only conceivable ones. The key issue is whether we consider feedbacks via altering resource distributions (our first hypothesis), altering resource levels (our second hypothesis) or altering the interaction with the only other fish in the system (our third hypothesis), it appears that some form of eco-evo feedback loop is necessary to propel the evolution of the LP guppies.

We are assessing these hypotheses through two approaches. First, we have used a series of short-term, factorial experiments to assess some of the potential mechanisms underlying each one. These experiments are *retrospective* studies of evolution that compare the qualities of fish adapted to HP and LP environments in a common setting. Because HP and LP fish represent the end points of adaptive evolution, the differences we observe in how they affect parameters of their ecosystems presumably quantify how their evolution has moulded their impact on their environment. These experiments also offer insights into whether those impacts can change the selective milieu for guppies. Second, we are performing *prospective* studies of evolution that reveal the correlated time-course of guppy evolution and guppy impacts on their ecosystem. Specifically, we perturbed a natural system by introducing guppies from a single HP locality into four headwater tributaries that previously had no guppies. We have spent the last 7–8 years in characterizing the evolutionary and ecological dynamics that followed these introductions.

4.2. Artificial streams: Retrospective studies of guppy evolution

We have performed several experiments in artificial streams that replicate the natural environment. We built 16 artificial streams alongside a natural stream in Trinidad. Part of the outflow from a spring that normally flows into the natural stream was diverted to a collecting tank, then gravity-fed to the 16 streams. Each stream had its own inlet valve, enabling us to equalize flow rates. The outflow from the streams returns to the natural stream. The artificial streams were lined with fresh natural substrate and seeded with invertebrates (minus invertebrate predators on guppies) at the outset of each experiment. They were then colonized and invertebrate populations were naturally replenished by adults emerging from the nearby stream. The ecosystems in each stream proved stable and reasonable facsimiles of natural streams (Bassar et al., 2010).

The fish in our experiments were collected from HP or LP localities found within the same river, in close proximity to one another. We duplicated these experiments by running them twice, each time with fish from a different river. Replicating across HP and LP fish from different rivers increases the robustness of our results and permits us to draw more general conclusions. Our duplicates always yielded qualitatively similar results and the data never described a significant interaction between replicate and any of our manipulated factors in the experimental designs. Here, we summarize three sets of results from these experiments.

4.2.1 LP and HP exert different direct and indirect effects on their ecosystems and the indirect effects create eco-evo feedbacks (Bassar et al., 2010, 2012)

We manipulated guppy phenotype (HP vs. LP), density (low vs. high, with high equal to a doubling of low) and access to algae (full vs. restricted). The high-density treatment approximated to the density in LP localities while the low-density treatment approximated the density in HP localities (Reznick et al., 1996a, 2001). The densities of the guppies were on the higher side of what is observed in natural streams. However, these numbers were chosen because preliminary studies in the mesocosms showed that these densities produced somatic growth rates that were comparable to guppy somatic growth rates in the natural streams (Bassar et al., 2013). In this sense, the numbers were calibrated to the availability of resources in the mesocosms.

Some mesocosms had no guppies, enabling us to evaluate the impact of guppies on ecosystem structure by comparing the 'no-guppy' treatment with the average of the treatments that include guppies. A requisite for eco-evo feedbacks is that the target species must alter some feature of its eco-system. Guppies exerted strong effects on the mesocosm ecosystem, decreasing algal standing stocks by nearly 80% (Fig. 1.1) and reducing invertebrate abundance (Bassar et al., 2010). More strikingly, the presence of guppies increased mass-specific gross primary productivity in the experimental eco-system and altered the rates of several other ecosystem processes like decomposition and nutrient flux rates (Bassar et al., 2010).

However, HP and LP guppies were not interchangeable in their effects on the experimental ecosystems (Fig. 1.1; Bassar et al., 2010). For example, LP guppies decreased algal standing stocks to a greater extent than did HP guppies. HP guppies depleted invertebrate abundance more than LP guppies. These results align closely with the observations that LP fish are more herbivorous than HP fish (Zandonà et al., 2011) and were confirmed when we evaluated the stomach contents of the mesocosm fish at the end of the experiment (Bassar et al., 2010). Another important consequence of the dietary difference is that HP guppies ate more invertebrate decomposers, which suppressed leaf decomposition rates in artificial streams with HP guppies (Bassar et al., 2010). Overall, the magnitude of the effects of exchanging phenotypes was large and for some variables, the HP–LP effect

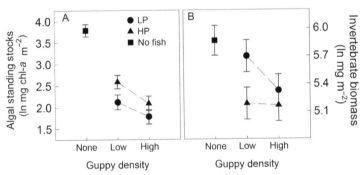

Figure 1.1 Standing stock of (A) algae and (B) invertebrates after 28 days from meso-cosm experiment with no guppies and guppy phenotype crossed with guppy density. Error bars are ±1 SE. *Redrawn from Bassar et al. (2010).*

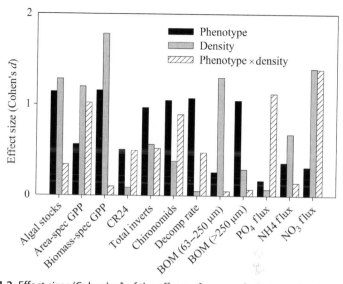

Figure 1.2 Effect sizes (Cohen's *d*) of the effects of guppy phenotype, density, and their interaction on multiple components of the ecosystem: algal stocks (mg chlorophyll-*a* per square metre), area-specific gross primary productivity (mg oxygen per square metre per day), biomass-specific gross primary productivity (mg oxygen per mg chlorophyll-*a* per day), community respiration over 24-h periods (mg oxygen per litre per day), total invertebrate biomass (mg per square metre), biomass of chironomids (mg per square metre), leaf decomposition rate (slope of the relationship between log leaf pack biomass and day), biological organic matter in the 63–250 µm range (g per square metre), biological organic matter larger than 250 µm (g per square metre), flux of phosphate ions (µg per hour per square metre), flux of ammonium ions (µg per hour per square metre), and flux of nitrate (µg per hour per square metre). *From Bassar et al. (2010).*

was equal to or larger than the effects of doubling guppy densities (Fig. 1.2). These results imply that the impact of guppies on the structure of their ecosystem changes as they adapt to LP environments.

On first inspection, the lower standing crop of algae in the LP treatment seems a simple consequence of their eating more algae, but consideration of the food web suggests the indirect effects of guppies on algal biomass might also differ between phenotypes. HP guppies have higher consumption rates of invertebrate grazers, which could have a top–down effect that enhances algal standing crops. HP guppies also excrete nutrients at higher rates (Bassar et al., 2010; Palkovacs et al., 2009), which could stimulate the growth of algae if algae are nutrient-limited. But because HP guppies are also eating more invertebrate decomposers thereby slowing leaf decomposition rates, they could be exerting a negative indirect effect on algae by lowering the rate at which nutrients become available for algal uptake (see Woodward et al., 2008).

The challenge then was to discern whether we could detect indirect effects of exchanging phenotypes and, if possible, distinguish the sources of those effects. By manipulating guppies' access to algae, we partitioned the net effect of each phenotype of guppy on algal biomass into direct and indirect effects. We did so by including electric enclosures (Connelly et al., 2008; Pringle and Blake, 1994) in each mesocosm (Bassar et al., 2012). We formed small quadrats out of copper wire and placed two in each mesocosm. One was attached to a fence charger that created a pulsed electric field that deterred guppies from entering the quadrat but did not deter invertebrates. The other quadrat was not electrified and served as a control. The electrified quadrat represented a patch of habitat that excluded guppy grazing, but exposed the rest of the ecosystem to the indirect consequences of guppy activities.

The experimental results were surprising (Bassar et al., 2012). We found no evidence that exchanging phenotypes would change the indirect effects of guppies on algal biomass; in fact, the sign of the effect estimated from the data was negative, which is in the opposite direction from what a trophic or nutrient cascade would predict. Not only was the estimated change in the indirect effect not statistically significant, its magnitude was quite small. The failure to detect the effect as statistically significant did not seem likely to be caused by low statistical power; had our residual degrees of freedom for the statistical test gone from 25 (the actual value) to 3000, the F-statistic would have remained insignificant (assuming that a larger sample size would not have appreciably changed the residual variance).

This result was paradoxical: we had expected a strong trophic cascade to emerge from the increased feeding of HP guppies on invertebrates, and indeed there was a substantial direct effect of exchanging phenotypes on invertebrate density. Why was there no evidence of an indirect consequence of invertebrate removal?

We resolved this paradox by exploring a mathematical model of this ecosystem (Bassar et al., 2012). The model described the dynamics through time of the concentration of dissolved nutrients (nitrogen) through compartments defined by the biomass of benthic organic material, primary producers, invertebrates and fish (Fig. 1.3A). We parameterized the model from the results of this experiment along with data from other experiments. The model reproduced the net effect on algae of exchanging the guppy

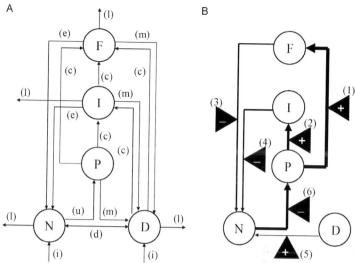

Figure 1.3 Graphical depictions of the model for the mesocosm ecosystem as presented in Bassar et al. (2012). (A) The direct flow of nitrogen, representing nutrients, among compartments is illustrated with arrows indicating the direction of flow. Compartments are fish (F), aquatic invertebrates (I), periphyton (P), detritus (D), and inorganic nutrients (nitrogen, N). Fluxes of matter and energy are driven by consumption (c), production of dead organic matter (egestion or mortality, m), excretion (e), decomposition (mineralization and immobilization, d) external in-flow (i), and losses (l). (B) Direct (1) and indirect ((2–6) ecological effects of phenotypic difference between high-predation (HP) guppies and low-predation (LP) guppies on algal biomass, measured in mesocosms. Plus or minus sign (inside triangles) next to ecosystem flow (arrows) indicates the effect of HP compared to LP that the corresponding flow mediates. Arrow thickness is proportional to effect size. (1) The P-to-F flow (consumption of periphyton by guppies) mediates a positive direct effect of exchanging HP for LP phenotype on periphyton biomass (because HP consume less periphyton than LP). (2) The P-to-I flow (consumption of periphyton by invertebrates) mediates a positive indirect effect of exchanging HP for LP phenotype on periphyton biomass (because HP consume more invertebrates that also feed on periphyton). (3) and (4) The F-to-N and I-to-N flows (excretion of fish and invertebrates) mediate negative indirect effects of exchanging HP for LP phenotype on periphyton biomass (because HP disrupt nutrient cycling compared to LP). (5) The indirect effect of decomposition (D-to-N flow) is positive but small. The negative effects (3) and (4) and small positive effect (5) result in a negative influence of HP phenotype on nutrient uptake by algae (N-to-P flow, 6). Precise quantification of the effects shows that the indirect 'trophic cascade' effect (2) is offset by the other indirect effect of nutrient cycling (3–6).

phenotypes as well as the strong direct effect of HP guppies on invertebrate abundance. It also reproduced the paradox of seeing no net indirect effect of exchanging guppy phenotypes. However, the model allowed us to estimate fluxes through pathways we could not separate experimentally. The model not only confirmed the signature of a large trophic cascade but also revealed an overwhelming bottom–up effect: fewer invertebrates in the presence of the HP phenotype resulted in much reduced nutrient cycling (due to a combination of reduced invertebrate excretion and slower decomposition of benthic organic matter). The two indirect effects almost entirely cancelled out each other's influence (Fig. 1.3B).

The results from combining our experimental and mathematical modelling approaches are important for two reasons. First, they indicate that a failure to find statistically significant net indirect effects though an experimental manipulation does not mean that individual indirect effects are not large and significant. Our experimental ecosystem was comparatively simple; more complex ecosystems with more paths through which indirect effects can flow would seem much more prone to demonstrate this phenomenon. If so, our results suggest that experimental studies of net indirect effects should always be complemented by mathematical models of those systems before definitive ecological conclusions are drawn.

Second, they suggest that sometimes what ecologists may find uninteresting, evolutionary biologists may find critically important (and of course vice versa). An ecologist might argue that because the net indirect effect is small, the individual indirect effects are of little interest or importance. However, in this case, the evolutionary biologist may think differently because the individual effects can generate selection pressures of their own. In particular, while exchanging HP for LP phenotypes may have no net indirect effect on algae, the fact that each phenotype affects the ecosystem in a different, complex manner can have evolutionary significance. In particular, we asked whether HP guppies affect the ecosystem in a way that facilitates an adaptive advantage for LP guppies. Were this to be the case, then some form of eco-evo feedback would be necessary to explain the evolution of the LP phenotype.

We were able to investigate this possibility with our mathematical analysis of the experimental ecosystems (Bassar et al., 2012). In particular, we asked whether the effects of the phenotypes on algal productivity would alter the selective landscape on herbivory itself. That is, how would selection *for* increased herbivory in LP environments change through the ways that the HP phenotype affects invertebrate abundance and algal productivity. We approximated selection on guppy herbivory by the change in algal biomass, ΔP, resulting from a hypothetical small change in the fish attack rate on algae. A negative change in ΔP indicates a diminishing return on fitness of increasing herbivory and thus a reduction in the force of selection for increased herbivory. Of course, ΔP itself is the sum of direct and indirect effects, ΔP_{dir} and ΔP_{indir} and so we can ask if changes in the selective landscape are driven more by direct or indirect effects.

We found that the feedback loop increased the intensity of selection favouring increased herbivory in LP guppies (Bassar et al., 2012). This increase would accelerate their divergence from the HP feeding pattern. Moreover, this increase was driven by the *indirect* effects of guppies, not their direct effects. When LP herbivory increased, the change in ΔP_{dir} was negative, indicating a 'diminishing return' on fitness of further increasing herbivory. However, the change in ΔP_{indir} was positive and actually larger than the negative change in ΔP_{dir}, which implies that the total effect of increasing LP herbivory was to 'increase fitness returns' and thus reinforce selection for herbivory. Thus, the indirect effects of each phenotype on algal productivity act to favour divergence in diet and would close the 'eco-evo loop' to be at least partially responsible for the evolution of some of the traits that characterize the LP phenotype.

4.2.2 The fitness advantage of HP 'superguppies' evaporates at high densities (Bassar et al., 2013)

We applied integral projection matrices to the performance of the HP and LP phenotypes at low and high densities in these experiments to estimate the population growth rates of each phenotype under low- and high-density conditions. HP guppies have higher population growth rates than LP guppies when the two were compared at low population densities. These differences disappeared at high population densities (Fig. 1.4), suggesting that density-dependent selection through intra-specific competition may level the evolutionary playing field between LP and HP phenotypes. However, LP guppies never showed higher population growth rates than HP guppies at the densities we employed. Density-dependent selection may thus be necessary to explain the evolution of the LP phenotype, but it is not sufficient.

We used an additional modelling effort to test these ideas more fully. We drew on the data from these experiments, plus additional data from our manipulative studies of natural populations, to parameterize a series of invasion analyses that asked how density-dependent vital rates affected the ability of the LP phenotype to invade a population of HP phenotypes. We found

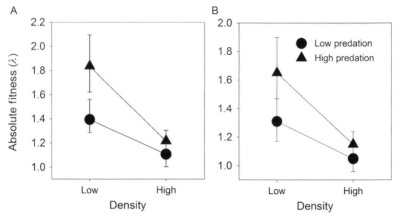

Figure 1.4 Growth rates of LP and HP guppies at low and high densities in mesocosms. Panel (A) are the results from LP and HP populations from the Guanapo river and panel (B) are the results from LP and HP populations from the Aripo river. Error bars are 95% confidence intervals. *Redrawn from data in Bassar et al. (2013).*

that, in the absence of any density-dependence, there would be no invasion of HP by LP. If we included the density-dependent effects and made the survival of HP guppies more sensitive to density than the survival of LP guppies, then LP phenotypes were more likely to invade a population of HP phenotypes. However, we could never find parameter values in which LP would supplant HP, which suggests, in agreement with our experiments, that density-dependent selection may be necessary but is not sufficient to drive the complete replacement of HP by LP.

4.2.3 The interactions between guppies and Rivulus can help drive the evolution of the LP phenotype (Palkovacs et al., 2009)

If the effects of guppies on their ecosystem play a pivotal role in driving the evolution of the LP phenotype, we should see changes in ecosystem structure caused by the invasion of an RO environment by a HP guppy. We should see further changes as the LP phenotype evolves and the resident *Rivulus* adapts to life with guppies. We examined this possibility with an experiment that simulated the temporal stages of the invasion of an RO environment by HP guppies. One treatment represents the pre-invasion phase: no guppies and *Rivulus* from an RO environment. The second treatment represents the early phase of the invasion; we paired guppies from an HP environment with *Rivulus* from an RO environment. The third treatment represents local adaptation, in which we paired guppies from an LP environment with *Rivulus* from an RO environment. The fourth treatment represents *Rivulus* adaptation to the guppy invasion, in which we paired *Rivulus* from LP locations with LP guppies. The contrast between treatments three and four examines how *Rivulus* evolution might alter the ecosystem.

The results not only confirmed the distinctive effects of HP and LP phenotypes on algal and invertebrate standing crops but also demonstrated an additional, striking evolutionary effect on the ecosystem. The abundance of invertebrates in the combination of LP guppies and *Rivulus* from an LP location was half of the abundance in the combination of LP guppies and *Rivulus* from an RO location. This result indicates that RO and LP *Rivulus* have different diets. It may be that this dietary shift either enhances the ability of guppies and *Rivulus* to co-exist or reflects the outcome of selection imposed by guppies on *Rivulus* feeding. Whether this shift is also part of an eco-evo feedback onto guppies that affects their selective milieu and perhaps even facilitates the refinement of the LP phenotype's evolution remains to be determined.

We have repeated this experimental design twice, each time with fish from a different river (Ronald Bassar et al., unpublished data). Our IPM analysis of the results shows that LP guppies have higher population growth rates, and presumably higher fitness, than HP guppies when both are exposed to the combination of high population densities and *Rivulus* from an RO locality. The fitness differences between HP and LP guppies are higher still when LP guppies are kept with *Rivulus* from an LP locality. This latter result suggests the possibility of some type of ecological divergence between LP guppies and LP *Rivulus* and supports the hypothesis that the interaction between guppies and *Rivulus* is one of the factors that propels the evolution of the LP phenotype.

A clear consequence of these results is that if we are to understand this system, we have to understand whether and how *Rivulus* may have evolved in the presence of guppies and, in particular, what their resource base is.

4.3. Interactions between guppies and *Rivulus*

Since density alone cannot account for the evolution of the LP phenotype, we turned to an additional important feature of the LP environment, which is the co-occurrence of guppies with *Rivulus*. Guppies have been the perennial victims in Trinidadian streams. The story to date has been all about how they accommodate the assault of different predators, including *Rivulus*, but guppies can also be the aggressor as they prey on newborn *Rivulus* and compete with *Rivulus* juveniles (Fraser and Lamphere, 2013). Guppies also shape the evolution of *Rivulus* life histories (Walsh and Reznick, 2008, 2009, 2010, 2011; Walsh et al., 2011). The remarkable aspect of this work is that the way guppies appear to shape the evolution of *Rivulus* is not a direct consequence of intraguild predation by guppies on newborn *Rivulus*, but an indirect consequence of their impact on resource availability and, through that, on *Rivulus* population density and dynamics (Walsh and Reznick, 2010, 2011).

We often find headwater streams in which barriers exclude all species of fish save *Rivulus*. Below such barriers lie LP environments that contain guppies and *Rivulus* as the only fish species. These settings offer us the opportunity to examine evolutionary interactions between two strongly interacting species via comparative and experimental studies of *Rivulus* from RO and LP environments.

Wild-caught *Rivulus* that co-occur with guppies are smaller at maturity and produce smaller eggs than their counterparts from RO localities. They also invest more in reproduction (Furness et al., 2012; Walsh and Reznick, 2009). If *Rivulus* life histories were shaped by guppy predation on juvenile *Rivulus* alone, then the early life history theory that models evolution without density regulation predicts that *Rivulus* should evolve delayed maturity and reduced reproductive allocation (Charlesworth, 1994; Gadgil and Bossert, 1970; Law, 1979; Michod, 1979). Since we see the opposite result—earlier maturity and increased reproductive investment, as opposed to delayed maturity and reduced reproductive investment—there must be some other explanation for how guppies shape *Rivulus* life histories.

Walsh and Reznick performed a common garden experiment on the grandchildren of wild-caught *Rivulus* from paired RO and LP localities in two different river systems and found that all of these differences in life histories persist, suggesting that they are genetic differences (Walsh and Reznick, 2010). They then added experimental evolution to the study by evaluating the life histories of *Rivulus* from localities where guppies had been introduced approximately 25–30 years earlier (Walsh and Reznick, 2011) and compared *Rivulus* from the sites where we had introduced guppies to study guppy evolution (Endler, 1980; Reznick et al., 1990, 1997) with those from upstream, above barriers that excluded the introduced guppies. They found the same differences in life histories as seen in natural LP–RO comparisons, thus showing that the *Rivulus* had evolved in response to the guppy introduction.

Walsh et al (2011) then considered aspects of the comparative population biology of *Rivulus* that lived with and without guppies to seek clues for why their life histories had evolved in an unexpected fashion. They found that the abundance of *Rivulus* in RO sites was twice as high in the LP sites immediately downstream from the barrier waterfall that excluded guppies. They also found that the *Rivulus* from LP sites had growth rates more than three times faster than those from RO sites via translocation experiments in which growth rates of the transplanted *Rivulus* accelerated to match the high growth rates of the resident *Rivulus* within the first month after transplantation. Plausible explanations for all of these patterns are that guppies reduce *Rivulus* abundance, either via predation on newborn or competition with juveniles (Fraser and Lamphere, 2013). An indirect consequence of their reduced density is that per capita food availability is higher, causing the higher individual growth rates. This effect (and perhaps the availability of juvenile guppies as a food source) may help explain why *Rivulus* life histories evolved as they did.

These common garden experiments provided additional information about how indirect effects may have shaped *Rivulus* evolution. In 'high food' treatments growth rates were higher and comparable to those seen in LP environments, whereas 'low food' treatments generated lower growth rates, comparable to those seen in RO environments. A compelling feature of the results is that there were significant interactions between populations and food availability in both experiments. The *Rivulus* from LP localities had earlier maturity and higher fecundity than those from RO localities only when food was abundant. These differences were either disappeared or reversed when they were compared at low food levels. Estimates of population growth rates derived from those differences in life history suggest that the LP *Rivulus* would have higher population growth rates, and hence higher fitness, when food is abundant—but that this advantage over RO *Rivulus* would be reversed at low food. Such interactions in the performance of two populations in one another's respective environments are a signature of local adaptation (Schluter, 2000). This suggests that *Rivulus* has adapted to the higher per capita food availability that was an indirect consequence of their interaction with guppies. This reinforces the conclusion from our mesocosm experiments comparing HP and LP guppies that indirect effects in trophic webs can be important sources of selection pressures (Walsh, 2012).

4.4. Focal streams: Prospective studies of evolution
4.4.1 Experimental introductions of guppy populations
Experiments in artificial streams characterize the ecological consequences of the end point of guppies adapting to life with or without predators, and also

of *Rivulus* adapting to life with or without guppies. Feedback between ecology and evolution is inferred from these results; for example, the evolution of the LP phenotype is suggestive of adaptation to high population densities. We also infer that the more general diet of LP guppies is an adaptation to their depletion of food availability.

A virtue of our study system is that it is also possible to study eco-evo feedbacks as a dynamic process by transplanting guppies from a HP environment into a previously guppy-free portion of stream that contains only *Rivulus*. We can then quantify the joint dynamics of guppy evolution and the changes that guppies impose on the environment. The time-course of guppy evolution is a general way of making inferences about eco-evo feedbacks. If the release from predation were the only reason for the evolution of the LP phenotype, then selection would be the most intense when the guppies were first introduced, because that is when they are furthest from the optimal phenotype. The intensity of selection and rate of evolution, as measured by the change in average phenotype per generation, should decline monotonically even as population densities increase (Fig. 1.5). If guppies are instead adapting to their own impacts on the ecosystem, the intensity of selection and rate of evolution would initially be small when guppy population densities are low because the guppies will have had only a small effect on the ecosystem. But as guppies increase in population density, their effects on the ecosystem should multiply and the eco-evo feedback loop should commence, with the result being an increase in the intensity of selection and rate of evolution (Fig. 1.5). As the system approaches a new ecological and evolutionary steady state at higher population densities (which we know is the case for LP populations), the intensity of selection and rate of evolution will decline.

We transplanted guppies from a single HP locality into four previously guppy-free tributaries to create an ecological and evolutionary disequilibrium that would select for the LP phenotype. The added dimensions to this work include (a) monitoring ecosystem responses to guppy introduction and guppy evolution, (b) experimental manipulation of primary productivity independently of predation risk by thinning canopy in two streams with two adjacent streams as controls and (c) monitoring the guppy populations with high resolution, monthly mark–recapture to enable us to reconstruct the time-course of selection and evolution. In monitoring the ecosystem, we periodically estimated standing crops of algae, invertebrates and benthic organic matter plus primary productivity and rates of algal growth (methods described in Kohler et al., 2012). The intention of canopy thinning was to

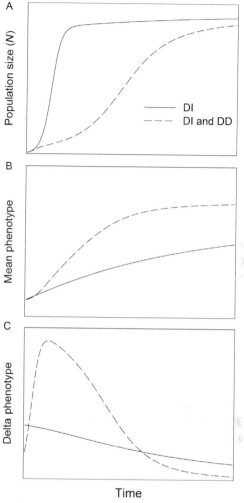

Time

Figure 1.5 Conceptual diagram illustrating dynamics of population size and phenotypic values. Two scenarios are envisioned. The first (DI) is the case where there are no eco-evo feedbacks and evolution of the LP phenotype is driven entirely by density-independent selection generated by the release from predation. In this case, fitness is variation in the intrinsic rate of growth among individuals in the population. The second case (DI and DD) reflects the action of density-dependent selection, in which the evolution of the LP phenotype is propelled by the release from predation plus the selection pressures generated at higher population densities. The joint trajectory of phenotype and population density should be different in the two cases and the joint trajectory of the change in mean phenotype and population density will be very different. Panels show the temporal course of population size (A), mean phenotype (B), and the change in mean phenotype (C).

generate an independent assessment of the importance of productivity and resource availability as factors that can shape adaptation by guppies and eco-evo feedbacks. We expected streams with thinned canopies to receive more light and have higher rates of primary productivity because we expected that light was limiting (see below and Kohler et al., 2012 for verification). If guppy evolution were to be driven by their high population densities and their depletion of food resources, then the strength of this feedback should depend on the extent of resource depletion. Were this to be the case, then the intensity of this feedback should be greater and the rate of evolution of guppies should be higher under an intact canopy.

Each stream has two types of control to provide a frame of reference for quantifying change. First, all streams have barriers that define the up- and downstream borders of the study site and hence the portion of stream occupied by the experimental population of guppies. These regions vary from 65 to 165 m in length. Above each upstream barrier is a section of stream that guppies cannot invade. This upstream region serves as a contemporary control for each experimental reach. We can analyse contemporary differences between control and experimental reaches to assess guppy impacts. Second, we initiated the monitoring of the ecosystem and a mark–recapture study of the *Rivulus* populations in the control and experimental reaches 1 year before the introduction. Having before and after data enables us to analyse the time series of changes in each site. We can evaluate time × stream reach interactions as a second measure of the impact of guppies on the ecosystem.

We initiated one set of introductions in 2008 (Upper and Lower Lalaja) and another in 2009 (Taylor and Caigual). Each set of introductions consisted of introductions into one stream with an intact canopy (Lower Lalaja or Caigual) and one with a thinned canopy (Upper Lalaja or Taylor). The introduced guppies were collected as juveniles from the single HP locality, then reared to adulthood in single sex groups. After they attained maturity, we mated them in groups of five males and five females. After 3 weeks, these fish were individually marked and photographed. We also removed three scales from each individual to provide a source of DNA. The males from a given breeding group were introduced into one of the streams and the females into the other. This means that females entered the stream carrying the sperm from one group of males but were introduced with a different group of males. Doing so increased the effective population size of the introduced populations and creates a broad overlap in their genetic composition.

We began monitoring these populations immediately after the introduction and have been doing so monthly ever since. At each census, marked fish are identified, photographed and weighed. All new recruits are given an individual mark, photographed, weighed and scales were collected to provide DNA. Fish are then returned to their site of capture. Our average probability of seeing a fish on any given census if it is alive is greater than 0.80. Fish are scored at 10 highly variable microsatellite loci to enable reconstruction of the pedigree and quantify each individual's lifetime reproductive success. For every individual that survives to reproduce, we have a greater than 0.98 probability of capturing, photographing and obtaining DNA from that fish at least once in its life (Lopez-Sepulcre et al., 2013).

Each year, beginning with the start of the introduction, we assess the life histories of the four experimental populations and the HP control site in a laboratory common garden. We collect juveniles from all five sites then rear them to a second generation in a common laboratory environment. The second-generation offspring are reared one per aquarium on controlled food availability, as in prior studies (Reznick, 1982). In this setting, we quantify age and size at maturity in males and females and many other aspects of female reproduction (offspring number, offspring size, frequency of reproduction). Differences that persist among populations after two generations in a common environment are interpreted as genetic differences.

4.4.2 Ecological consequences of canopy manipulations

Our first concern was establishing whether or not canopy thinning had a significant effect on the ecosystem and, if so, how large it was. In the first pair of streams (Upper and Lower Lalaja), canopy thinning increased light availability by approximately 55% and in the second pair (Taylor and Caigual), by 180% (Kohler et al., 2012). The primary productivity was higher by an order of magnitude (unpublished result) and the abundance of alpha chlorophyll was higher (Kohler et al., 2012). The thinning of the canopy increased the total abundance of all invertebrates in the pair of streams that saw the 180% increase in light availability (Taylor and Caigual) but did not have this effect in the pair that saw the more moderate increase in light availability (Upper and Lower Lalaja; Fig. 1.6; Thomas Heatherly et al., unpublished data).

The introduced guppy populations had a strongly seasonal cycle of abundance (Fig. 1.7). Each year, the populations increased during the dry seasons (February–May, September–October) and decline in the intervening wet seasons (June–September, October–January). The peak population sizes increased each year for the first 3 years. Some populations have shown

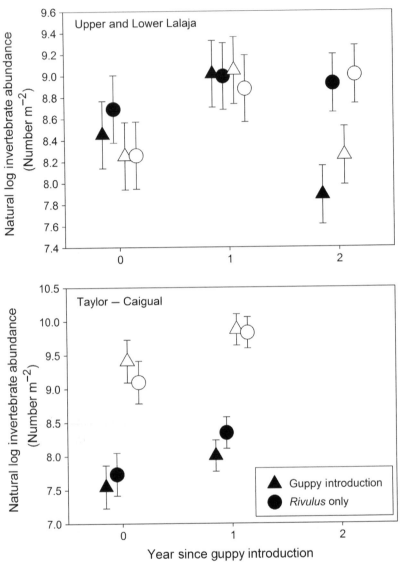

Figure 1.6 Count of macro-invertebrates in the four guppy introduction streams just prior to the introduction of guppies (years since = 0) and the first 2 years after in the Upper and Lower Lalaja system (top panel) and the first year after in the Taylor–Caigual System (lower panel). Each stream pair has one stream that had the canopy thinned (open symbols) and one stream with the canopy intact (closed symbols). Within each stream an upstream reach contains no guppies (RO reaches) and a downstream reach where guppies were introduced (guppy introduction). Data are from the pools of the streams in the dry season. *Figure drawn from unpublished data courtesy of Thomas Heatherly.*

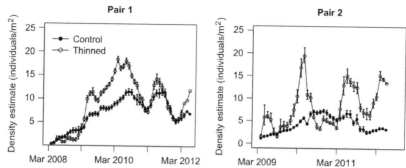

Figure 1.7 Estimated guppy densities in the four introduction streams. Points represent maximum likelihood estimates of density derived from mark–recapture statistics; bars represent standard errors. Data shown for the Upper and Lower Lalaja system (pair 1) and the Taylor–Caigual system (pair 2). Points in open symbols denote data from streams over which the canopy was thinned (Upper Lalaja or Taylor) and points in closed symbols denote data from control streams without any canopy thinning (Lower Lalaja and Caigual).

persistent declines from their third year peak, sometimes associated with identifiable habitat change. One such cause was the sediment outflow from a treefall that filled in a deep, well-populated pool, causing a decline in density.

4.4.3 The impact of guppies on Rivulus

Fraser and Lamphere (2013) compared the population densities and size structures of the *Rivulus* populations in the Taylor–Caigual pair of focal streams for the first 15 months after the introduction of guppies and showed that *Rivulus* were rarer but larger in the guppy introduction site relative to the control reaches. This pattern reflected the reduced recruitment rate of young of the year after guppies had been introduced. Fraser and Lamphere complemented these observations with experiments that showed that the most likely cause of this shift in size distribution is guppy predation on newborn *Rivulus*, rather than competition between adult guppies and juvenile *Rivulus*.

4.4.4 The impact of guppies on Invertebrates

The guppy introductions caused a decrease in the abundance and biomass of predatory and gathering invertebrates 2 years post-introduction compared with the upstream controls in the Upper and Lower Lalaja system (Fig. 1.6; unpublished data of Thomas Heatherly et al.). This 2-year delay between the guppy introduction and a measurable effect of guppies on invertebrate abundance might have occurred because the peak abundances of the guppies during the longer dry season (February–May) was progressively higher each year for the first 3 years. The higher population densities of guppies may be required for there to be a measurable impact of guppies on invertebrate abundance. Further work will test this hypothesis.

4.4.5 Do guppies change the structure of natural ecosystems?

There is emerging evidence that guppies have a lasting effect on the abundance and size structure of *Rivulus* populations and cause *Rivulus* life histories to evolve. There are hints that ecological character divergence evolves between the two species. We also know that guppies reduce invertebrate abundance both in our artificial stream experiments (Bassar et al., 2010) and after they are introduced into natural streams (Fig. 1.6; unpublished data of Thomas Heatherly et al.). We will continue to quantify their impact and will be able to evaluate it in two ways, first via contemporary comparisons between the control and the introduction reaches in each of the four streams and second via the time series of changes in the guppy introduction sites.

4.4.6 Guppy evolution

We will assess the trajectory of guppy evolution by integrating the laboratory common garden assessment of genetic differences between the introduced population of guppies and the ancestral population with the ongoing mark–recapture study of the introduced populations and pedigree reconstruction. The mark–recapture study provides the trends in morphological change, and the pedigrees from our genetic samples provide estimates of character heritability in the field.

The dramatic seasonal cycles in guppy abundance and recruitment, perhaps driven by cycles in resource abundance, present an interesting challenge to the assessment of how selection on guppies might change over time. Guppy generation time is sufficiently short for there to be multiple generations within a year, or even within a long dry season. It is conceivable that cycles in resource abundance cause selection for very different attributes over the course of a year. Such variation could cause an irregular path for the evolutionary change from a HP ancestor to an LP descendant rather than a regular, progressive one; while the path may show the net change from one to another, it may also display short periods of stasis or even periods in which the direction of evolution is reversed. Our idealized depiction of alternative scenarios of guppy evolution without and with eco-evo feedbacks (Fig. 1.5) may dramatically understate the potential complexity of this process.

4.4.7 Future work on guppies and their ecosystem

The data in hand from the focal streams, along with the data we are continuing to gather, will ultimately allow us to test additional hypotheses about the interplay of evolution and ecology. For example, our canopy treatments offer the opportunity to see how ecological context affects the importance of eco-evo feedbacks. If, as results to date indicate, density-dependent selection and effects of guppies on resources are critical for generating eco-evo feedbacks, then we expect those feedbacks to be less important in the open canopy streams where the productivity of the habitat is higher.

The canopy treatment will also allow us to examine the role of seasonality on both population dynamics and the evolutionary process. While wet and dry seasons affect both types of streams, results to date indicate that the seasonal fluctuations in guppy populations have much greater amplitude in the open canopy streams. We suspect that primary and secondary productivity will also show greater amplitude in the open canopy streams because of their more rapid increases in the dry seasons.

With sufficient long-term observation, we can also ask if the evolution of the LP phenotype in turn affects population dynamics, which would bring the eco-evo feedback loop around another half-turn. We can use IPM models based on the earliest mark–recapture data to predict what longer term population dynamics should show in the absence of any subsequent evolution. Comparing these predictions to the actual patterns, and to patterns predicted by IPM models of the last stages of guppy evolution, can tell us if adaptive evolution might stabilize population dynamics (Mueller et al., 2000; Stokes et al., 1988).

Finally, we will develop our approach to ask more general questions about how, why and when evolution occurs in nature via statistical analyses of the time-course of selection and evolution using well-established methods currently applied to single, unmanipulated populations. The large majority of such long-term mark–recapture studies reveal evidence for substantial directional selection on heritable traits but little evidence of the predicted directional evolution. The goal of such analyses has thus mostly been to explain why evolution has not occurred when we otherwise should expect it to. Such results have been reported by Charmantier et al. (2008) for egg-laying date in the great tit population in Wytham Woods, Kruuk et al. (2002) for antler size in the red deer population at Rhum, Merila et al. (1997)

for body size in the collard flycatcher population on the island of Gotland and Ozgul et al. (2009) for body weight in the sheep on St. Kilda Island, to cite a few examples.

Our study differs from these because it is a replicated experiment in a system in which we expect to see directional evolution. More to the point, we have manipulated a factor (canopy cover) that should influence the rate of directional evolution in the near term. We can use these well-established methods in the same way as have others, but in circumstances under which we expect the rate of evolution to be quite different. Our approach thus enables us to, for the first time, detail the dynamics of evolution and characterize the relative importance of factors that have nullified or masked evolution in prior studies, such as confounding environmental factors (Garant et al., 2004; Ozgul et al., 2009), phenotypic plasticity (Charmantier et al., 2008), genetic correlations (Morrissey et al., 2012; Sheldon et al., 2003) or sexual conflict (Foerster et al., 2007). With this approach, we should be able to find different factors propelling different rates of evolution under either open or closed canopies and ascertain whether they have their origins in eco-evo feedbacks.

5. CONCLUSIONS

We began by defining eco-evo feedbacks as occurring when an organism modifies some feature of its environment and, by doing so, changes the nature of selection it experiences (Ferriere et al., 2004). We then described how a research programme could be developed to investigate whether such feedbacks have been important in moulding ecosystem properties and trait distributions in natural populations. Our work with guppies illustrates such a research programme. It identifies effects of guppies on various facets of their ecosystem and, through those effects, focuses attention on three possible sources of feedback as major drivers of the evolution of guppy life history. Those potential sources are the direct and indirect effects of guppies on their ecosystems, the role of density-dependent selection through intra-specific competition and density- and frequency-dependent interactions with R. hartii. We have evidence indicating that each of these sources of feedback contributes to the evolution of the LP phenotype and our goal in the years ahead is to estimate their relative importance.

Our introduction experiment will ultimately show us the joint trajectories of guppy evolution and ecosystem change but those trajectories alone cannot demonstrate causation. There is an important role for additional experiments to describe the relative importance of each source of eco–evo feedback in this system, along identifying the ecological mechanisms that create those feedbacks. There is a considerable need for new mathematical modelling to assess the likely outcomes of whichever mechanisms appear most pervasive. More specifically, we expect that only through the combined results of experiments, models and the monitoring of our introduction experiment we will be able to really understand the role of eco–evo feedbacks in guppy evolution.

Our studies of guppies offer a unique template for studying the genetics of adaptation. A full understanding of the genetic control of adaptive evolution in nature must be built on all three parts: genetics, the nature of adaptation and natural history context of that adaptation. Our observations, experiments and modelling are building the knowledge of 'adaptation' and 'nature'. We know that our introduced fish will eventually leave descendants of the LP phenotype and we will know, ultimately, the why and how. But we also have tissue samples from all guppies we have studied in our introduction experiments so the DNA describing all phases of the evolutionary trajectory will be available. These samples can offer a unique and invaluable look at how nature moulds adaptations from start to finish and along which genetic paths it travels along the way. It would be exciting if we could match key moments in the ecological trajectories of population density and ecosystem change with key shifts in DNA sequences. Such a match would be the ultimate verification of eco–evo feedbacks as a prevailing force in adaptive evolution.

Finally, we can speculate where else our approach might be applied and how general our results might be in other ecosystems. The first requirement for the empirical study of eco–evo feedbacks is identifying a species that exerts strong and readily quantifiable effects on the ecosystem in which it is found. The ecological literature suggests dozens of such systems: every time we see the characterization of a keystone species or the induction of trophic cascades we encounter a candidate species for the study of eco–evo interactions (e.g. Woodward et al., 2008). While the ecology of such strong interactions have been studied for decades, there has been virtually no consideration of their evolutionary consequences. For example, Estes et al. (2011) illustrate the dramatic consequences of trophic downgrading, or the elimination of apex predators. Likewise, Paine (1966, 1974) characterized the consequences of the natural presence or absence of a keystone predator. What we need to imagine in such circumstances is the consequences of the resulting ecosystem changes for the nature of selection on the species instigating these changes and for the selective milieu of the rest of the species in the system. We could ask, for instance, if the elk of Yellowstone Park with or without wolves or the sea urchins on the seafloor around Amchitka island before or after the extirpation of otters (Estes et al., 2011) have anything in common with LP guppies in the way they deplete their environment of resource, restructure their ecosystems and, perhaps, impose selection on themselves, and cause themselves to adapt to what they have done to their ecosystem.

REFERENCES

Abrams, P.A., Matsuda, H., 1997. Prey adaptation as a cause of predator–prey cycles. Evolution 51, 1742–1750.
Alexander, H.J., Taylor, J.S., Wu, S., Breden, F., 2006. Parallel evolution and vicariance in the guppy (*Poecilia reticulata*) over multiple spatial and temporal scales. Evolution 60, 191–208.
Barson, N.J., Cable, J., Oosterhout, C.V., 2009. Population genetic analysis of microsatellite variation of guppies (*Poecilia reticulata*) in Trinidad and Tobago: evidence for a dynamic source-sink metapopulation structure, founder events, and population bottlenecks. J. Evol. Biol. 22, 485–497.
Bashey, F., 2008. Competition as a selective mechanism for larger offspring size in guppies. Oikos 117, 104–113.
Bassar, R.D., Marshall, M.C., Lopez-Sepulcre, A., Zandonà, E., Auer, S.K., Travis, J., Pringle, C.M., Flecker, A.S., Thomas, S.A., Fraser, D.F., Reznick, D.N., 2010. Local adaptation in Trinidadian guppies alters ecosystem processes. Proc. Natl. Acad. Sci. U.S.A. 107, 3616–3621.
Bassar, R.D., Ferriere, R., Lopez-Sepulcre, A., Marshall, M.C., Travis, J., Pringle, C.M., Reznick, D.N., 2012. Direct and indirect ecosystem effects of evolutionary adaptation in the Trinidadian guppy (*Poecilia reticulata*). Am. Nat. 180, 167–185.
Bassar, R.D., López-Sepulcre, A., Reznick, D.N., Travis, J., 2013. Experimental evidence for density-dependent regulation and selection on Trinidadian guppy life histories. Am. Nat. 181, 25–38.
Cameron, T.C., Plaistow, S., Mugabo, M., Piertney, S.B., Benton, T.G., 2014. Eco-evolutionary dynamics: experiments in a model system. In: Moya-Laraño, J., Rowntree, J., Woodward, G. (Eds.), Advances in Ecological Research, Vol. 50: Eco-evolutionary Dynamics. Elsevier, Amsterdam, pp. 171–206.
Charlesworth, B., 1971. Selection in density-regulated populations. Ecology 52, 469–474.
Charlesworth, B., 1994. Evolution in Age-Structured Populations. Cambridge University Press, Cambridge.
Charmantier, A., McCleery, R.H., Cole, L.R., Perrins, C., Kruuk, L.E.B., Sheldon, B.C., 2008. Adaptive phenotypic plasticity in response to climate change in a wild bird population. Science 320, 800–803.
Clarke, B., 1972. Density-dependent selection. Am. Nat. 106, 1–13.
Cockerham, C.C., Burrows, P.M., Young, S.S., Prout, T., 1972. Frequency-dependent selection in randomly mating populations. Am. Nat. 106, 493–515.
Connelly, S., Pringle, C.M., Bixby, R.J., Brenes, R., Whiles, M.R., Lips, K.R., Kilham, S., Huryn, A.D., 2008. Changes in stream primary producer communities resulting from large-scale catastrophic amphibian declines: can small-scale experiments predict effects of tadpole loss? Ecosystems 11, 1262–1276.

Coulson, T., 2012. Integral projections models, their construction and use in posing hypotheses in ecology. Oikos 121, 1337–1350.

Coulson, T., Tuljapurkar, S., Childs, D.Z., 2010. Using evolutionary demography to link life history theory, quantitative genetics and population ecology. J. Anim. Ecol. 79, 1226–1240.

de Villemereuil, P.B., Lopez-Sepulcre, A., 2010. Consumer functional responses under intra- and inter-specific interference competition. Ecol. Model. 222, 419–426.

Duffy, M.A., Sivars-Becker, L., 2007. Rapid evolution and ecological host-parasite dynamics. Ecol. Lett. 10, 44–53.

Duffy, M.A., Hall, S.R., Caceres, C.E., Ives, A.R., 2009. Rapid evolution, seasonality, and the termination of parasite epidemics. Ecology 90, 1441–1448.

Dunson, W., Travis, J., 1991. The role of abiotic factors in community organization. Am. Nat. 138, 1067–1091.

Easterling, M.R., Ellner, S.P., Dixon, P.M., 2000. Size-specific sensitivity: applying a new structured population model. Ecology 81, 694–708.

Ellner, S.P., 2013. Rapid evolution: from genes to communities, and back again? Funct. Ecol. 27, 1087–1099.

Endler, J.A., 1978. A predator's view of animal color patterns. Evol. Biol. 11, 319–364.

Endler, J.A., 1980. Natural selection on color patterns in *Poecilia reticulata*. Evolution 34, 76–91.

Estes, J.A., Terborgh, J., Brashares, J.S., Power, M.E., Berger, J., Bond, W.J., Carpenter, S.R., et al., 2011. Trophic downgrading of planet Earth. Science 333, 301–306.

Ferriere, R., Dieckmann, U., Couvet, D., 2004. Evolutionary Conservation Biology. Cambridge University Press, Cambridge.

Foerster, K., Coulson, T., Sheldon, B.C., Pemberton, J.M., Clutton-Brock, T.H., Kruuk, L.E.B., 2007. Sexually antagonistic genetic variation for fitness in red deer. Nature 447, 1107–1109.

Fowler, N.L., Pease, C.M., 2010. Temporal variation in the carrying capacity of a perennial grass population. Am. Nat. 175, 504–512.

Fowler, N.L., Overath, R.D., Pease, C.M., 2006. Detection of density dependence requires density manipulations and calculation of lambda. Ecology 87, 655–664.

Fraser, D.F., Lamphere, B.A., 2013. Experimental evaluation of predation as a facilitator of invasion success in a stream fish. Ecology 94, 640–649.

Furness, A.I., Walsh, M.R., Reznick, D.N., 2012. Convergence of life-history phenotypes in a Trinidadian killifish (*Rivulus hartii*). Evolution 66, 1240–1254.

Gadgil, M., Bossert, W.H., 1970. Life historical consequences of natural selection. Am. Nat. 104, 1–24.

Garant, D., Kruuk, L.E.B., McCleery, R.H., Sheldon, B.C., 2004. Evolution in a changing environment: a case study with great tit fledging mass. Am. Nat. 164, E115–E129.

Ghalambor, C.K., Reznick, D.N., Walker, J.A., 2004. Constraints on adaptive evolution: the functional trade-off between reproduction and fast-start swimming performance in the Trinidadian guppy (*Poecilia reticulata*). Am. Nat. 164, 38–50.

Haskins, C.P., Haskins, E.F., McLaughlin, J.J., Hewitt, R.E., 1961. Polymorphism and population structure in *Lebistes reticulata*, a population study. In: Blair, W.F. (Ed.), Vertebrate Speciation. University of Texas Press, Austin, TX, pp. 320–395.

Hiltunen, T., Ellner, S.P., Hooker, G., Jones, L.E., Hairston, N.G., 2014. Eco-evolutionary dynamics in a three-species food web with intraguild predation: intriguingly complex. In: Moya-Laraño, J., Rowntree, J., Woodward, G. (Eds.), Advances in Ecological Research, Vol. 50: Eco-evolutionary Dynamics. Elsevier, Amsterdam, pp. 41–74.

Houde, A.E., 1997. Sex, Color and Mate Choice in Guppies. Princeton University Press, Princeton, NY.

Jayakar, S.D., 1970. A mathematical model for the interaction of gene frequencies in a parasite and its host. Theor. Popul. Biol. 1, 140–164.

Jorgensen, C., Auer, S.K., Reznick, D.N., 2011. A model for optimal offspring size in fish, including live-bearing and parental effects. Am. Nat. 177, E119–E135.

Kirkwood, T.B.L., 1993. The disposable soma theory: evidence and implications. Neth. J. Zool. 43, 359–363.

Kohler, T.J., Heatherly, T.I., El-Sabaawi, R., Zandonà, E., Marshall, M.C., Flecker, A.S., Pringle, C.M., Reznick, D.N., Thomas, S.A., 2012. Flow, nutrients, and light availability influence Neotropical epilithon biomass and stoichiometry. Freshw. Sci. 31, 1019–1034.

Kokko, H., López-Sepulcre, A., 2007. The ecogenetic link between demography and evolution: can we bridge the gap between theory and data? Ecol. Lett. 10, 773–782.

Kot, M., 2001. Elements of Mathematical Ecology. Cambridge University Press, Cambridge.

Kruuk, L.E.B., Slate, J., Pemberton, J.M., Brotherstone, S., Guinness, F., Clutton-Brock, T.H., 2002. Antler size in red deer: heritability and selection but no evolution. Evolution 56, 1683–1695.

Langerhans, R.B., DeWitt, T.J., 2004. Shared and unique features of evolutionary diversification. Am. Nat. 164, 335–349.

Lauridsen, R.B., Edwards, F.K., Cross, W.F., Woodward, G., Hildrew, A.G., Jones, J.I., 2014. Consequences of inferring diet from feeding guilds when estimating and interpreting consumer-resource stoichiometry. Freshw. Biol.. http://dx.doi.org/10.1111/fwb.12361.

Law, R., 1979. Optimal life histories under age-specific predation. Am. Nat. 114, 399–417.

Leon, J.A., 1974. Selection in contexts of interspecific competition. Am. Nat. 108, 739–757.

Levin, S.A., Udovic, J.D., 1977. A mathematical model of coevolving populations. Am. Nat. 111, 657–675.

Lopez-Sepulcre, A., Gordon, S.P., Patterson, I.G., Bentzen, P., Reznick, D.N., 2013. Beyond lifetime reproductive success: the posthumous reproductive dynamics of male Trinidadian guppies. Proc. Biol. Sci. 280, 20131116.

MacArthur, R.H., 1962. Some generalized theorems of natural selection. Proc. Natl. Acad. Sci. U.S.A. 48, 1893–1897.

MacArthur, R.H., Wilson, E.O., 1967. The Theory of Island Biogeography. Princeton University Press, Princeton.

Mattingly, H.T., Butler IV, M.J., 1994. Laboratory predation on the Trinidadian guppy: implications for the size-selective predation hypothesis and guppy life history evolution. Oikos 69, 54–64.

Medawar, P.B., 1952. An Unsolved Problem in Biology. H. K. Lewis & Co., London

Merila, J., Sheldon, B.C., Ellegren, H., 1997. Antagonistic natural selection revealed by molecular sex identification of nestling collared flycatchers. Mol. Ecol. 6, 1167–1175.

Michod, R.E., 1979. Evolution of life histories in response to age-specific mortality factors. Am. Nat. 113, 531–550.

Morrissey, M.B., Walling, C.A., Wilson, A.J., Pemberton, J.M., Clutton-Brock, T.H., Kruuk, L.E.B., 2012. Genetic analysis of life history constraint and evolution in a wild ungulate population. Am. Nat. 179, E97–E114.

Moya-Laraño, J., Bilbao-Castro, J.R., Barrionuevo, G., Ruiz-Lupión, D., Casado, L.G., Montserrat, M., Melián, C.J., Magalhães, G., 2014. Eco-evolutionary spatial dynamics: rapid evolution and isolation explain food web persistence. In: Moya-Laraño, J., Rowntree, J., Woodward, G. (Eds.), Advances in Ecological Research, Vol. 50: Eco-evolutionary Dynamics. Elsevier, Amsterdam, pp. 75–144.

Mueller, L.D., Joshi, A., Borash, D.J., 2000. Does population stability evolve? Ecology 81, 1273–1285.

O'Steen, S., Cullum, A.J., Bennett, A.F., 2002. Rapid evolution of escape ability in Trinidadian guppies (Poecilia reticulata). Evolution 56, 776–784.

Otto, S.P., Day, T., 2007. A Biologist's Guide to Mathematical Modeling in Ecology and Evolution. Princeton University Press, Princeton.

Ozgul, A., Tuljapurkar, S., Benton, T.G., Pemberton, J.M., Clutton-Brock, T.H., Coulson, T., 2009. The dynamics of phenotypic change and the shrinking sheep of St. Kilda. Science 325, 464–467.

Paine, R.T., 1966. Food web complexity and species diversity. Am. Nat. 100, 65–75.

Paine, R.T., 1974. Intertidal community structure: experimental studies on the relationship between a dominant competitor and its principal predator. Oecologia 15, 93–120.

Palkovacs, E.P., Marshall, M.C., Lamphere, B.A., Lynch, B.R., Weese, D.J., Fraser, D.F., Reznick, D.N., Pringle, C.M., Kinnison, M.T., 2009. Experimental evaluation of evolution and coevolution as agents of ecosystem change in Trinidadian streams. Philos. Trans. R. Soc. Lond. B Biol. Sci. 364, 1617–1628.

Palkovacs, E.P., Wasserman, B.A., Kinnison, M.T., 2011. Eco-evolutionary trophic dynamics: loss of top predators drives trophic evolution and ecology of prey. PLoS ONE 6, e18879.

Pimentel, D., 1961. Animal population regulation by the genetic feed-back mechanism. Am. Nat. 95, 65–79.

Pimentel, D., 1968. Population regulation and genetic feedback. Science 159, 1432–1437.

Pringle, C.M., Blake, G.A., 1994. Quantitative effects of atyid shrimp (Decapoda: Atyidae) on the depositional environment in a tropical stream: use of electricity for experimental exclusion. Can. J. Fish. Aquat. Sci. 51, 1443–1450.

Pulliam, H.R., 1974. On the theory of optimal diets. Am. Nat. 108, 59–74.

Reznick, D.N., 1982. The impact of predation on life history evolution in Trinidadian guppies: genetic basis of observed life history patterns. Evolution 36, 1236–1250.

Reznick, D., Bryant, M., 2007. Comparative long-term mark-recapture studies of guppies (*Poecilia reticulata*): differences among high and low predation localities in growth and survival. Ann. Zool. Fenn. 44, 152–160.

Reznick, D.N., Bryga, H., 1987. Life-history evolution in guppies (*Poecilia reticulata*). 1. Phenotypic and genetic changes in an introduction experiment. Evolution 41, 1370–1385.

Reznick, D.N., Bryga, H.A., 1996. Life-history evolution in guppies (*Poecilia reticulata*: Poeciliidae). 5. Genetic basis of parallelism in life histories. Am. Nat. 147, 339–359.

Reznick, D., Endler, J.A., 1982. The impact of predation on life history evolution in Trinidadian guppies (*Poecilia reticulata*). Evolution 36, 160–177.

Reznick, D.N., Bryga, H., Endler, J.A., 1990. Experimentally induced life-history evolution in a natural population. Nature 346, 357–359.

Reznick, D.N., Butler, M.J., Rodd, F.H., Ross, P., 1996a. Life-history evolution in guppies (Poecilia reticulata).6. Differential mortality as a mechanism for natural selection. Evolution 50, 1651–1660.

Reznick, D.N., Rodd, F.H., Cardenas, M., 1996b. Life-history evolution in guppies (*Poecilia reticulata*: Poeciliidae). IV. Parallelism in life-history phenotypes. Am. Nat. 147, 319–338.

Reznick, D.N., Shaw, F.H., Rodd, F.H., Shaw, R.G., 1997. Evaluation of the rate of evolution in natural populations of guppies (*Poecilia reticulata*). Science 275, 1934–1937.

Reznick, D., Butler, M.J., Rodd, H., 2001. Life-history evolution in guppies. VII. The comparative ecology of high- and low-predation environments. Am. Nat. 157, 126–140.

Reznick, D.N., Bryant, M.J., Roff, D., Ghalambor, C.K., Ghalambor, D.E., 2004. Effect of extrinsic mortality on the evolution of senescence in guppies. Nature 431, 1095–1099.

Reznick, D.N., Bassar, R.D., Travis, J., Rodd, F.H., 2012. Life history evolution in guppies VIII: the demographics of density regulation in guppies (*Poecilia reticulata*). Evolution 66, 2903–2915.

Rodd, F.H., Reznick, D.N., 1997. Variation in the demography of guppy populations: the importance of predation and life histories. Ecology 78, 405–418.

Roughgarden, J., 1971. Density-dependent natural selection. Ecology 52, 453–468.

Roughgarden, J., 1976. Resource partitioning among competing species – co-evolutionary approach. Theor. Popul. Biol. 9, 388–424.

Schluter, D., 2000. The Ecology of Adaptive Radiation. Oxford University Press, Oxford.

Schoener, T.W., 2011. The newest synthesis: understanding the interplay of evolutionary and ecological dynamics. Science 331, 426–429.

Seghers, B.H., 1973. An analysis of geographic variation in the antipredator adaptations of the guppy, Peocilia reticulata. (Ph.D. thesis). University of British Columbia.

Seghers, B.H., 1974. Schooling behavior in the guppy (*Poecilia reticulata*): an evolutionary response to predation. Evolution 28, 486–489.

Seghers, B.H., Magurran, A.E., 1995. Population differences in the schooling behavior of the Trinidad guppy, *Poecilia reticulata*—adaptation or constraint. Can. J. Zool. 73, 1100–1105.

Sheldon, B.C., Kruuk, L.E.B., Merila, J., 2003. Natural selection and inheritance of breeding time and clutch size in the collared flycatcher. Evolution 57, 406–420.

Smallegange, I., Coulson, T., 2013. Towards a general, population-level understanding of eco-evolutionary change. Trends Ecol. Evol. 28, 1–6.

Smallegange, I.M., Deere, J.A., 2014. Eco-evolutionary interactions as a consequence of selection on a secondary sexual trait. In: Moya-Laraño, J., Rowntree, J., Woodward, G. (Eds.), Advances in Ecological Research, Vol. 50: Eco-evolutionary Dynamics. Elsevier, Amsterdam, pp. 145–170.

Smouse, P., 1976. The implications of density-dependent population growth for frequency- and density-dependent selection. Am. Nat. 110, 849–860.

Stokes, T.K., Gurney, W.S.C., Nisbet, R.M., Blythe, S.P., 1988. Parameter evolution in a laboratory insect population. Theor. Popul. Biol. 34, 248–265.

Travis, J., Lotterhos, K.E., 2013. Using experiments and models to untangle direct and indirect effects in a "simple" food web: is there hope for understanding fishery systems? Bull. Mar. Sci. 89, 317–336.

Wallace, B., 1975. Hard and soft selection revisited. Evolution 29, 465–473.

Walsh, M.R., 2012. The evolutionary consequences of indirect effects. Trends Ecol. Evol. 28, 23–29.

Walsh, M.R., Reznick, D.N., 2008. Interactions between the direct and indirect effects of predators determine life history evolution in a killifish. Proc. Natl. Acad. Sci. U.S.A. 105, 594–599.

Walsh, M.R., Reznick, D.N., 2009. Phenotypic diversification across an environmental gradient: a role for predators and resource availability on the evolution of life histories. Evolution 63, 3201–3213.

Walsh, M.R., Reznick, D.N., 2010. Influence of the indirect effects of guppies on life-history evolution in *Rivulus hartii*. Evolution 64, 1583–1593.

Walsh, M.R., Reznick, D.N., 2011. Experimentally induced life-history evolution in a killifish in response to introduced guppies. Evolution 65, 1021–1036.

Walsh, M.R., Fraser, D.F., Bassar, R.D., Reznick, D.N., 2011. The direct and indirect effects of guppies: implications for life-history evolution in *Rivulus hartii*. Funct. Ecol. 25, 227–237.

Williams, G.C., 1957. Pleiotropy, natural selection and the evolution of senescence. Evolution 11, 398–411.

Woodward, G., Papantoniou, G., Edwards, F., Lauridsen, R.B., 2008. Trophic trickles and cascades in a complex food web: impacts of a keystone predator on stream community structure and ecosystem processes. Oikos 117, 683–692.

Zandonà, E., Auer, S.K., Kilham, S.S., Howard, J.H., López-Sepulcre, A., O'Connor, M.P., Bassar, R.D., Osorio, O., Pringle, C.M., Reznick, D.N., 2011. Diet quality and prey selectivity correlate with life histories and predation regime in Trinidadian guppies. Funct. Ecol. 25, 964–973.

CHAPTER TWO

Eco-Evolutionary Dynamics in a Three-Species Food Web with Intraguild Predation: Intriguingly Complex

Teppo Hiltunen[*,1,2], **Stephen P. Ellner**[*], **Giles Hooker**[†], **Laura E. Jones**[*], **Nelson G. Hairston Jr.**[*,‡]

[*]Department of Ecology and Evolutionary Biology, Cornell University, Ithaca, New York, USA
[†]Department of Biological Statistics and Computational Biology, Cornell University, Ithaca, New York, USA
[‡]Swiss Federal Institute of Aquatic Science and Technology, Eawag, Dübendorf, Switzerland
[1]Present address: Department of Food and Environmental Sciences/Microbiology, University of Helsinki, Helsinki, Finland.
[2]Corresponding author: e-mail address: teppo.hiltunen@helsinki.fi

Contents

Abstract

We explore the role of rapid evolution in the community dynamics of a three-species planktonic food web with intraguild predation. Previous studies of a two-species predator–prey system showed that rapid evolution of an anti-predator defence trait in the prey results in long-period antiphase predator–prey cycles (predator maxima coinciding with prey minima and vice versa) that are virtually diagnostic of eco-evolutionary dynamics. Here, we ask if there exist diagnostic population dynamics for a food web where algae are consumed by two predators (flagellates and rotifers), while rotifers also consume flagellates. With genetically homogeneous non-evolving prey, we previously predicted theoretically, and confirmed experimentally, that population cycles exhibit short-period oscillations with peaks in prey density followed by peaks in flagellates and then rotifers. In contrast, when prey defence can evolve, theory predicts a wide diversity of possible dynamics depending upon the trade-off between defences against the two predators. When defence against one predator implies vulnerability to the other, the predicted pattern is that predators "take turns": one predator peaks at each prey minimum, while the other remains rare because prey are defended against it. There is strong selection for prey to evolve defence against the abundant predator (losing defence against the rare one); once this happens, predator dominance reverses rapidly. This pattern is what we generally observed in seven separate microcosms (sampled daily for 130–330 days). Cycles in which predator abundances alternate between stasis and rapid change may be explained using the concept of canards from dynamical systems theory. Nevertheless, details differed among experimental runs, making patterns diagnostic of eco-evolutionary dynamics difficult to identify.

1. INTRODUCTION

Eco-evolutionary dynamics has emerged as a distinct field of study within ecology, the focus of journal issues, meetings and many conference sessions and symposia. Although there is not complete agreement about the meaning of "eco-evolutionary dynamics", an essential component is a feedback loop between ecological and evolutionary dynamics. The ecology → evolution link represents ecological change resulting in natural selection that produces a change in some ecologically important trait. The evolution → ecology link represents the evolved change in the trait causing a change in the ecological dynamics. The cycle then continues with ecological change further altering the direction or strength of selection, which causes further trait change, and so on. A diagram or verbal depiction of the feedback loop prefaces most talks and many papers about eco-evolutionary dynamics.

In the last few years, both the eco → evo links and the evo → eco links (and in rare cases both) have been documented in a variety of systems including laboratory microcosms (Becks et al., 2010; Hiltunen et al., 2014; Yoshida et al., 2003, 2007); mesocosms and enclosures (Agrawal et al., 2013; Bassar et al., 2010; Harmon et al., 2009; Wymore et al., 2011); and field studies of Darwin's finches (Grant and Grant, 2002), fence lizards (Sinervo et al., 2000), freshwater copepods (Ellner et al., 1999; Hairston et al., 1996), Soay sheep (Pelletier et al., 2007) and the Glanville fritillary butterfly (Hanski, 2011). In some cases, the system remains dynamic because of variability in the external environment (e.g. finches, copepods) or because transient dynamics are reinitiated each year by seasonality (e.g. Gallagher, 1982); in other cases, the variability is the result of internal processes and intra- or inter-specific interactions (fence lizards, Glanville fritillary, predator–prey microcosms). However, only in a few cases has a complete feedback loop between evolutionary and ecological *dynamics* been demonstrated within a single system (Becks et al., 2012). The key feature distinguishing eco-evolutionary dynamics from other "subspecies" of rapid evolution is that the reciprocal feedbacks alter the pattern of temporal variation in the interacting organismal traits and ecological variables.

Understanding how rapid evolution in ecologically important traits can interact with populations, community and ecosystem dynamics requires experimental manipulation of both heritable trait variation and the environment, to see what changes as a function of presence versus absence of one or both of the evo-eco links. One approach to this challenge is to bring the complexity of nature into more controlled environments, such as mesocosms or field enclosures. But to date, published experiments have all measured effects at a single point in time (Bassar et al., 2010; De Meester et al., 2011; Van Doorslaer et al., 2009), rather than the ongoing reciprocal effects on temporal dynamics that are a defining feature of eco-evolutionary dynamics (Fussmann et al., 2007; Post and Palkovacs, 2009; Wymore et al., 2011).

In this chapter, we take a second approach: adding complexity to simple laboratory predator–prey microcosms, to explore how evolution in one of the key players affects temporal dynamics. We have previously shown that cyclical, endogenously driven eco-evolutionary dynamics occur over a broad range of experimental conditions in rotifer–algal chemostat microcosms. When only a single predator and a single prey species are present, theory predicts (Jones and Ellner, 2007) and our empirical studies have confirmed (Becks et al., 2012; Yoshida et al., 2003), the existence of antiphase cycles between predator and prey abundance: prey are most abundant when predators are least abundant, and vice versa, so peaks in predator abundance lag peaks in prey abundance by half a cycle period (e.g. Fig. 2.1A). These long-period, antiphase cycles are diagnostic for the presence of eco-evolutionary cycling in our microcosms (e.g. Becks et al., 2012; Fussmann et al., 2000; Jones and Ellner, 2007; Yoshida et al., 2003) and in other similar systems (Hiltunen et al., 2014). Classic predator–prey models

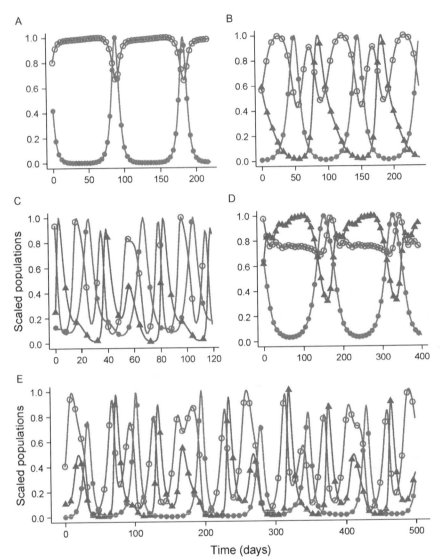

Figure 2.1 Simulations of model (1) showing various kinds of population dynamics that can occur, in theory, with rapid evolution of prey defence traits. Open circles (green in online version) are prey abundance; solid circles (red in online version) are top predator abundance; solid triangles (purple in online version) are intermediate predator abundance. All curves are scaled relative to their maximum value over the time period plotted. (A) Food web with only the top predator and two prey genotypes (defended and undefended), showing the long-period antiphase cycles characteristic of eco-evolutionary dynamics. (B) The simplest pattern of predator species taking turns: each prey peak is followed by a peak in one or the other predator. (C, D) Two examples of more complicated cycles with predators taking turns. In (C), the intermediate predator completes two cycles before prey with effective defence against it become dominant, allowing the top predator to increase. In (D), the two predators are antiphase with each other, but one is nearly in phase and the other nearly antiphase with the predator. (E) Chaotic dynamics, with a variable number of intermediate predator peaks in between top predator peaks.

give a quarter-period lag between prey and predator peaks, and the only other mechanism known to produce antiphase cycles under some conditions (i.e. predator maturation delays: de Roos and Persson, 2003) cannot produce cycles like those in Fig. 2.1A, whose period is orders of magnitude longer than the maturation time of the organisms (Hiltunen et al., 2014).

But, what happens in more complex food webs? Does any diagnostic temporal pattern of cycling exist in more complex systems that would allow an investigator to infer from long-term population data the presence of eco-evolutionary feedbacks? To answer this question, we explore here a food web more complex by one species than the single predator–single prey system we have studied to date: one with intraguild predation, where a top predator species consumes an intermediate predator species which consumes a basal prey species, and the top predator also consumes the intermediate predator.

We have already studied the effects of prey evolution in this system theoretically (Ellner and Becks, 2011), finding a remarkably broad array of possible dynamical outcomes (some of these are shown in Fig. 2.1B–E). Which outcome occurs depends upon whether the prey can evolve a single defence against both predators, or instead separate defences are required against each predator, and it also depends on the costs for each defence alone and in combination (Ellner and Becks, 2011). With two possible trait states for the prey (defended or not against each predator type), there are 11 different possible combinations of two or more prey genotypes, and some combinations can have different dynamics depending on the cost and effectiveness of defence and other model parameters. However, perhaps the most likely of the 11 scenarios is an evolutionary trade-off such that prey can be defended against either predator (at relatively low cost) but the two defences are incompatible; that is, it is either impossible to be defended against both predator species at the same time (e.g. prey cannot be small and large the same time), or it is prohibitively costly. In this case, the predicted dynamics typically involve the predators "taking turns" (Fig. 2.1B–D), with the top and intermediate predators roughly antiphase with each other because the prey population is alternately well-defended against one predator and then against the other (whereas without prey evolution, the two predators can be nearly in phase with each other: Fig. 2.2A). But apart from this situation, there are a wide variety of possibilities for the dynamics of the three populations.

Nevertheless, we can still ask if the potential dynamics with evolution of prey defence traits differ in some way from those that occur without prey evolution. Here, theory is more encouraging. When prey are not evolving, the theoretical prediction is that when cycles occur, they will have relatively short

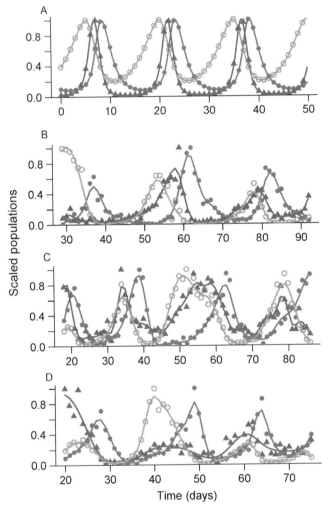

Figure 2.2 Theoretical and experimental results for three-species food web with intraguild predation *without* defence evolution by prey: alga *Chlorella autotrophica* (prey), rotifer *Brachionus plicatilis* (top predator) and flagellate *Oxyrrhis marina* (interme-diate predator), replotted from Hiltunen et al. (2013). Symbols (and colours in online version) are the same as in Fig. 2.1. (A) The theoretical predictions from a model for this food web with a single prey genotype, and (B–D) experimental results; the symbols are the data, and the curves are spline smooths of the data (see Sections 2 and 2.4).

periods (comparable to single prey–single predator cycles), with the peak in prey preceding that of the intermediate predator, followed then by the peak in the top predator. Experimental results with the rotifer–flagellate–algal chemostat system were remarkably consistent with the predicted dynamics (Fig. 2.2, replotted results reported by Hiltunen et al., 2013). With prey evolution, the predicted dynamics, although of great variety, generally have a longer cycle period than without evolution and in some cases a much longer period (Ellner and Becks, 2011). However, this is not an absolute prediction, and in any case, it is comparative: all else being equal (but *only* if all else is equal), prey defence evolution tends to lengthen cycle period. A second distinctive property is that in a purely "ecological" interaction, the growth rate of a predator population should be predictable from only the abundances of its prey, and of its own natural enemies, but when prey are rapidly evolving defences, the predator population growth rate cannot be predicted from population abundances alone because prey quality is changing. Thus, at least in the present simple food web, the inability to predict predator dynamics based only on species abundances should also be diagnostic of evo-evolutionary dynamics (though not necessarily of evolution in prey defence traits).

Our objective in the research reported here was to test these hypotheses in our three-species, intraguild predation, chemostat microcosms, and further to see if the patterns we observe are consistent with the patterns we predicted before the experiments were conducted (Ellner and Becks, 2011). We find that, as predicted, cycle periods observed with evolving, genetic variable prey were substantially longer with prey evolution (ca. 100 days) than without (ca. 20 days). Furthermore, the observed dynamics were often consistent with our prior predictions for situations in which defences against the two predators are incompatible (Fig. 2.1B and C). As expected, predator population growth rates were more predictable from population abundances alone when prey evolution was not occurring. Several patterns in our data suggest that the observed eco-evolutionary dynamics may be a *canard*, a recently discovered phenomenon in dynamical systems theory involving trajectories that spend a long time near unstable objects (Wechselberger, 2005). We explain how canards readily occur in models for eco-evolutionary dynamics and how a canard scenario can explain the observed long periods of stasis (dominance by just one predator) followed by rapid shift to dominance by the other, with predator population growth rates being most predictable during the shifts in dominance.

2. METHODS

2.1. Study species and setting up the experimental community

We used a three-species, intraguild predation, marine planktonic food web as our study system. It consisted of a prey alga, *Chlorella autotrophica*, a flagellate intermediate predator, *Oxyrrhis marina*, and a rotifer top predator, *Brachionus plicatilis*. Cultures of *C. autotrophica* (CCMP 243) and *O. marina* (CCMP604) were obtained from the National Center for Marine Algae and Microbiota (NCMA); *B. plicatilis* was obtained from Florida Aqua Farms (FL, USA). In choosing a three-species system, we tested several candidate combinations of taxa with the goal of identifying one for which all three species would coexist in our chemostats. Although we already had two well-tested freshwater predator–prey systems available (*Brachionus calyciflorus–Chlorella vulgaris*, Fussmann et al., 2000; Yoshida et al., 2003; *B. calyciflorus–Chlamydomonas reinhardtii*, Becks et al., 2010, 2012), we found it challenging to find an intermediate protist predator that would coexist in either. Initial attempts to use *Coleps hirtus* or *Ochromonas danica* ultimately failed because neither would coexist with *B. calyciflorus*. Indeed, *Ochromonas* proved to be extremely toxic to the rotifers (Hiltunen et al., 2012).

We eventually chose a marine system because salinity provided, along with temperature and chemostat dilution rate, an additional controllable parameter. This made it possible to find conditions (detailed below) that constrained rotifer growth rate, preventing these consumers from driving the flagellates extinct from the combination of predation and competition. Holt and Polis (1997) observed in their theoretical models that the stable periodic oscillations produced by intraguild predation were often of large amplitude with a high likelihood of extinction by demographic stochasticity. Consistent with this expectation, in preliminary runs of our microcosms, we found that the flagellates sometimes went extinct at the population-density minimum of an oscillation. To avoid this problem, we supplemented the abundance of flagellates by continuously pumping into our experimental chemostat a low number of *O. marina* ($\sim 10^4$ cells day^{-1}) from a separate chemostat. This external source amounted to only 0.8–5.6% of the maximum concentration of the flagellate cells in our chemostat runs, but was sufficient to prevent stochastic extinction.

2.2. Controlling the initial genetic variation in prey populations

In previously reported results from two-species rotifer–algal chemostats (both for *Brachionus–Chlorella* and for *Brachionus–Chlamydomonas*), we have documented rapid evolution of algal defence traits that changed predator–prey cycle dynamics substantially (Becks et al., 2010; Yoshida et al., 2003, 2007). In those studies, chemostats were started either with a single prey genotype in order to prevent prey evolution by eliminating all heritable variation or with multiple prey genotypes upon which selection could act. To allow the same kind of comparison for this study, we produced genetic diversity for the presence or absence of prey defence by exposing three separate source populations of *C. autotrophica* to either predation only by *O. marina*, predation only by *B. plicatilis*, or no predation, in continuous culture for 6 months. These chemostats were run at a dilution rate of 0.1 day^{-1}, but otherwise under the same conditions as our three-species dynamics experiments (see below).

Manipulating the potential for algal evolution in the three-species experiments proved to be difficult (Hiltunen et al., 2013). Several experimental runs were started with only *C. autotrophica* from the chemostat without predators in order to minimize genetic variation for defence against predation, but evidence of prey defence evolution quickly appeared as seen by the formation of cell clumping, shown previously to be effective against flagellate consumers in chemostat microcosms (Boraas et al., 1998; and see Section 2.3). This was accompanied by additional indirect evidence of prey evolution: a marked change in community dynamics away from the patterns expected when eco-evolutionary processes are absent. Either genetic variation for defence was initially present despite many generations of selection for competitive ability rather than defence, or else defended genotypes were quickly produced by mutation; in previous experiments where the initial algal population was monoclonal (Yoshida et al., 2003), algal defence nonetheless appeared within several weeks. Conversely, some experimental runs were started with algae from all three source populations so that genetic variation for defence was present from the outset, but clumping quickly became rare and remained rare, and there was no evidence of prey defence evolution (direct or indirect) for several months, or until after a subsequent inoculation with algae from multiple source populations.

2.3. Community dynamics experiment

To test the effect of prey evolution on community dynamics, experiments were conducted in 380-mL continuous culture chemostats maintained in growth chambers held at 21 °C (±1 °C) with constant illumination (120 $\mu Em^{-2} s^{-1}$ from Sylvania Gro-lux wide spectrum fluorescent lamps). We used a modified *f*/2 culture medium (Guillard and Ryther, 1962) with nitrogen at 80 $\mu M\ L^{-1}$ as the limiting nutrient (all other nutrients were in excess of algal requirements) and salinity adjusted to 35 g L^{-1} with commercial salt mix (Tropic Marin, MA, USA). The dilution rate of the chemostats was 0.25 day^{-1}. Chemostats were bubbled with air to maintain thorough mixing, with bubbling cycled on (0.5 min) and off (9.5 min) because flagellates were unable to capture the algae in the turbulent conditions created by bubbling. Each chemostat was sampled daily or every second day for between 130 and 330 days, following the same general methodology as in our previous studies (Becks et al., 2010; Fussmann et al., 2000; Hiltunen et al., 2013; Yoshida et al., 2003). Duplicate samples of all species were taken daily, one each through ports near the top and bottom of the chemostat. Algal and flagellate densities were counted under a compound microscope using a haemocytometer for algae (Improved Neubauer, Hausser Scientific, Horsham, PA, USA) and a Sedgewick-Rafter cell for flagellates (Wildlife Supply Co., FL, USA). Rotifers were enumerated under a dissecting microscope. All reported abundances are the means of the values from the top- and bottom-port samples.

In addition to population abundances, we also monitored algal cell clumping (number of cells/clump) as a potential prey defence trait. Boraas et al. (1998) found that *C. vulgaris* evolved a clumped growth form (not really "colonies" since the mucilaginous agglomeration had no regular arrangement of cells) in the presence of the predatory (phagotrophic) flagellate *Ochromonas vallescia*, and Becks et al. (2010, 2012) found similar evolution of clumping by *C. reinhardtii* in the presence of *B. calyciflorus*. In both cases, the clumped growth form was found to be highly heritable, though others have found it to be inducible both for *Chlamydomonas* (Lurling and Beekman, 2006) and for another chlorophyte alga, *Scenedesmus subspicatus* (Hessen and Van Donk, 1993). Previous research did not reveal clumping by *Chlorella* as a defence against rotifers, presumably because the cells are too small for aggregation to prevent ingestion. However, *C. vulgaris* did evolve another effective defence (Yoshida et al., 2003) in which single cells passed through the rotifer gut unharmed (Meyer et al., 2006). Both this defence and cell clumping in *Chlamydomonas* came with a cost in reduced growth rate under conditions of nutrient limitation (Becks et al., 2012; Yoshida et al., 2004). The experimental results we report here indicate that, in addition to *Chlorella* clumping, these prey evolved some form of defence against the two predator species (resulting in low or negative predator growth rate even at high algal density) that was not visible under the microscope and that we therefore could not quantify directly.

Many of the experiments required an initialization phase for establishment of rotifer populations, involving changes in chemostat dilution rate, additional rotifer inoculations or both because the predator population sometimes failed to establish after inoculation. Successful rotifer introductions frequently began with several days of population decline or slow growth, despite high abundance of algae. The data that we present and analyze here omit these initial periods. We removed all data prior to the first day on which both the dilution rate was 0.25 day^{-1} (the nominal experimental condition) and the measured rotifer density had doubled from that on the day following inoculation.

2.4. Data smoothing

Some of our analyses and our plots of data from the community dynamics experiments used estimated continuous-time population trajectories; in particular, estimated continuous-time population trajectories were used to estimate instantaneous rates of population growth. We describe in this section the technical details of how these trajectory estimates were obtained. In all cases, we used B-splines fitted by maximum penalized likelihood (Wahba, 1990). Data on mean clump size were smoothed with cubic splines, using the R function smooth.spline with default values of all parameters. For population data, we used quartic splines with third-derivative roughness penalty because these higher-order splines give better estimates of the trajectory's derivative (Wahba, 1990). The higher-order splines were fitted using the smooth.basis function in R's fda library (Ramsay et al., 2013), with smoothing parameter selected by generalized cross-validation (gcv). To avoid overfitting, model degrees of freedom in the gcv criterion were overweighted by a factor of 1.2; overweighting by a factor of 1.4 is sometimes recommended (Gu, 2013), but a lower overweighting factor reduces the bias of the fitted curve and 1.2 was sufficient to eliminate all apparent instances of overfitting in our data. To reduce heteroskedasticity, smoothing was done on square-root transformed data, and the fitted smooth trajectory was then squared. The derivative of population size $N(t)$ was estimated as $dN/dt = 2y(t \, dy/dt$, where $y(t)$ is the estimated trajectory and dy/dt is the estimated time derivative for the square-root transformed data (so that $N = y^2$).

2.5. Estimating predictability of predator dynamics

The instantaneous growth rates for rotifers, $W_R = (1/R) \, dR/dt$, and flagellates, $W_F = (1/F) \, dF/dt$, were computed from smoothed trajectories as described in the preceding paragraph (Section 2.4). To estimate how well each predator population's instantaneous growth rate could be predicted purely from the abundance of algae and of the other predator, we fitted a generalized additive spline (GAM) model for W_R with smoothed algal and flagellate abundance as the two predictors and for W_F with smoothed algal and rotifer abundance as the two predictors, using R's mgcv package (Wood, 2006, 2011). Model degrees of freedom were again overweighted by a factor of 1.2 to avoid overfitting. Model specifications were of the form

```
fitR <- gam(WR ~ s(algae)+s(flagellates),gamma=1.2)

fitF <- gam(WF ~ s(algae)+s(rotifers),gamma=1.2)
```

where `algae`, `rotifers` and `flagellates` are the smoothed population trajectories described in Section 2.4. The adjusted r^2 value of the fitted model (as reported by summary.gam) is our measure of predictability. By omitting each predator's density as a covariate in predicting its own population growth, we assume that there is no within-species predator interference. We know of no evidence for this kind of interference in either predator, and because the system is kept well mixed and the predators are not able to actively search for prey, we believe that this is a reasonable assumption.

For all of the experimental runs with three species, the procedure described in the previous paragraph was applied to the entire data series, giving one r^2 value for each predator species. For all of the longer runs with prey defence evolution, we also applied the procedure to subsets of the data comparable in length to the shorter no-evolution data sets. W_R, W_F and smoothed population trajectories were estimated using the entire data series. GAM models for W_R and W_F were fitted for a series of overlapping data windows of 60 days duration, with windows starting at 30-day intervals. For example, if a data series ran from days 30 to 280, we fitted GAM models and obtained r^2 values for the time windows from days 30 to 89, days 60 to 119, days 90 to 149 and so on. The last data window was allowed to be over 60 days long so that it could run to the end of the data set (rather than omitting the end of the data set, or having the final window be shorter than 60 days). The 60-day window width was chosen because it approximates the full length of the no-evolution data sets. Consequently, the predictability estimates for the 60-day windows of experimental runs with prey evolution are directly comparable to the corresponding values for the full no-evolution data sets, which thus provide the "null expectation" for predator growth rate predictability in the absence of prey defence evolution. The 30-day overlap (rather than using non-overlapping 60-day windows) was done to increase the likelihood of having some windows correspond well to high-predictability periods. We did not also consider shorter overlaps (and therefore more windows) because of the risk that this would lead to high-predictability values arising as "false positives" in a large number of trials.

2.6. Models for community and eco-evolutionary dynamics

The theoretical predictions that we test here (Figs. 2.1 and 2.2) were derived previously (Ellner and Becks, 2011) using a small-fluctuations analysis of models for the three-species food web with one or more prey clones. Because the analysis was based on Taylor-series approximations for near-equilibrium dynamics, the qualitative predictions are robust to model details such predator functional responses. For numerical simulations in this paper, we used a model very similar to that in Hiltunen et al. (2012), except that multiple prey genotypes can be present. For the case of two prey genotypes, the model equations are as follows:

$$
\begin{aligned}
\frac{dS}{dt} &= \delta(1-S) - rS\left(\frac{A_1}{k_1 + S} + \frac{A_2}{k_2 + S}\right) \\
\frac{dA_i}{dt} &= A_i\left[\frac{rS}{k_i + S} - \frac{p_i gR}{k_R + E_R + \alpha_F F} - \pi_i hF - \delta\right], \quad i = 1, 2 \\
\frac{dR}{dt} &= R\left[\frac{gQ}{k_R + E_R + \alpha_F F} + \frac{\eta F}{k_R + E_R + \alpha_F F} - \delta\right] \\
\frac{dF}{dt} &= F\left[hE_F - \frac{\eta R}{k_R + Q + \alpha_F F} - \delta\right] + I_F
\end{aligned}
\tag{2.1}
$$

The state variables are S = limiting substrate, A = algae (of two types $i = 1$, 2), R = rotifers and F = flagellates, with all parameters positive. The equations represent a well-mixed chemostat with constant inflow of the limiting substrate and constant outflow of all species at dilution rate δ, with all populations measured in units of limiting substrate (and all scaled relative to the concentration of substrate in the inflowing culture medium). The algae and rotifers have type-II functional responses (as is commonly assumed), and for simplicity, the flagellates have a type-I functional response. The parameters p_i and π_i are the probability of capture and consumption by rotifers and flagellates, respectively, for algal genotype i. E_R and E_F are the total amount of "edible" prey available to rotifers and algae, respectively, $E_R = p_1 A_1 + p_2 A_2$, $E_F = \pi_1 A_1 + \pi_2 A_2$. I_F is the rate of flagellate input from the external source. Our system operated at $\delta = 0.25/d$ and in the units of model (1) the flagellate input rate is approximately $0.001/d$ (Hiltunen et al., 2013).

3. RESULTS

3.1. Two-species (single predator, single prey) experiments

Communities with algal prey and only one predator species were studied to verify that predator–prey cycles would occur, and that algae could evolve defences against both predators.

With only rotifers as predator (Fig. 2.3A–D), once a rotifer population was successfully established after their second introduction on day 50, the community quickly settled into the eco-evolutionary cycles of predator and prey abundance (Fig. 2.3A) previously observed in other rotifer–algal pairs (Becks et al., 2010; Yoshida et al., 2003). This was accompanied by small oscillations in prey clumping (Fig. 2.3B). In Fig. 2.3C, a plot of rotifer population growth rate ($1/R\ dR/dt$) against algal abundance exhibits counterclockwise cycles, again matching the results for our previously rotifer–algal systems with evolution of prey defence (Becks et al., 2010; Shertzer et al., 2002). The dashed horizontal line is at $dR/dt=0$, emphasizing that across a wide range of intermediate algal densities, rotifer growth rate can be either positive or negative, revealing that algal quality as well as algal abundance is determining rotifer population growth rate. The small oscillations in prey clumping (Fig. 2.3B) suggest that prey clumping is not an important component of the *C. autotrophica* defence against rotifers (in contrast to our observation for *C. reinhardtii* where clumping was the primary defence; Becks et al., 2010, 2012) because *Chlorella* clumps are rare and too small to be an obstacle to ingestion by rotifers. However, Fig. 2.3D shows that algal population growth rate was positively correlated with mean clump size, as if clumping actually did provide defence against rotifer predation. Taken together, these results suggest that clumping is somehow associated with the algal defence trait in this experiment, perhaps as a side effect of some change in cell wall chemistry that prevents rotifers from digesting the algae (as we found previously for *C. vulgaris*; Meyer et al., 2006).

With only flagellates as predator (Fig. 2.3E–H), there is again clear evidence of defence evolution but the connection to cell clumping is inconsistent. The first replicate (Fig. 2.3E and F) began with classic predator–prey cycles (days 0–90), with short periods and quarter-period lag between prey and predator peaks. Following that, flagellate density remained low for

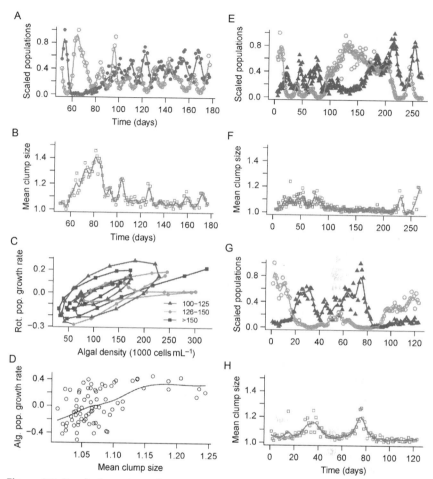

Figure 2.3 Results from three chemostat experiments with algae and a single predator (rotifers: A–D, flagellates: E–F and G–H). (A) Rotifer and algal abundance, as in Fig. 2.2. (B) Mean algal clump size; symbols are the data, the curve is a spline smooth (see Section 2). (C) Rotifer per-capita population growth rate $(1/R)$ dR/dt plotted as a function of algal density. Symbols (and colours in online version) indicate different time periods in the panel (A) data. The first data point is marked by a large triangle (purple in online version). (D) Algal population growth rate $(1/A)$ dA/dt as a function of algal mean clump size. The trend line is local linear regression using the R function scatter.smooth. (E, F) and (G, H) Two experiments with flagellates and algae, plotted as in (A) and (B).

several months (days ~90–150), while algae increased to a level that earlier had been sufficient for rapid flagellate population growth (e.g. on day 40). This indicates that the algae had developed defence against consumption by flagellates, but there was no concomitant increase in cell clumping until several months later (Fig. 2.3F), when flagellate growth resumed and algal density dropped (day 230 and later). In contrast, in the second replicate (Fig. 2.3G and H), algal clumping increased after both of the two large flagellate peaks. The time delay between flagellate increase and elevated clumping suggests that the clumping response was evolved rather than induced because plastic algal defences appear within a single algal generation (Hessen and Van Donk, 1993; Lurling and Beekman, 2006). At the end of this experiment, algal defence appeared to evolve (day 90) while clumping was almost entirely absent.

These single-predator experiments confirm that algal defences evolve against each predator and moreover suggest that the defences are different in nature: defence against rotifers was associated with cell clumping, while defence against flagellates was effective for several months while clumping was entirely absent.

3.2. Three-species (two predators, single prey) experiments with prey defence evolution

We turn now to the dynamics with both predators present, when prey defence evolution occurred and the predicted dynamics (Fig. 2.1) are very different from the dynamics predicted and observed in the absence of prey defence evolution (Fig. 2.2). The results in Fig. 2.2 establish the baseline against which the results in this section can be compared, to see how prey defence evolution affects the dynamics of the three-species system.

Two extended chemostat runs (Fig. 2.4A–D) were initialized with prey from a single source population, and therefore low heritable variation so that prey evolution would not occur until heritable variation was generated by mutations. In both cases, the initial dynamics aligned with the theoretical predictions for the dynamics when prey are not evolving (Fig. 2.2A). In Fig. 2.4A and B, the initial period is the same data as Fig. 2.2B. Then, following the introduction of algae from multiple source populations on days 102–104, there was an increase in prey clumping and the dynamics quickly transitioned to the "predators taking turns" pattern (Fig. 2.1B and C) that was predicted to occur when defences against the two predators are incompatible, with two periods of rotifer dominance separated by a long period of flagellate dominance. Similarly in Fig. 2.4C and D, the rise in algal clumping was followed by a transition from short-period cycles like those predicted for non-evolving prey, to a "predators taking turns" pattern.

Figure 2.4E–H shows chemostat runs that were initialized with genetically variable prey from multiple source populations, including algae exposed to either rotifer or flagellate predation. In Fig. 2.4E and F, the

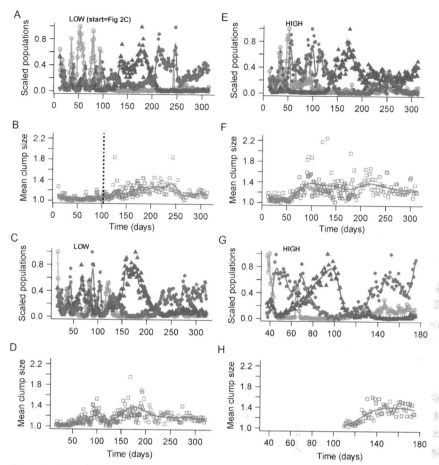

Figure 2.4 Results from four chemostat experiments (A–B, C–D, E–F and G–H) with algae and both predators. In (A), (C), (E) and (G), the symbols (and colours in online version) are the same as in Fig. 2.1, and "high" or "low" indicate the initial level of algal genetic variability (see Section 2). The initial period of the data in (A) is the same as was plotted in Fig. 2.2C; the dashed vertical lines in B indicate the inoculation of additional algal genotypes into that chemostat on days 102,103 and 104 after the experiment was initiated.

transition to long-period "predators taking turns" dynamics occurred quickly, after two short cycles that align with the prediction for non-evolving prey; in Fig. 2.4G and H, the long-period "predators taking turns" pattern was present from the start.

In these four runs, it was generally the case that prey abundance was higher when rotifers were dominant than when flagellates were dominant. The relationship between prey clumping and predator abundances was inconsistent. For example, in Fig. 2.4B and C, clumping seems to be a response to flagellate density (increasing after flagellates increase and decreasing after flagellates increase), but in Fig. 2.4A and B, clumping remains at its highest level in that run long after flagellates have declined (days 210–250), and in Fig. 2.4E and F, clumping arises before the first substantial flagellate peak.

The other three extended runs with prey evolution (Fig. 2.5A–F) are more difficult to characterize. Like Fig. 2.4A and B, Fig. 2.5A and B was

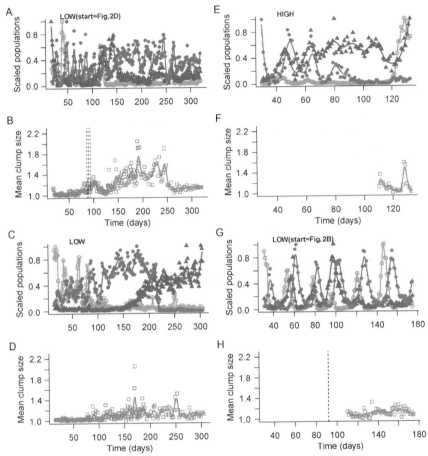

Figure 2.5 Results from four additional chemostat experiments (A–B, C–D, E–F and G–H) with algae and both predators. In (A), (C), (E) and (G), the symbols (and colours in online version) are the same as in Fig. 2.1, and "high" or "low" indicate the initial level of algal genetic variability (see Section 2). The initial period of the data in (C) is the same as was plotted in Fig. 2.2D; the dashed vertical lines in (D) indicate the inoculation of the chemostat with additional algal genotypes on days 87, 90 and 94. The initial period of the data in (G) is the same as was plotted in Fig. 2.2B; the dashed vertical lines in (H) indicate the inoculation of the chemostat with additional algal genotypes on day 92.

initiated with low prey genetic diversity and initially followed the predicted pattern for non-evolving prey (the initial phase of Fig. 2.5A and B is the same data as Fig. 2.2D). Also as in Fig. 2.4A and B, introduction of genetically variable prey (days 87 and 94) was soon followed by an upturn in clumping and a marked change in the dynamics. However, instead of predators "taking turns", the rotifers remained dominant for over 150 days, until the end of the run, even though clumping dropped substantially after day 250. The downturn in clumping was coincident with a brief dip in rotifer abundance (which we cannot explain) that allowed a brief spike in algal density.

In Fig. 2.5C and D, short-period cycles with rotifers dominant gave way to an extended period of rotifer dominance, possibly with longer-period cycles, followed by flagellate dominance and near extinction of the rotifers until the end of the run. These dramatic changes in the predator populations were not clearly associated with any substantial changes in algal clumping, which remained low relative to other runs. In Fig. 2.5E and F, though the initial prey genetic variance was high, the run started with short-period cycles matching the pattern for non-evolving prey, followed by a period of flagellate dominance. However, the run did not continue long enough for us to know whether a "predators taking turns" pattern would have developed. Finally, for completeness, Fig. 2.5G and H shows our one additional extended three-species run. In this case, addition of algal genetic diversity (day 92) did not cause a shift away from the expected pattern for non-evolving prey (the initial period of Fig. 2.5G and H, prior to the addition of algal genetic diversity, is the same data as Fig. 2.2B).

3.3. Prey evolution and the predictability of population dynamics

We turn now to our second general prediction that rapid prey defence evolution should decrease the ecological predictability of population dynamics. By "ecological predictability", we mean the extent to which population growth rates at a given time can be predicted purely from the abundances of the other species present at the same time, as in standard (non-evolutionary) multispecies community models with constant parameters. The structure of our experimental food web implies that ecological predictability should be high when species' traits are not evolving because each predator's birth and death rates are determined by the abundances of algae and of the other predator. Note that "predictability" here is about predicting the current rate of change, not about predicting the system's future state directly (though of course the rate of change has implications for the future).

We tested this prediction by smoothing the predator population time series to estimate instantaneous rates of population growth and then fitting generalized additive models for rotifer and flagellate population growth rates as a function of algal abundance and the abundance of the other predator (see Section 2 for details). The adjusted r^2 values for these models are our measure of predictability (Table 2.1). Because algal population growth also depends on available nitrogen concentration, which was not measured, we did not assess algal growth rate in this analysis.

As expected, the predictabilities were generally a good deal higher for the data without prey defence evolution than for data with prey defence evolution (Table 2.1, column "All data"). For both flagellates and rotifers, all but one of the data sets with prey evolution had lower predictability than

Table 2.1 Predictability of rotifer and flagellate per-capita population growth rates, as a function of algal abundance and the abundance of the other predator

	Rotifers (r^2)		Flagellates (r^2)	
Figure (days)	All data	60 days (median)	All data	60 days (median)
No prey evolution				
2.2B (29–92)	0.80		0.89	
2.5G (93–175)	0.74		0.74	
2.2C (18–86)	0.52		0.59	
2.2D (20–75)	0.91		0.70	
Prey evolution				
2.4A (102–313)	0.30	0.73	0.24	0.45
2.4C (70–320)	0.39	0.75	0.42	0.67
2.4E (70–313)	0.50	0.69	0.50	0.53
2.4G (40–175)	0.67	0.79	0.40	0.72
2.5A (95–320)	0.31	0.45	0.43	0.67
2.5C (75–305)	0.03	0.63	0.25	0.58
2.5E (29–133)	0.04	0.44	0.60	0.62

Tabulated values are the adjusted r^2 value of the fitted generalized additive model (GAM) as reported by the summary.gam function in R. "All data" means that a single GAM was fitted to the entire span of days given in the leftmost column. "60 days (median)" means that a series of GAMs were fitted to overlapping data windows of 60 days duration, and the tabulated value is the median adjusted r^2 of that set of models. See Section 2.5 for details.

the *minimum* predictability of the data sets without prey evolution (rotifers: 4G, $r^2 = 0.67$; flagellates: 5E, $r^2 = 0.60$).

However, this comparison is confounded by the fact that the data sets with prey evolution are generally much longer than the no-evolution data sets. We therefore also computed the predictability for 60-day-long windows of the data sets with evolution (days 1–60, 31–90 and 61–120 of the time period analyzed), and in Table 2.1, we report the median r^2 values for these windows (column "60-day median"). Because the 60-day windows are comparable in length to a complete no-evolution data set, the predictabilities for the no-evolution data sets provide the "null expectation" for predictability in the absence of prey defence evolution. The 60-day predictabilities are typically much higher than the complete data set predictabilities, and the median predictability within 60-day windows is, for all of the data sets with prey evolution, comparable to predictability values of the no-evolution data sets. We explain in the following section why this difference between short- and long-term predictability is an expected consequence of the eco-evolutionary dynamics in our experimental system.

3.4. Canard cycles and regime shifts in eco-evolutionary dynamics

We now suggest a simple mechanism that can explain two features of the experimental results in Figs. 2.4–2.5 and Table 2.1: (1) long periods of dominance by one predator followed by a rapid switch to a long period of dominance by the other; (2) when prey defence evolves rapidly, predator population growth rates remain predictable within short time windows, but predictability over long time windows is much lower than when prey are not evolving.

The mechanism illustrated by simulations of model (1) in Fig. 2.6 is that the eco-evolutionary dynamics of prey defence evolution can readily produce a kind of dynamics known as *canards*. A canard is a trajectory of a dynamical system that spends a long time near an unstable object (e.g. Wechselberger, 2005). Figure 2.6 shows simulation results from model (1) with two genotypes, each well defended against one predator and completely vulnerable to the other. Changes in the predator populations (Fig. 2.6A and B) drive and are in turn driven by changes in the frequency f of the prey type defended against flagellates (Fig. 2.6C).

As in our experiments with prey evolution, most of the time, one of the predators is much more common than the other (Fig. 2.6B), and the transitions between periods of rotifer and flagellate dominance are relatively

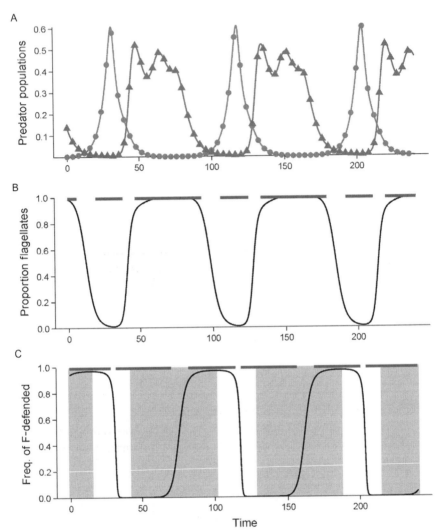

Figure 2.6 Canard cycling in model (1) for the three-species food web. (A) Population densities of the two predators (rotifers, circles – red in online version; flagellates, triangles – purple in online version) measured in units of the limiting resource, scaled relative to its concentration in the inflowing nutrient medium. (B) Proportion of flagellates, i.e., flagellate abundance divided by the total abundance of flagellates and rotifers. Solid bars at the top of the panel indicate times when the proportion is above 0.9 or below 0.1. (C) Frequency of the flagellate-defended genotype, $f(t)$. Solid bars at the top of the panel indicate times when the frequency is above 0.9 or below 0.1. Grey shading indicates periods of time when selection favours the flagellate-defended algal genotype. Parameter values: $\delta=0.25$, $r=0.8$, $g=0.5$, $h=1$, $k_R=0.1$, $\eta=0.01$, $h=1$, $\alpha_F=0.2$, $l_F=0.001$, $k_1=k_2=0.05$. Algal type 1 was fully edible to rotifers and 95% defended against flagellates; for algal type 2 the reverse was true.

short. In the model, this results from similar dynamics in the two prey genotypes (Fig. 2.6C). Most of the time, one of the genotypes is at very high frequency and the other is at very low frequency, with brief periods when the
algal population rapidly goes from flagellate defended ($f=1$) to rotifer defended ($f=0$) or vice versa. The bursts of rapid evolutionary change do not
occur when the direction of selection changes, but rather they occur after a
substantial delay. Grey shading in Fig. 2.6C indicates periods of time when
selection "points up" (the flagellate-defended genotype is favoured) because
flagellate predation is more intense than rotifer predation. But at the start of
each such period, the flagellate-defended genotype is very rare so the
response to selection is delayed. The prey population is "stuck" near $f=0$
because genetic variance in defence was eroded to nearly zero by the prior
period of strong selection in favour of the rotifer-defended genotype.
Because the rate of evolution is proportional to trait variance (Falconer,
1960; Fisher, 1930), evolution is extremely slow, while variance of the
defence trait is extremely low, and trait variance grows very slowly because
the rate of evolution is so low. Eventually, the prey population "breaks
loose" and then evolves rapidly to the point that the rotifer-defended genotype is nearly absent. This allows the rotifer population to increase, but by
then the prey population is "stuck" near $f=1$ and the rotifer-defended genotype remains rare long after rotifers greatly outnumber flagellates.

Figure 2.6 was generated by model (1), which has five state variables
when there are two types of algae, but the qualitative dynamics are robust
to model details and we can examine them in a simpler model with only
three state variables. Suppose (unrealistically) that the total algal abundance
is held constant (at 1, without loss of generality) by some process unrelated to
predation. The only state variables are rotifer abundance R, flagellate abundance F and the frequency f of the flagellate-defended genotype. Specifically,
we consider the following "minimal" model for our experimental food web:

$$\frac{dR}{dt} = \frac{gRf}{k_R + f} + \frac{vRF}{k_I + F} - m_R R(1 + \alpha_R R)$$

$$\frac{dF}{dt} = \frac{hF(1-f)}{k_F + (1-f)} - \frac{vRF}{k_I + F} - m_F F(1 + \alpha_F F) \qquad (2.2)$$

$$\frac{df}{dt} = f(1-f)\left[\frac{hF}{k_F + (1-f)} - \frac{gR}{k_R + f}\right]$$

The model assumes that defence is perfect: flagellates eat only rotifer-
defended algae, and rotifers eat only flagellate-defended algae. The three

terms on the right-hand side of the dR/dt equation say, respectively, that rotifers eat flagellate-defended algae (and only those algae) with a type-II functional response; that rotifers eat flagellates with a type-II functional response; and that rotifers have density-independent mortality at per-capita rate m_R plus density-dependent mortality. The dF/dt equation is similar, except that rotifer consumption results in loss rather than gain. The df/dt equation is the standard continuous-time model for haploid selection without mutation or drift in a population of constant size because the factor in square brackets is the fitness difference between flagellate-defended and rotifer-defended prey.

Figure 2.7A confirms that this model, with suitable parameters, behaves just like our more realistic model: long periods of dominance by one predator interrupted by quick switches in dominance, and long periods of evolutionary stasis punctuated by rapid evolution from one defence trait to the other. Parameter values affect cycle shape and period, but the qualitative behaviour occurs so long as there are population cycles because it reflects two consistent features: (1) which allele is favoured depends only on predator relative abundances, so large-amplitude population cycles push the allele frequencies to the extremes of 0 and 1; (2) these extreme allele frequencies lead to a slow response when the direction of selection changes because genetic variance is depleted. Intraguild predation does not play an essential role; in our experimental system and models, it is important only because it facilitates coexistence of two predators on one prey species in the homogeneous chemostat environment.

Because the simplified model is three dimensional, we can visualize its dynamics (Fig. 2.7B). In Fig. 2.7B, the solid black curve is the stable cycle, and the thin grey curve is a trajectory that starts off the stable cycle and converges onto it. Symbols on the curve indicate the direction of selection: triangles (purple in online version) indicate that flagellate-defended genotype is favoured, and circles (red in online version) indicate that the rotifer-defended genotype is favoured. Time $t=20$ in Fig. 2.7A corresponds to the trajectory being near the origin (lower left corner) in Fig. 2.7B: both predators are scarce, and the prey are defended against rotifers. The flagellates can eat most of the prey and so start to increase in abundance, so then defence against flagellates is favoured (the trajectory segment with triangles moving "into the page" in Fig. 2.7B). But because genetic variance is low, there is little prey evolution even while flagellate abundance increases by three orders of magnitude. This is where the trajectory qualifies as a canard. It remains very near a region in the plane $f=0$ which is unstable: trajectories

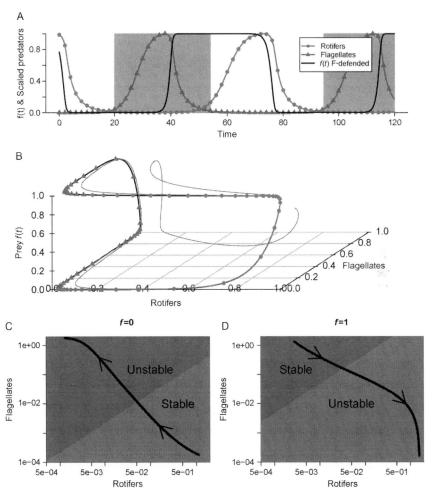

Figure 2.7 Canard cycle in model (2) for two predators feeding on two competing prey genotypes, each perfectly defended against one predator but undefended against the other. (A) Trajectories of predator abundance (rotifers, solid circles – red in online version; flagellates, solid triangles – purple in online version) and the frequency f of the flagellate-defended prey type (black). Grey shading indicates periods when selection favours the flagellate-defended genotype. (B) Three-dimensional plot of model solution trajectories. The solid black curve is the stable limit cycle; the thin grey curve is a trajectory starting off the limit cycle. Symbols on the limit cycle indicate the direction of selection (circles – red in online version: rotifer-defended algal genotype is favoured; triangles – purple in online version: flagellate-defended genotype is favoured). (C, D) Projections of the limit cycle trajectory onto the planes $f=0$ and $f=1$. Periods when $f<0.05$ and $f>0.95$, respectively, are plotted in (C) and (D).

that start near that region move further away from the plane, not closer to it, because the term in square brackets in the df/dt equation above is positive. Gradually, trait variance builds up, and the trajectory jumps up to the plane $f=1$. The flagellates starve and die, so the trajectory comes back to ($R=0$, $F=0$) only now f is near 1 rather than near 0. As a result, the sequence of events since $t=20$ repeats, except that this time the rotifers increase, and there is a delayed response to selection in favour of rotifer defence followed by a quick jump back to $f=0$. Again, the trajectory qualifies as a canard by staying near $f=1$ while selection is pushing it off that plane.

Another way to view the dynamics is by projecting them onto the planes $f=0$ and $f=1$ where they spend most of their time (Fig. 2.7C and D; note the log-scale axes). The plotted curves are the trajectory segments on which f is below 0.05 or above 0.95. Starting at bottom right in Fig. 2.7C, the trajectory has just jumped onto $f=0$ at a location where rotifers are common and flagellates scarce, so that region of the plane is locally stable (it is better to be rotifer-defended than flagellate-defended). As flagellates increase and rotifers decrease, the trajectory enters an unstable region of the plane (where it is better to be flagellate defended). It remains there until genetic variance builds up enough for the population to jump over to the top-left corner of Fig. 2.7D, where the process repeats in reverse.

Ecologists are often told that an unstable equilibrium in a model, or an unstable periodic orbit, is biologically irrelevant. Model solutions move away from unstable objects like that, so we expect not to observe them in the real world (if the model describes the real world). Surprisingly, that idea is not entirely correct. As in Figs. 2.6 and 2.7, model solutions can get very close to an unstable object (by coming in from a special direction or, as in Figs. 2.6 and 2.7, by approaching a locally stable part of the object). After that, solutions can take a very long time to move away, so that the unstable object is a recurrent and important feature of the dynamics. Canards are not typical for dynamical systems in general, so their existence and significance were not recognized until recently. But Figs. 2.6 and 2.7 illustrate that canards can very easily arise in eco-evolutionary dynamics. The biological mechanism is simply that strong selection erodes genetic variance, so the response to selection is delayed until variance accumulates again.

The canard mechanism is a deterministic explanation for the delayed response to a change in the direction of selection. The trajectory never gets all the way to $f=0$ where the flagellate-defended genotype is completely absent. When selection shifts so that flagellate defence is again favoured in Fig. 2.7, the allele frequencies in the simulated population correspond (for the algal population sizes in our experiments) to about a dozen flagellate-defended individuals being still present. The population is not waiting for random mutation to reintroduce that genotype, it is waiting for the progeny of the few flagellate-defended cells to take over the population. In smaller populations, stochastic loss of the rarer allele might imply that there would also be a wait for random mutation to reintroduce the currently favoured genotype.

Because we were unable to obtain unambiguous measurements of prey defence, we have no direct evidence that canard cycles occurred in this experimental system. However, some indirect evidence is provided by the predator predictabilities within 60-day time windows (Table 2.1). During "slow evolution" periods when the defence trait is relatively constant (indicated by the thick lines at the top of Fig. 2.6C), rotifer and flagellate population growth rates should be predictable from the abundances of prey and of the other predator, just as predictable as in the "no-evolution" runs (Fig. 2.2). In the canard scenario, the system spends most of its time in a "slow evolution" state. A typical 60-day window from an experimental run with prey evolution should exhibit predator predictabilities comparable to a complete (roughly 60 days) no-evolution experimental run, and that is what we observe (Table 2.1, the 60-day (median) column).

Another general feature of the canard scenario is that "slow evolution" periods centre on the times when selection changes direction because the system is flipping from rotifer dominance to flagellate dominance or vice versa. That is, the borders between shaded and unshaded regions in Fig. 2.6C align with the brief periods when the proportion of flagellates (in Fig. 2.6B) is changing rapidly. Therefore, predator population growth rates should be most predictable (as a function of algal and other-predator abundances) during times when the flagellate:rotifer ratio is changing quickly.

To test this second prediction, for each of the 60-day time windows, we measured the rate of change in the ratio of predator abundances by fitting a linear regression line to log(rotifers/flagellates) values during that time window and recording the slope of the line. We then performed a one-tailed test for the predicted positive association between the absolute value of the slope and the predator predictability (r^2), using Kendall's nonparametric correlation coefficient τ because there is no *a priori* basis for the assumptions of statistical tests using the Pearson correlation coefficient. As predicted, there was a positive correlation for both rotifers ($\tau = 0.25$, $P = 0.02$, $n = 35$) and flagellates ($\tau = 0.20$, $P = 0.04$, $n = 35$), though neither correlation was very strong.

In summary, we observe two patterns in the experimental population data that are predicted under the hypothesis that the long periods of single–predator dominance are explained by long periods of trait stasis (with one genotype dominant) separated by short periods of rapid trait evolution, and we have shown that evolutionary trajectories of that kind are a natural and robust theoretical prediction in models for prey evolving defences against multiple predators. It remains to be seen whether other hypotheses can also explain those patterns, and more critically, we need to develop multi-predator experimental systems in which the prey defence traits are known and can be measured directly rather than inferred.

4. DISCUSSION AND CONCLUSIONS

Before the start of our experiments, we predicted that the temporal community dynamics with prey evolution should be markedly different from those without prey genetic variation and so without evolution (Ellner and Becks, 2011): this is what we observed. Although there is no unique diagnostic eco-evolutionary pattern comparable to that documented for a single predator–single prey system, it is nevertheless clear that the feedbacks resulting from rapid evolution of prey defence traits radically changed the dynamics of our three-species experimental system. In particular, if defences against the two predators are incompatible, theory predicts that the predators should "take turns" (i.e. only one of the predators is abundant at a time, with each predator numerically dominant for an extended period), and that is the pattern we observed most often in the data presented here.

The long cycles in our experiments here are distinct from the short-period non-evolutionary three-species cycles we previously reported (Hiltunen et al., 2013). However, theory does not predict a distinct qualitative break in cycle period between non-evolutionary and evolutionary cycles, analogous to the distinct gap in cycle periods (scaled relative to maturation time) predicted between consumer–resource cycles and cycles driven by intraspecific competition (Murdoch et al., 2002). In addition, there has been to date no exhaustive study of model and parameter space that would permit a definitive prediction of how cycle period is affected by evolution.

With experimental single predator–single prey systems, the presence of prey evolution can often be detected from a qualitative change in the temporal consumer–resource dynamics from a normal quarter-period phase lag between predator and prey to antiphase cycles (Becks et al., 2012; Hiltunen et al., 2014; Yoshida et al., 2003, 2007). However, with the very slightly increased food web complexity studied here, such signature dynamics are much less evident. For example, based on theory (Ellner and Becks, 2011), there are at least 11 different qualitatively distinct types of three-species food webs with prey evolution (i.e. with and/or without defence against either the top predator, or intermediate predator, or both), and in some cases even a single type of these food webs can exhibit several different dynamic outcomes. With this array of possible patterns, it is very difficult to infer with confidence the presence of the prey evolution merely from observed temporal community dynamics. The only general prediction for the presence of evolution is that the resulting dynamics can be very different from the unique pattern that is predicted when neither predator nor prey are evolving (Fig. 2.2A).

For our three-species community, we assessed, as a possible route to detecting the presence or absence of prey trait evolution, estimating the predictability of the top predator's growth rate based purely on the abundance of each of its two prey populations. The rationale for this approach is the fact that in the absence of evolution, on average, all prey individuals within a species should be of equal food quality to the predator so that only the combined abundances of the two prey species are important in determining predator growth rate. Because evolution can radically change the quality of prey as food from one time interval to another, predator dynamics should become much less predictable when prey quality is evolving but only prey quantity is used to predict predator dynamics. For our data, as expected, the predictability of the predator growth rate was substantially greater in runs without prey evolution than for those with prey evolution (Table 2.1).

One feature of many of our chemostat runs with prey evolution was a pattern of extended periods of numerical dominance by only one of the two predator species, while the other remained at very low density, followed by a rapid reversal with the rare species becoming dominant and the other becoming scarce. We suggest that this dynamical pattern, consistent with the outcome expected when prey defences are incompatible, may represent the canard dynamics described in the previous section. In a population and community context, canards may provide a previously unappreciated mechanism for patterns such as pest or disease outbreaks. For example, canard "explosions" have been employed to explain insect–pest outbreaks in forests (Brøns and Kaasen, 2010). We suggest that in our system, alternating periods of relative stasis followed by rapid shifts in predator domination may be a canard, wherein the prey very slowly evolve defence against the dominant predator (slow because the defended genotype is extremely rare), until the favoured allele is frequent enough for selection to elicit a response. Defence against the dominant predator then evolves quickly, resulting in a rapid switch in which predator is dominant (Figs. 2.6 and 2.7). To our knowledge, this would be the first example of canards in a controlled experimental system. Canard "explosions" (Brøns and Kaasen, 2010) occur in a narrow range of parameters but eco-evolutionary canards are robust in our models and occur for a broad range of parameters such that selection for defence against the dominant predator is strong. However, all our models describe haploid single-locus selection, and other interesting phenomena might be found in diploid or quantitative trait models where dominance, epistasis and recombination would all play a role in the response to strong selection favouring a rare allele or combination of alleles.

In general, when new species are added to a community, the number of ecological interactions multiplies and each species may face an increased array of selection pressures: ultimately, the number of possible evolutionary outcomes must also increase (Strauss, 2013). The results presented here illustrate clearly an explosion in dynamical complexity that occurs with only a modest increase in food web complexity from the two-species (single predator–single prey, rotifer–algal) system we have studied previously with and without prey evolution (Becks et al., 2010, 2012; Yoshida et al., 2003, 2007) to this three-species (rotifer–flagellate–algal) system, first without (Hiltunen et al., 2013) and now with evolution. Our experimental results confirm theory (Ellner and Becks, 2011) suggesting just such a marked expansion of possible outcomes. Whereas increasing food web complexity with the addition of an intermediate predator in the absence of evolution produced temporal dynamics that were still very straightforward to interpret (Hiltunen et al., 2013), the addition of prey genetic diversity and hence their capacity to evolve leads to an intriguingly complex array of dynamical outcomes.

Our results lead us to question the extent to which the patterns we observed in simple two-species food webs can be expected to also occur in more complex natural food webs. We suggest that it will be at least very challenging to infer the mechanisms underlying the dynamics of natural communities from temporal patterns of population abundances alone, especially if the potential for rapid and reversible evolution of the players is not taken into account. Furthermore, even when the potential for this evolution is recognized—and we note that rapid contemporary evolution is likely the norm in natural communities (Hairston et al., 2005; Hendry and Kinnison, 1999; Post and Palkovacs, 2009)—the range of theoretically possible outcomes is so large that inferring causal mechanisms by matching observed to predicted patterns will be extremely difficult. Traits and their dynamics will have to be studied directly and in parallel with population dynamics.

ACKNOWLEDGEMENTS

We thank K. Blackley, T. Hermann, A. Looi, D. Rosenberg and C. Zhang, for help with experimental set-up and sampling, and A. Barreiro, C. M. Kearns and L. R. Schaffner for laboratory assistance. This research was primarily supported by Grant No. 220020137 from the James S. McDonnell Foundation. US NSF Grant DEB-1256719 partially supported the research of S. P. E, N. G. H and G. H., and N. G. H. was supported by Eawag, the Swiss Federal Institute of Aquatic Science and Technology, while this chapter was being prepared for publication.

REFERENCES

Agrawal, A.A., Johnson, M.T.J., Hastings, A.P., Maron, J.L., 2013. A field experiment demonstrating plant life-history evolution and its eco-evolutionary feedback to seed predator populations. Am. Nat. 181, S35–S45.

Bassar, R.D., Marshall, M.C., Lopez-Sepulcre, A., Zandona, E., Auer, S.K., Travis, J., et al., 2010. Local adaptation in Trinidadian guppies alters ecosystem processes. Proc. Natl. Acad. Sci. U.S.A. 107, 3616–3621.

Becks, L., Ellner, S.P., Jones, L.E., Hairston Jr., N.G., 2010. Reduction of adaptive genetic diversity radically alters eco-evolutionary community dynamics. Ecol. Lett. 13, 989–997.

Becks, L., Ellner, S.P., Jones, L.E., Hairston Jr., N.G., 2012. The functional genomics of an eco-evolutionary feedback loop: linking gene expression, trait evolution, and community dynamics. Ecol. Lett. 15, 492–501.

Boraas, M.E., Seale, D.B., Boxhorn, J.E., 1998. Phagotrophy by a flagellate selects for colonial prey: a possible origin of multicellularity. Evol. Ecol. 12, 153–164.

Brøns, M., Kaasen, R., 2010. Canards and mixed-mode oscillations in a forest pest model. Theor. Popul. Biol. 77, 238–242.

De Meester, L., Van Doorslaer, W., Geerts, A., Orsini, L., Stoks, R., 2011. Thermal genetic adaptation in the water flea *Daphnia* and its impact: an evolving metacommunity approach. Integr. Comp. Biol. 51, 703–718.

De Roos, A.M., Persson, L., 2003. Competition in size-structured populations: mechanisms inducing cohort formation and population cycles. Theor. Popul. Biol. 63, 1–16.

Ellner, S.P., Becks, L., 2011. Rapid prey evolution and the dynamics of two-predator food webs. Theor. Ecol. 4, 133–152.

Ellner, S.P., Hairston Jr., N.G., Kearns, C.M., Babaï, D., 1999. The roles of fluctuating selection and long-term diapause in microevolution of diapause timing in a freshwater copepod. Evolution 53, 111–122.

Falconer, D.S., 1960. Introduction to Quantitative Genetics. Oliver and Boyd Press, London, UK.

Fisher, R.A., 1930. The Genetical Theory of Natural Selection. Oxford University Press, Oxford, UK.

Fussmann, G.F., Ellner, S.P., Shertzer, K.W., Hairston Jr., N.G., 2000. Crossing the Hopf bifurcation in a live predator-prey system. Science 290, 1358–1360.

Fussmann, G.F., Loreau, M., Abrams, P.A., 2007. Eco-evolutionary dynamics of communities and ecosystems. Funct. Ecol. 21, 465–477.

Gallagher, J.C., 1982. Physiological variation and electrophoretic banding patterns of genetically different seasonal populations of *Skeletonema costatum* (Bacillariophyceae). J. Phycol. 18, 148–162.

Grant, P.R., Grant, B.R., 2002. Unpredictable evolution in a 30-year study of Darwin's finches. Science 296, 707–711.

Gu, C., 2013. Smoothing Spline ANOVA Models, second ed. Springer, New York.

Guillard, R.R.L., Ryther, J.H., 1962. Studies of marine planktonic diatoms. I. *Cyclotella nana* Hustedt and *Detonula confervacea Cleve*. Can. J. Microbiol. 8, 229–239.

Hairston Jr., N.G., Kearns, C.M., Ellner, S.P., 1996. Phenotypic variation in a zooplankton egg bank. Ecology 77, 2382–2392.

Hairston Jr., N.G., Ellner, S.P., Geber, M.A., Yoshida, T., Fox, J.A., 2005. Rapid evolution and the convergence of ecological and evolutionary time. Ecol. Lett. 8, 1114–1127.

Hanski, I.A., 2011. Eco-evolutionary spatial dynamics in the Glanville fritillary butterfly. Proc. Natl. Acad. Sci. U.S.A. 108, 14397–14404.

Harmon, L.J., Matthews, B., Des Roches, S., Chase, J.M., Shurin, J.B., Schluter, D., 2009. Evolutionary diversification in stickleback affects ecosystem functioning. Nature 458, 1167–1170.

Hendry, A.P., Kinnison, M.T., 1999. Perspective: the pace of modern life: measuring rates of contemporary microevolution. Evolution 53, 1637–1653.

Hessen, D.O., Van Donk, E., 1993. Morphological changes in *Scenedesmus* induced by substances released from *Daphnia*. Archiv. Hydrobiol. 127, 129–140.

Hiltunen, T., Barreiro, A., Hairston Jr., N.G., 2012. Mixotrophy and the toxicity of *Ochromonas* in a pelagic food web. Freshw. Biol. 57, 2262–2271.

Hiltunen, T., Jones, L.E., Ellner, S.P., Hairston Jr., N.G., 2013. Temporal dynamics of a simple community with intraguild predation: an experimental test. Ecology 94, 773–779.

Hiltunen, T., Hairston Jr., N.G., Hooker, G., Jones, L.E., Ellner, S.P., 2014. A newly discovered role of evolution in previously published consumer–resource dynamics. Ecol. Lett. 17 (8), 915–923.

Holt, R.D., Polis, G.A., 1997. A theoretical framework for intraguild predation. Am. Nat. 149, 745–764.

Jones, L.E., Ellner, S.P., 2007. Effects of rapid evolution on predator-prey cycles. J. Math. Biol. 55, 541–573.

Lurling, M., Beekman, W., 2006. Palmelloids formation in *Chlamydomonas reinhardtii*: defence against rotifer predators? Ann. Limnol. Int. J. Limnol. 42, 65–72.

Meyer, J.R., Ellner, S.P., Hairston Jr., N.G., Jones, L.E., Yoshida, T., 2006. Evolution on the time scale of predator-prey dynamics revealed by allele-specific quantitative PCR. Proc. Natl. Acad. Sci. U.S.A. 103, 10690–10695.

Murdoch, W.W., Kendall, B.E., Nisbet, R.M., Briggs, C.J., McCauley, E., Bolser, R., 2002. Single-species models for many-species food webs. Nature 417, 541–543.

Pelletier, F., Clutton-Brock, T., Pemberton, J., Tuljapurkar, S., Coulson, T., 2007. The evolutionary demography of ecological change: linking trait variation and population growth. Science 315, 1571–1574.

Post, D.M., Palkovacs, E.P., 2009. Eco-evolutionary feedbacks in community and ecosystem ecology: interactions between the ecological theatre and the evolutionary play. Philos. Trans. R. Soc. B 364, 1629–1640.

Ramsay, J.O., Wickham, H., Graves, S., Hooker, G., 2013. fda: functional data analysis. R package version 2.3.8. http://CRAN.R-project.org/package=fda.

Shertzer, K.W., Ellner, S.P., Fussmann, G.F., Hairston Jr., N.G., 2002. Predator-prey cycles in a live aquatic microcosm: testing hypotheses of mechanism. J. Anim. Ecol. 71, 802–815.

Sinervo, B., Svensson, E., Comendant, T., 2000. Density cycles and an offspring quantity and quality game driven by natural selection. Nature 406, 985–988.

Strauss, S., 2013. Ecological and evolutionary responses in complex communities: implications for invasions and eco-evolutionary feedbacks. Oikos 123, 257–266.

Van Doorslaer, W., Stoks, R., Duvivier, C., Bednarska, A., De Meester, L., 2009. Population dynamics determine genetic adaptation to temperature in *Daphnia*. Evolution 63, 1867–1878.

Wahba, G., 1990. Spline Models for Observational Data. SIAM, Philadelphia, PA.

Wechselberger, M., 2005. Existence and bifurcation of canards in R^3 in the case of a folded node. SIAM J. Appl. Dyn. Syst. 4, 101–139.

Wood, S.N., 2006. Generalized Additive Models: An Introduction with R. Chapman and Hall/CRC, New York.

Wood, S.N., 2011. Fast stable restricted maximum likelihood and marginal likelihood estimation of semiparametric generalized linear models. J. R. Stat. Soc. B 73, 3–36.

Wymore, A.S., Keeley, A.T.H., Yturralde, K.M., Schroer, M.L., Propper, C.R., Whitham, T.G., 2011. Genes to ecosystems: exploring the frontiers of ecology with one of the smallest biological units. New Phytol. 191, 19–36.

Yoshida, T., Jones, L.E., Ellner, S.P., Fussmann, G.F., Hairston Jr., N.G., 2003. Rapid evolution drives ecological dynamics in a predator-prey system. Nature 424, 303–306.

Yoshida, T., Hairston Jr., N.G., Ellner, S.P., 2004. Evolutionary tradeoff between defense against grazing and competitive ability in a simple unicellular alga, *Chlorella vulgaris*. Proc. R. Soc. Lond. B 271, 1947–1953.

Yoshida, T., Ellner, S.P., Jones, L.E., Bohannan, B.J.M., Lenski, R.E., Hairston Jr., N.G., 2007. Cryptic population dynamics: rapid evolution masks trophic interactions. PLoS Biol. 5, 1868–1879.

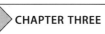

CHAPTER THREE

Eco-Evolutionary Spatial Dynamics: Rapid Evolution and Isolation Explain Food Web Persistence

Jordi Moya-Laraño*,1, José Román Bilbao-Castro†,
Gabriel Barrionuevo†, Dolores Ruiz-Lupión*, Leocadio G. Casado†,
Marta Montserrat‡, Carlos J. Melián§, Sara Magalhães¶

*Department of Functional and Evolutionary Ecology, Estación Experimental de Zonas Áridas, EEZA–CSIC, Carretera de Sacramento s/n., Almería, Spain
†Department of Informatics, University of Almeria, Cañada de San Urbano S/N, Almería, Spain
‡Instituto de Hortofruticultura Subtropical y Mediterránea "La Mayora" (IHSM–UMA–CSIC), Consejo Superior de Investigaciones Científicas, Algarrobo–Costa, Málaga, Spain
§Department of Fish Ecology and Evolution, Center for Ecology, Evolution and Biogeochemistry, Swiss Federal Institute of Aquatic Science and Technology, Switzerland
¶Centro de Biologia Ambiental, Faculdade de Ciências da Universidade de Lisboa, Lisbon, Portugal
1Corresponding author: e-mail address: jordi@eeza.csic.es

Contents

Advances in Ecological Research, Volume 50
ISSN 0065-2504
http://dx.doi.org/10.1016/B978-0-12-801374-8.00003-7

Abstract

One of the current challenges in evolutionary ecology is understanding the long-term persistence of contemporary-evolving predator–prey interactions across space and time. To address this, we developed an extension of a multi-locus, multi-trait eco-evolutionary individual-based model that incorporates several interacting species in explicit landscapes. We simulated eco-evolutionary dynamics of multiple species food webs with different degrees of connectance across soil-moisture islands. A broad set of parameter combinations led to the local extinction of species, but some species persisted, and this was associated with (1) high connectance and omnivory and (2) ongoing evolution, due to multi-trait genetic variability of the embedded species. Furthermore, persistence was highest at intermediate island distances, likely because of a balance between predation-induced extinction (strongest at short island distances) and the coupling of island diversity by top predators, which by travelling among islands exert global top-down control of biodiversity. In the simulations with high genetic variation, we also found widespread trait evolutionary changes indicative of eco-evolutionary dynamics. We discuss how the ever-increasing computing power and high-resolution data availability will soon allow researchers to start bridging the *in vivo–in silico* gap.

The more ambitious plan may have more chances of success... This sounds paradoxical... [but] the paradox disappears if we look closely at a few examples... provided [these are] not based on a mere pretension but on some vision of the things beyond those immediately present.

György Pólya (1957)

1. INTRODUCTION

1.1. Food webs and eco-evolutionary dynamics

Although the notion that ecology and evolution operate at similar timescales was put forward nearly 50 years ago (Pimentel, 1968), the first use of the term eco–evolutionary dynamics dates to Savill et al. (1997). The reciprocal effect of ecology and evolution on biological interactions is central to this concept, and eco–evolutionary dynamics have mainly been studied in this context. This is illustrated by the fact that the most prominent reviews (e.g. Carroll et al., 2007; Schoener, 2011) and the seminal studies on eco-evolutionary dynamics (Reznick et al., 1990; Yoshida et al., 2003) concern species interactions. At least two main factors may account for this focus on biological interactions: first, the selection pressure posed by (mainly) antagonistic interactions is very strong (Benkman, 2013), so evolutionary responses should be rapid, merging ecological and evolutionary timescales. Indeed, several studies have documented fast responses to the selection pressure posed by predators (e.g. Orsini et al., 2012; Reznick et al., 1990) or parasites (e.g. Kraaijeveld and Godfray, 1997; Martins et al., 2013). Second, ecological dynamics are best described for species interactions, thus providing a solid foundation for incorporating predictions of evolutionary feedbacks in ecology.

Most eco-evolutionary studies have addressed antagonistic interactions between two species (Losos et al., 2004; Palkovacs and Post, 2009; Terhorst et al., 2010). Beginning with Darwin's tangled bank (Darwin, 1959), however, ecologists have long been aware of the far more complex web of interactions in which individuals are embedded (Elton, 1927; Winemiller and Polis, 1996). Hence, incorporating ecological networks in the framework of eco-evolutionary dynamics is a natural extension, as highlighted in a series of reviews and opinion pieces (Agrawal et al., 2006; Bolnick et al., 2011; Fontaine et al., 2011; Fussmann et al., 2007; Olesen et al., 2010; Stewart et al., 2013; Thompson, 1998).

Most available experimental studies of eco-evolutionary dynamics in food webs have tested how individuals with different evolutionary histories differentially affect an ecological community (Bassar et al., 2012; Chislock et al., 2013; Farkas et al., 2013; Harmon et al., 2009; Ingram et al., 2012; Lau, 2012; Palkovacs and Post, 2009; Urban, 2013; Walsh et al., 2012). For example, Bassar et al. (2012; see also Travis et al., 2014, chapter 1 of this volume) showed that guppies that had evolved in ponds with or without predators had different effects on the prey invertebrate community. These examples show how the product of a rapid evolutionary process (i.e. organisms that have evolved in one environment or another) affect ecosystems, yet the strength and sign of prey selection (i.e. positive or negative frequency-dependent selection) remain to be tested in experiments that also account for prey abundance (see Melian et al., 2014).

Another possible way by which evolution can affect ecology on a contemporary timescale is if the evolution of the genetic composition of populations alters ecological dynamics (Becks et al., 2010; Johnson et al., 2009; Rowntree et al., 2011; Yoshida et al., 2003). For instance, predator–prey dynamics in genetically homogeneous prey populations changed dramatically when the prey population included two clones with different resistance properties (Yoshida et al., 2003). This led to rapid evolution of the prey population, which, in turn, affected the ecological dynamics of predator–prey cycles. A similar approach was used in host–parasite webs by Lennon and Martiny (2008), in which the introduction of viruses in a community of algae caused rapid evolution of resistance in the algal host, following a dampening of the initial effect of viruses on nutrient cycling. Similarly, Johnson et al. (2009) showed rapid evolution of plant traits and used a model to predict the impact of such changes in the arthropod community inhabiting the plants. Such an approach requires either populations with different standing genetic variations at the outset, in order to have a control for the evolutionary rate, or a strong modelling approach to generate testable predictions on the relative role of ecology and evolution in the dynamics of prey and predators.

In summary, standing genetic variation and the strength and sign of selection can alter population dynamics of species in ways that may be difficult to anticipate. One question in particular that remains to be addressed is how genetic variation at the outset of dynamics and the strength and sign of selection may affect food web structure and dynamics. Earlier results, using a modelling approach, suggested that higher genetic variability for traits contributes to food web stability by increasing connectance and variability in interaction strengths (Moya-Larano, 2011, see also Melián et al., 2011). This issue is particularly relevant in "metanetwork" food webs that are scattered in a heterogeneous space, in which local interactions will shape the outcome of each subpopulation.

1.2. Space, the next frontier

Although the literature on eco-evolutionary dynamics has been growing at an extraordinary pace, the incorporation of space has lagged behind somewhat (Urban et al., 2008), which is surprising given its prominence in other ecological and evolutionary fields (Levins, 1968). Actually, early attempts at combining ecological changes in population densities with evolutionary changes in gene frequencies have been done in the context of spatially heterogeneous environments (Levene, 1953), and some researchers have even produced spatial ecological and evolutionary data in the same study system (e.g. Singer and Thomas, 1996; Thomas et al., 1996).

Dispersal is a very powerful trait that links ecology and evolution: for instance, migration among patches changes both the density and the allele frequency of populations. In turn, both the connectivity among patches and their genetic composition can affect the sign and strength of selection for dispersal. For example, traits from rare migrants can become dominant (i.e. the advantage of the rare or negative frequency-dependent selection) or quickly go extinct in a new patch (i.e. the advantage of the common or positive-frequency-dependent selection). The role of spatial heterogeneity in eco-evolutionary dynamics has only been addressed in few experimental systems (Farkas et al., 2013; Hanski, 2011; Kerr et al., 2006; Singer and McBride, 2012), with a particularly well-documented exception being that of the Glanville fritillary butterfly, where the allele frequency in the *pgi* dispersing gene is driven by spatial heterogeneity (Hanski, 2011).

Given the links between dispersal and the dynamics of interacting populations and allele frequencies, several avenues of research are worthy of further exploration to help us understand how systems in spatially heterogeneous landscapes are shaped by eco-evolutionary dynamics.

1.3. Merging space, food webs and evolution

Recently, researchers (mainly ecologists) have focused on the study of multi-trophic metacommunities (i.e. communities linked by dispersal and trophic interactions—Haegeman and Loureau, 2014; Holyoak et al., 2005; Melian et al., 2014; Pillai et al., 2011; Wilson, 1992). The composition of such multi-trophic metacommunities reflects that of simple communities, in which species may be linked by a linear food chain or embedded in a complex food web (Bohan and Woodward, 2013; Tamadoni-Nezhad et al., 2013; Winemiller and Polis, 1996).

Spatial heterogeneity plays an important role in structuring food webs. Indeed, McCann et al. (2005) and Rooney et al. (2006) showed that the high mobility of predators can couple different food web energy channels across space and contribute to global food web stability (the "bird feeder effect") (McCann et al., 2005; Rooney et al., 2006). Moreover, dispersal of omnivores may contribute to the robustness of the metacommunity, thereby enhancing food web complexity and species diversity (Pillai et al., 2011). None of these examples, however, considers the evolution of any of the players involved in the food web.

In a subsequent study, a mathematical model predicted that local extinction patterns in a predator–prey metacommunity had differential effects on the evolution of dispersal in predators and prey (Pillai et al., 2012). This promising result suggests that emergent properties stem from the combination of ecology and evolution in metacommunities of two species. Hence, eco-evolutionary dynamics in metacommunities composed of more complex food webs could provide the foundation from which emergent patterns are generated. Still, both empirical and theoretical studies on this topic are conspicuously lacking (Thuiller et al., 2013; Urban et al., 2008): we aim to merge empirical observations and theory by extending food webs in heterogeneous space using invertebrate soil food webs as a model system (Moya-Laraño et al., 2012).

1.4. Soil food webs as a model system

Soil food webs are diverse systems that underpin the decomposition of organic material and nutrient recycling in terrestrial ecosystems (André et al., 2002; Decaëns, 2010; Hättenschwiler et al., 2005; Swift et al., 1979; Wardle, 2006). In recent years, both laboratory (Brose et al., 2008; Schneider et al., 2012) and field experiments (Chen and Wise, 1999; McLaughlin et al., 2010; Moya-Larano and Wise, 2007; Scheu and Schaefer, 1998; Wise and Chen, 1999) have been conducted to address relevant ecological questions related to food web theory, including how water affects decomposition processes indirectly via its effects on the food web (Lensing and Wise, 2006). Soils are highly heterogeneous ecosystems (Moore et al., 2004), with strong spatial heterogeneity in water content (Schume et al., 2003). This can drive the spatial structuring of leaf-litter food webs at micro-, local and regional scales (Melguizo-Ruiz et al., 2012). At the micro-environmental scale, water can accumulate in soil patches due to the micro-topography of the area in which the leaf litter sits (e.g. the base of slopes) or other landmarks which affect the micro-environment (e.g. underneath shrubs at the base of tree trunks). Since soil moisture affects soil fauna, by attracting them to moist areas during dry conditions (Verdeny-Vilalta and Moya-Laraño, 2014), these moisture pockets may work as islands of productivity during drought conditions. This provides an ideal scenario for studying metacommunity dynamics, using micro-environmental patches connected by migration.

1.5. Aims: A few examples of hypothesis testing using Weaver

Here, we present Weaver, an Individual-Based Modelling computer program that aims to fill the gap between empirical observations of individual-based food webs in heterogeneous space with the theoretical predictions coming from eco-evolutionary multi-trophic metacommunity dynamics. This program is an extension of a former simpler platform (mini-Akira, Moya-Laraño et al., 2012) which, by having greatly increased computing performance, allows the exploration of individual-based eco-evolutionary dynamics in multi-species food webs across space. This framework links genes to ecosystems through space, reaching an unprecedented level of comprehensiveness that provides insight into ecological and evolutionary dynamics at the gene, individual, population and community levels at different spatial scales. All of these can also be linked to ecosystem processes, such as top-down control of predators inducing trophic cascades affecting basal resources at different temperatures (Moya-Laraño et al., 2012), or the role of predators for maintaining biodiversity across space under different food web, island and genetic configurations. In addition, rather than being a "black box", the present framework, and as in empirical studies, produces several detailed outputs (including gene spatio-temporal dynamics) that can be used for understanding the various mechanisms behind eco-evolutionary dynamics. The researcher can track everything that each gene and each individual has done in the simulation. As far as we know, no other framework is currently capable of providing such detail. However, although we develop some new hypotheses in Section 4, here we are focused primarily on exploring food web persistence and trait evolution under different genetic and ecological scenarios: one of the great advantages is that one can perform additional simulation experiments (e.g. eliminating particular species or knocking down variability in a particular trait) to uncover the mechanisms responsible for the patterns emerging from simulations.

The overall purpose of the simulations presented here is to illustrate the usefulness of our IBM framework through a few examples, addressing some of the main open questions in the field of eco-evolutionary dynamics and ecological networks in space. We ask six questions: (1) Does connectance affect food web persistence? (2) Does standing intraspecific genetic variation alter food web persistence by triggering rapid evolution? (3) In a multi-trophic metacommunity context, how does the spatial structure (i.e. degree of spatial isolation) alter persistence and rapid evolution in food webs? (4) In all of the above scenarios, do predators inhibit prey populations in rapid evolving predator–prey systems and, if so, do predators contribute to maintain prey diversity? (5) How does standing genetic variation alter the evolutionary rate and the persistence of predator–prey systems? (6) Do traits evolve differently in prey under purely competitive environments (without predators) versus those where both predation and competition are at play?

Of course, the results derived from this approach are merely digital approximations of nature: to take full advantage of this simulation platform, the results must be tested in real systems and fed back to one each other iteratively. We propose how to integrate simulations with real systems to link, step by step, the *in vivo–in silico* gap, with the potential of generating an unprecedented level of understanding about how real ecosystems work. For instance, if one output is produced in nature that the simulations are not reproducing, the new estimates and parameters found in the experiment can be included in further simulations. We have included a subsection in Section 4 to explain how to implement this Feedback Research Program (FRP). However, we stress again that performing studies at the digital level only, as we do here, can provide important clues and hypotheses that need to be empirically tested in future experiments.

2. MATERIALS AND METHODS

Unless stated otherwise, we here use the same approach as in the former paper (Moya-Laraño et al., 2012), in which we provided a level of detail that is beyond the scope of this chapter. We therefore refer to the reader who wants to fully grasp all the underpinnings and details of this modelling framework to the above reference. However, we have made an effort to explain the most relevant parameters necessary to follow the approach and make this article as self-explanatory as possible. When necessary, we have actually replicated some of the information in Moya-Laraño et al. (2012; Appendix).

We modelled 20-species beech forest soil food webs (Melguizo-Ruiz et al., 2012) with differing degrees of connectivity and genetic variation in the same 13 traits as in Moya-Laraño et al., 2012 (Appendix), namely: fixed body size at birth; amount of energy for maintenance and growth at birth; growth ratio; phenology (or genetically determined development time determining birth date beyond environmental constraints such as changes in temperature); searching area; voracity; sprint speed; metabolic rate; temperature plasticity for speed; voracity and searching area; and activation energy for metabolic rate. Estimates of ecological ranges for temperature-dependent traits were obtained from the literature (Dell et al., 2011; Ehnes et al., 2011). The amount of genetic trait variability is governed by the parameter φ (Moya-Laraño et al., 2012), which can be thought of as a genetic restriction parameter ranging between 0 and 1 (Appendix). Values of φ close to 0 indicate that the trait has the highest possible genetic variation (i.e. across the entire phenotypic range; Table 3.1) and

Table 3.1 Species and trait ranges included in simulations

Species ID	Class	Common name	Taxon[a]	Feeding guild	Trait ranges	
					Energy tank (%)[b]	Growth ratio
(a) Webs with connectance 0.1 or 0.55						
aca1	Arachnida	Mite	Mesostigmatida	Predator	0.25–0.5	1.35–1.45
aca2	Arachnida	Mite	Prostigmatida	Predator	0.25–0.5	1.35–1.45
aca3	Arachnida	Mite	Mesostigmatida	Predator	0.25–0.5	1.35–1.45
aca4	Arachnida	Mite	Prostigmatida	Predator	0.25–0.5	1.35–1.45
spd1	Arachnida	Spider	Agelenidae	Predator	0.25–0.5	1.20–1.30
spd1	Arachnida	Spider	Erigoninae	Predator	0.25–0.5	1.25–1.35
spd3	Arachnida	Spider	Dysderidae	Predator	0.25–0.5	1.15–1.25
spd4	Arachnida	Spider	Theridiidae	Predator	0.25–0.5	1.15–1.25
geo1	Chilopoda	Centipede	Geophilomorpha	Predator	0.25–0.5	1.15–1.25
lit1	Chilopoda	Centipede	Lithobiomorpha	Predator	0.25–0.5	1.15–1.25
opi1	Arachnida	Hartvestmen	Opilionida	Predator	0.25–0.5	1.15–1.25
col1	Insecta	Springtail	Collembola	Fungivore	0.25–0.5	1.25–1.35
col2	Insecta	Springtail	Collembola	Fungivore	0.25–0.5	1.35–1.45
col3	Insecta	Springtail	Collembola	Fungivore	0.25–0.5	1.15–1.25
enc1	Oligochaeta	Potworm	Enchytraeidae	Fungivore	0.25–0.5	1.35–1.45
enc2	Oligochaeta	Potworm	Enchytraeidae	Fungivore	0.25–0.5	1.25–1.35
enc3	Oligochaeta	Potworm	Enchytraeidae	Fungivore	0.25–0.5	1.25–1.35
ori1	Arachnida	Mite	Oribatida	Fungivore	0.25–0.5	1.35–1.45
ori2	Arachnida	Mite	Oribatida	Fungivore	0.25–0.5	1.15–1.25
ori3	Arachnida	Mite	Oribatida	Fungivore	0.25–0.5	1.25–1.35

Continued

Table 3.1 Species and trait ranges included in simulations—cont'd

Species ID	Trait ranges					
	Phenology (days)	Body size at birth (mg)	Assimilation efficiency	Voracity	Sprint speed	Search area
aca1	3–11	0.001–0.003	0.7–0.9	0.55–0.75	0.1–0.3	0.1–0.4
aca2	3–11	0.002–0.004	0.7–0.9	0.55–0.75	0.1–0.3	0.1–0.4
aca3	3–11	0.003–0.005	0.7–0.9	0.55–0.75	0.1–0.3	0.1–0.4
aca4	3–11	0.004–0.006	0.7–0.9	0.55–0.75	0.1–0.3	0.1–0.4
spd1	3–11	0.013–0.033	0.7–0.9	0.55–0.75	0.1–0.3	0.1–0.4
spd1	3–11	0.008–0.010	0.7–0.9	0.55–0.75	0.1–0.3	0.1–0.4
spd3	3–11	0.049–0.069	0.7–0.9	0.55–0.75	0.1–0.3	0.1–0.4
spd4	3–11	0.009–0.029	0.7–0.9	0.55–0.75	0.1–0.3	0.1–0.4
geo1	3–11	0.09–0.10	0.7–0.9	0.55–0.75	0.1–0.3	0.1–0.4
lit1	3–11	0.045–0.065	0.7–0.9	0.55–0.75	0.1–0.3	0.1–0.4
opi1	3–11	0.004–0.026	0.7–0.9	0.55–0.75	0.1–0.3	0.1–0.4
col1	3–11	0.001–0.003	0.7–0.9	0.55–0.75	0.1–0.3	0.1–0.3
col2	3–11	0.004–0.024	0.7–0.9	0.55–0.75	0.1–0.3	0.1–0.3
col3	3–11	0.002–0.003	0.7–0.9	0.55–0.75	0.1–0.3	0.1–0.3
enc1	3–11	0.0001–0.002	0.5–0.7	0.55–0.75	0.1–0.3	0.1–0.3
enc2	3–11	0.0001–0.002	0.5–0.7	0.55–0.75	0.1–0.3	0.1–0.3
enc3	3–11	0.001–0.003	0.5–0.7	0.55–0.75	0.1–0.3	0.1–0.3
ori1	3–11	0.003–0.005	0.7–0.9	0.55–0.75	0.1–0.3	0.1–0.3
ori2	3–11	0.0001–0.002	0.7–0.9	0.55–0.75	0.1–0.3	0.1–0.3
ori3	3–11	0.001–0.003	0.7–0.9	0.55–0.75	0.1–0.3	0.1–0.3

Trait ranges

Species ID	met_rate	Q10 voracity	Q10 speed	Q10 search area	Activation energy (eV)
aca1	0.6–0.8	3–4	1.5–2.5	2.0–2.5	0.33–0.42
aca2	0.6–0.8	3–4	1.5–2.5	2.0–2.5	0.36–0.46
aca3	0.6–0.8	3–4	1.5–2.5	2.0–2.5	0.33–0.42
aca4	0.6–0.8	3–4	1.5–2.5	2.0–2.5	0.36–0.46
spd1	0.5–0.7	3–4	1.5–2.5	2.0–2.5	0.65–0.75
spd1	0.5–0.7	3–4	1.5–2.5	2.0–2.5	0.65–0.75
spd3	0.5–0.7	3–4	1.5–2.5	2.0–2.5	0.65–0.75
spd4	0.5–0.7	3–4	1.5–2.5	2.0–2.5	0.65–0.75
geo1	0.46–0.66	3–4	1.5–2.5	2.0–2.5	0.75–0.85
lit1	0.45–0.65	3–4	1.5–2.5	2.0–2.5	0.75–0.85
opi1	0.46–0.76	3–4	1.5–2.5	1.5–2.5	0.65–0.75
col1	0.65–0.85	2–4	1.5–2.5	1.5–2.5	0.61–0.70
col2	0.65–0.85	2–4	1.5–2.5	1.5–2.5	0.61–0.70
col3	0.55–0.85	2–4	1.5–2.5	1.5–2.5	0.61–0.70
enc1	0.70–0.90	2–4	1.5–2.5	1.5–2.5	0.39–0.49
enc2	0.70–0.90	2–4	1.5–2.5	1.5–2.5	0.39–0.49
enc3	0.70–0.90	2–4	1.5–2.5	1.5–2.5	0.39–0.49
ori1	0.57–0.77	2–4	1.5–2.5	1.5–2.5	0.66–0.76
ori2	0.57–0.77	2–4	1.5–2.5	1.5–2.5	0.66–0.76
ori3	0.57–0.77	2–4	1.5–2.5	1.5–2.5	0.66–0.76

Continued

Table 3.1 Species and trait ranges included in simulations—cont'd

| | | | | | Trait ranges | |
Species ID	Class	Common name	Taxon[a]	Feeding guild	Energy tank (%)[b]	Growth ratio
(b) Web with connectance 0.3						
aca1	Arachnida	Mite	Mesostigmatida	Predator	0.25–0.5	1.35–1.45
aca2	Arachnida	Mite	Prostigmatida	Predator	0.25–0.5	1.35–1.45
aca3	Arachnida	Mite	Mesostigmatida	Predator	0.25–0.5	1.35–1.45
aca4	Arachnida	Mite	Prostigmatida	Predator	0.25–0.5	1.35–1.45
aca5	Arachnida	Mite	Mesostigmatida	Predator	0.25–0.5	1.35–1.45
aca6	Arachnida	Mite	Prostigmatida	Predator	0.25–0.5	1.20–1.30
spd1	Arachnida	Spider	Agelenidae	Predator	0.25–0.5	1.25–1.35
spd2	Arachnida	Spider	Erigoninae	Predator	0.25–0.5	1.15–1.25
spd3	Arachnida	Spider	Dysderidae	Predator	0.25–0.5	1.15–1.25
spd4	Arachnida	Spider	Theridiidae	Predator	0.25–0.5	1.15–1.25
spd5	Arachnida	Spider	Dysderidae	Predator	0.25–0.5	1.15–1.25
spd6	Arachnida	Spider	Erigoninae	Predator	0.25–0.5	1.15–1.25
geo1	Chilopoda	Centipede	Geophilomorpha	Predator	0.25–0.5	1.15–1.25
geo2	Chilopoda	Centipede	Geophilomorpha	Predator	0.25–0.5	1.15–1.25
lit1	Chilopoda	Centipede	Lithobiomorpha	Predator	0.25–0.5	1.15–1.25
lit2	Chilopoda	Centipede	Lithobiomorpha	Predator	0.25–0.5	1.15–1.25
col1	Insecta	Springtail	Collembola	Fungivore	0.25–0.5	1.25–1.35
col2	Insecta	Springtail	Collembola	Fungivore	0.25–0.5	1.35–1.45
enc1	Oligochaeta	Potworm	Enchytraeidae	Fungivore	0.25–0.5	1.35–1.45
ori1	Arachnida	Mite	Oribatida	Fungivore	0.25–0.5	1.35–1.45

Trait ranges

Species ID	Phenology (days)	Body size at birth (mg)	Assimilation efficiency	Voracity	Sprint speed	Search area
aca1	3–11	0.001–0.003	0.7–0.9	0.55–0.75	0.1–0.3	0.1–0.4
aca2	3–11	0.002–0.004	0.7–0.9	0.55–0.75	0.1–0.3	0.1–0.4
aca3	3–11	0.003–0.005	0.7–0.9	0.55–0.75	0.1–0.3	0.1–0.4
aca4	3–11	0.004–0.006	0.7–0.9	0.55–0.75	0.1–0.3	0.1–0.4
aca5	3–11	0.0028–0.0035	0.7–0.9	0.55–0.75	0.1–0.3	0.1–0.4
aca6	3–11	0.002–0.003	0.7–0.9	0.55–0.75	0.1–0.3	0.1–0.4
spd1	3–11	0.013–0.033	0.7–0.9	0.55–0.75	0.1–0.3	0.1–0.4
spd2	3–11	0.008–0.010	0.7–0.9	0.55–0.75	0.1–0.3	0.1–0.4
spd3	3–11	0.049–0.069	0.7–0.9	0.55–0.75	0.1–0.3	0.1–0.4
spd4	3–11	0.009–0.029	0.7–0.9	0.55–0.75	0.1–0.3	0.1–0.4
spd5	3–11	0.06–0.08	0.7–0.9	0.55–0.75	0.1–0.3	0.1–0.4
spd6	3–11	0.0035–0.0045	0.7–0.9	0.55–0.75	0.1–0.3	0.1–0.4
geo1	3–11	0.09–0.10	0.7–0.9	0.55–0.75	0.1–0.3	0.1–0.4
geo2	3–11	0.05–0.07	0.7–0.9	0.55–0.75	0.1–0.3	0.1–0.4
lit1	3–11	0.045–0.065	0.7–0.9	0.55–0.75	0.1–0.3	0.1–0.4
lit2	3–11	0.040–0.050	0.7–0.9	0.55–0.75	0.1–0.3	0.1–0.4
col1	3–11	0.001–0.003	0.7–0.9	0.55–0.75	0.1–0.3	0.1–0.4
col2	3–11	0.004–0.024	0.7–0.9	0.55–0.75	0.1–0.3	0.1–0.3
enc1	3–11	0.0001–0.002	0.5–0.7	0.55–0.75	0.1–0.3	0.1–0.3
ori1	3–11	0.003–0.005	0.7–0.9	0.55–0.75	0.1–0.3	0.1–0.3

Continued

Table 3.1 Species and trait ranges included in simulations—cont'd

Species ID	Trait ranges				
	Metabolic rate	Q10 voracity	Q10 speed	Q10 search area	Activation energy (eV)
aca1	0.6–0.8	3–4	1.5–2.5	2.0–2.5	0.33–0.42
aca2	0.6–0.8	3–4	1.5–2.5	2.0–2.5	0.36–0.46
aca3	0.6–0.8	3–4	1.5–2.5	2.0–2.5	0.33–0.42
aca4	0.6–0.8	3–4	1.5–2.5	2.0–2.5	0.36–0.46
aca5	0.6—0.8	3–4	1.5–2.5	2.0–2.5	0.33–0.42
aca6	0.6–0.8	3–4	1.5–2.5	2.0–2.5	0.36–0.46
spd1	0.5–0.7	3–4	1.5–2.5	2.0–2.5	0.45–0.55
spd2	0.5–0.7	3–4	1.5–2.5	2.0–2.5	0.45–0.55
spd3	0.5–0.7	3–4	1.5–2.5	2.0–2.5	0.45–0.55
spd4	0.5–0.7	3–4	1.5–2.5	2.0–2.5	0.45–0.55
spd5	0.5–0.7	3–4	1.5–2.5	2.0–2.5	0.45–0.55
spd6	0.5–0.7	3–4	1.5–2.5	2.0–2.5	0.45–0.55
geo1	0.46–0.66	3–4	1.5–2.5	2.0–2.5	0.45–0.55
geo2	0.46–0.66	3–4	1.5–2.5	2.0–2.5	0.45–0.55
lit1	0.45–0.65	3–4	1.5–2.5	2.0–2.5	0.45–0.55
lit2	0.45–0.65	3–4	1.5–2.5	2.0–2.5	0.45–0.55
col1	0.65–0.85	2–4	1.5–2.5	1.5–2.5	0.61–0.70
col2	0.65–0.85	2–4	1.5–2.5	1.5–2.5	0.61–0.70
enc1	0.70–0.90	2–4	1.5–2.5	1.5–2.5	0.39–0.49
ori1	0.57–0.77	2–4	1.5–2.5	1.5–2.5	0.66–0.76

[a]Taxonomic level specification differs depending on the group.

[b]Trait definitions: energy tank, percentage of body size devoted to maintenance and growth at birth; growth ratio, ratio between two instar body lengths; phenology, time between egg laying and birth; body size at birth, mass of the structural body size; assimilation efficiency, percentage of ingested food converted to own mass; voracity, sprint speed, search area, and metabolic rates are mass scaling coefficients for ingested mass, maximum speed, number of cells travelled per day, and metabolic rates, respectively; Q10s and activation energy for metabolic rate denote genetic variability in temperature plasticity for the same four traits (further information can be found in Moya-Laraño et al., 2012 and in Appendix).

a value of 1 means that all animals are genetically identical for that particular trait. For simplicity, we used the same value for all traits and did not play for different amounts of genetic variation in different traits. $\varphi = 0.99$ means that animals are almost genetically identical; hence, adaptation cannot occur from standing genetic variation. In the absence of genetic correlation among traits (Moya-Laraño et al., 2012), $\varphi = 0.01$ means that genetic constraints are minimal, so evolution can occur rapidly and in any direction. To narrow down the questions to be answered, we ran all simulations without genetic correlation among traits (Moya-Laraño et al., 2012). We therefore compared scenarios in which all species had maximum genetic variation in all traits ($\varphi = 0.01$) against others in which genetic variability was restricted to a minimum ($\varphi = 0.99$; Appendix).

One important feature of Weaver is that it can restrict which species are able to feed on each other by including a vector of edible species, allowing initializing simulations with particular food web structures. To manipulate connectance and to simulate realistic food webs, we used the program Network3D to build two random 20-species webs restricted to the niche model (Williams and Martinez, 2000): one with relatively low connectance (0.1) and another with relatively high connectance (0.3). To fit in the animals from the beech forest food web, top predators were assigned to the largest species (harvestmen, spiders and centipedes), intermediate size predators to predators of the smallest size (Mesostigmata and Prostigmata mites) and the smallest sizes to fungivores, namely, springtails (Collembola), oribatid mites (Oribatida) and enchytraeid worms (Clitellata). Within each taxon, we chose a diversity of offspring and adult body sizes to generate across-species diversity. The ranges of the other traits were similar among species, with the exception of that of enchytraeids. Indeed, due to their high starvation resistance, likely coming from their low activation energies for metabolic rates (Ehnes et al., 2011; see Appendix), this group grew to disproportionately large numbers in our simulations as compared to other fungivore species. We believe that some additional constraints, such as their low desiccation resistance (Lindberg et al., 2002; Maraldo et al., 2008), which is not yet incorporated in our framework, may make populations of these worms growing at lower rates in the wild despite their relatively low energy expenditure to activate metabolism. We thus decided to compensate this by decreasing their assimilation efficiencies relative to other taxa (Table 3.1). In addition, to further test the effect of connectivity on food web persistence, we simulated a hypergeneralist food web, in which all predators were able to feed on all the species, including themselves (connectance = 0.55; Fig. 3.1).

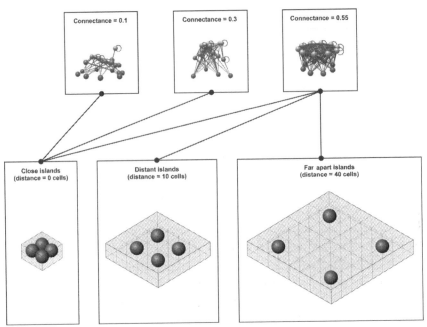

Figure 3.1 Food web and metacommunity structures included in the simulations. The joining segments indicate which food web structures were tested under which metacommunity spatial structures. The spheres in the lower panel correspond to microislands of moisture (moisture pockets) in the forest floor.

Food webs with connectance 0.1 and 0.55 had the exact same species: 11 species of predators and 9 species of fungivores. To allow its fitting to the niche model, the food web with intermediate connectance (0.3) necessarily included a different number of predators (16) and fungivores (4; Fig. 3.1 and Table 3.1). Hence, we created our hypergeneralist web by allowing all predators in the original niche model food web with connectance 0.1 to feed on all other prey and predators, therefore manipulating the latter to have a connectance of 0.55. Food web structure was built taken into account animals only, and for simplicity, we included in all webs a single fungus species upon which all fungivores fed. A simple basal resource makes these webs somehow unrealistic. However, given that we did not explicitly include trade-offs between traits in the simulations (see Section 4) nor resource-dependent assimilation efficiencies, both of which could explain the coexistence of different fungivore species, having different species of fungi or just one was functionally equivalent. As highly connected webs were the most persistent, the remaining simulations were performed using these hypergeneralist webs.

To incorporate a spatial component, we simulated a four-micro-island scenario mimicking water (moisture) pockets in the forest floor (e.g. Melguizo-Ruiz et al., 2012) in which fungi were able to grow (Fig. 3.1). The space surrounding these islands was drier (initialized at 0% RH), thus not allowing fungi to grow. Since animals did not sense water, nor did water directly affect them, this 0% RH had effect only on fungi (i.e. it could have been any number below that which allowed fungus growth in this scenario, <85% see Appendix). Therefore, the four micro-islands had equal basal productivity. These micro-islands were spheres of 5-cell (patch) units of radius (see Appendix) and were either close to each other (distance among centres 10 cells, distance among borders 0 cells) or at a distance (10- or 40-cell distance among borders). The dimensions of the Worlds containing these islands were (depth × width × length) $10 \times 20 \times 20$, $10 \times 50 \times 50$ or $10 \times 80 \times 80$ cells, respectively. For the two last scenarios, in which islands were far apart, we minimized edge effects by allowing a 10-cell space around islands. Note that this is a generous edge which matches the distance between islands in the simulations at intermediate distances. Since so far migration in Weaver depends merely on an animal's mobility which in turn depends on several state variables (e.g. fungi or prey availability, predatory threat, internal stage—condition c—and the trait searching area), here we did not consider long-distant dispersal (e.g. aerial dispersal in springtails and spiders).

To include additional realism in the simulations, we initialized the density of each species and instar following mass–abundance allometric constraints (Reuman et al., 2009), for which we used the equation $N = 74.8M^{-0.75}$ as in Schneider et al. (2012). As in other equations (see Appendix), we assumed 70% of water body content to calculate the number of individuals of each instar and species. To accommodate the output coming from the above allometric equation to our simulation, we applied the above equation to all instars and species and the absolute resulting number was then divided by the total number of individuals, therefore obtaining a fraction for each species and instar. To calculate the absolute number of each species and instar in the simulation, we then multiplied this fraction by the total number of individuals at initialization (i.e. the community size, which was set at 20,000) and rounded the resulting number. We tested the effect of predators on prey density and diversity (like in keystone predation— Paine, 1966) by running simulations with all species in the food web and comparing these with simulations in which predators were excluded, i.e., with fungivores only.

Because these simulations have a strong stochastic component, to assess the consistency of the results, we ran five replicates of each simulation. The deterministic effects were sufficiently strong as to allow enough statistical power to test the main hypotheses. However, as the main patterns are summarized in the statistical analyses, we only display a few of the dynamics as examples. All statistical analyses were performed in R (R development core team 2014). The main hypotheses stated above were tested using general linear models (GLMs) on all replicates, with proportion of prey, predators and all species remaining at the end as dependent variables. When necessary, we ran log–likelihood ratio tests to unravel potential differences among groups and post hoc Tukey tests to compare pairs of groups (package "multcomp"). All simulations were run for 200 days. For one of the parameter combinations that showed the strongest signs of stability, we ran one additional simulation for 500 days for two reasons: (1) to determine how many and which species would remain (long-term persistence) and to draw the final (persistent) web with species-to-species interaction strengths and (2) to explore trait evolutionary dynamics, for which we measured the change in constitutive traits (i.e. those genetically determined as opposed to plastic or environmentally determined traits) through time by fitting splines in a GLM (R library "splines") in which time (day) was the independent variable and the constitutive trait value the dependent variable. The results of the simulations were then plotted with 95% confidence bands using the library "effects" (Fox, 2003) in R. We further explored trait evolution in one of the fungivore species that persisted until the end in one of the webs (predators present), as well as in simulations that involved competition only. The purpose of this analysis was to explore whether trait evolution differed in a purely competitive environment relative to one in which both predation and competition occurred: we predicted that traits associated with competition or anti-predatory behaviour would evolve differently.

3. RESULTS
3.1. Connectance and food web persistence

An example of the dynamics can be found in Fig. 3.2. Connectance increased the proportion of species remaining at the end of the simulation by 1.7×. An example of the dynamics can be found in Fig. 3.2 (GLM, $b=0.34$, $t_{14}=2.66$, $P=0.0197$; Fig. 3.3). This effect was stronger and highly significant for predators (GLM, $b=1.02$, $t_{14}=6.79$, $P<0.0001$;

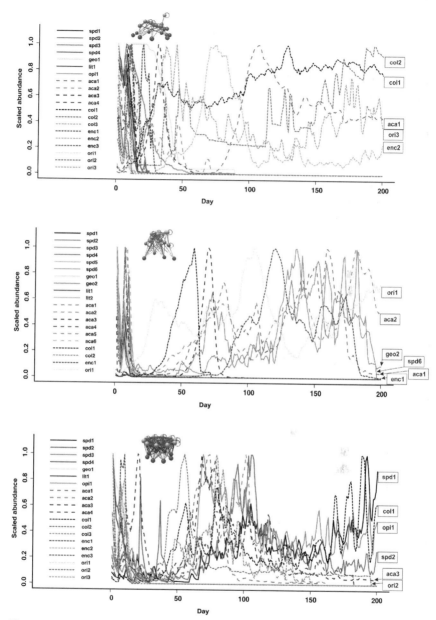

Figure 3.2 Dynamics of the population of each species (starting with 20 species) during 200 days, when the connectance of the food web was low (0.1, upper panel), relatively high (0.3, middle panel) or very high (0.55, lower panel). Figures depict one replicate out of the five ran per simulation. Abundances of each species (*Y*-axis) are shown scaled at 0–1. Dotted lines correspond to fungivores (oribatid mites, springtails and potworms), dashed lines to small predators (predatory mites) and solid lines to large predators (spiders, opilionids and centipedes). The codes on the right facilitate the identification of extant species in B/W printings. Names in the legend correspond to those in Table 3.1.

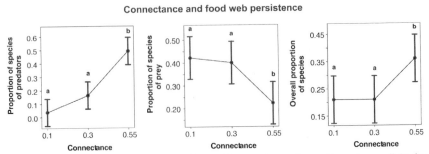

Figure 3.3 Proportion of predator (left panel), prey (central panel), and overall (right panel) species remaining at the end of the simulations ($N=5$) when the connectance of the food web was low (0.1), relatively high (0.3) or very high (0.55). Values are least-squares means $\pm$ SE. Letters on top of bars denote significant differences among groups (<0.05).

Fig. 3.3). However, increasing connectance increased the extinction rate for prey (GLM, $b=-0.46$, $t_{14}=-3.32$, $P=0.006$; Fig. 3.3). Note however that the trends are not linear, as the webs with connectance 0.1 and 0.3 are similar to each other (post hoc "Tukey" test, $P>0.13$ for both comparisons) and they both significantly differ from the web with connectance 0.55 (both $P<0.025$).

3.2. Genetic variation and food web persistence

High genetic variation allowed the persistence of $3.6\times$ more species than either no genetic variation at all, or an intermediate level. An example of the dynamics can be found in Fig. 3.4 (prey, GLM, $b=0.06$, $t_{14}=3.17$, $P=0.007$; predators, GLM, $b=0.23$, $t_{14}=5.54$, $P<0.0001$; overall, GLM, $b=0.15$, $t_{14}=5.3$, $P=0.0001$; Fig. 3.5). However, some patterns are clearly non-linear, and allowing the highest variability was only significantly different for prey between the two extremes (highest vs. lowest, post hoc "Tukey" test, $P=0.006$, remaining comparisons $P>15$), being the two lowest levels of genetic variation not significantly different for the proportion of predators that persisted ($P=0.662$) and both significantly lower than the scenario with high genetic variation (both $Ps<0.0001$). The pattern for overall diversity was similar to that of the predators (comparison between the two lowest levels, $P=0.443$; comparison between each of the two lowest levels vs. high genetic variation, both $P<0.0001$).

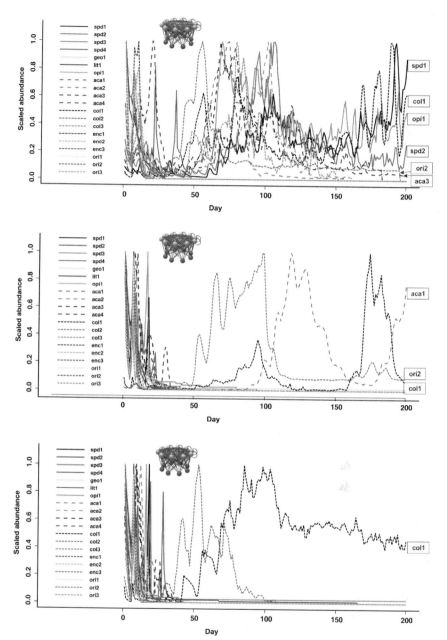

Figure 3.4 Dynamics of the population of each species during 200 days when the connectance of the food web was very high (0.55), and trait genetic variation was either high ($\varphi = 0.01$, superior panel), medium ($\varphi = 0.49$, middle panel) or almost zero ($\varphi = 0.99$, inferior panel). Figures depict one replicate out of the five ran per simulation. Abundances of each species (Y-axis) are shown scaled at 0–1. Dotted lines correspond to fungivores (oribatid mites, springtails and potworms), dashed lines to small predators (predatory mites) and solid lines to large predators (spiders, opilionids and centipedes). The codes on the right facilitate the identification of extant species in B/W printings. Names in the legend correspond to those in Table 3.1.

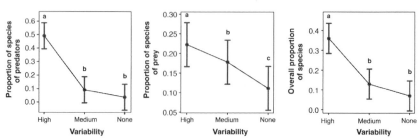

Figure 3.5 Proportion of predator (left panel), prey (central panel) and overall (right panel) species remaining at the end of the simulations ($N=5$) when genetic variation in the species embedded in the food web was high ($\varphi=0.01$), intermediate ($\varphi=0.49$), or almost zero ($\varphi=0.99$). Values are least-squares means ± SE. Letters on top of bars denote significant differences among groups (<0.05).

3.3. Island distance and food web persistence

The overall proportion of species was between $2.6\times$ and $3.5\times$ more persistent in islands that were contiguous to each other or at relatively shorter distances (minimum distance 10 cells), respectively, than when they were farther apart (40 cells). An example of the dynamics can be found in Fig. 3.6. In addition, the proportion of prey species was 2–$2.5\times$ more persistent at intermediate distances. The three models were highly significant (all $P<0.0001$; Fig. 3.7). Post hoc comparisons revealed no differences between the shortest and longest distances in the proportion of prey species that remained (Tukey test, $P=0.750$). However, the proportion of species remaining was significantly higher at intermediate distances when compared with the shortest ($P<0.0001$) or longest ($P<0.0001$) distances. Predator species, on the other hand, persisted equally well at the shortest and intermediate distances ($P=0.969$) and a $4.5\times$ higher proportion of predators persisted at these two shorter distances relatively to the longest distances (both $Ps<0.0001$). The pattern of overall diversity also showed the highest persistence of species at intermediate distances, with only the comparison between shortest and intermediate distances being marginally significant ($P=0.065$), being the other two comparisons highly significant (both $Ps<0.0001$). A total 50% of species remained at the end of the simulation at intermediate distances.

3.4. Predator top-down control on prey diversity

Overall, the presence of predators resulted in high extinction rates on prey. An example of the dynamics (Fig. 3.8) reveals that the presence of predators

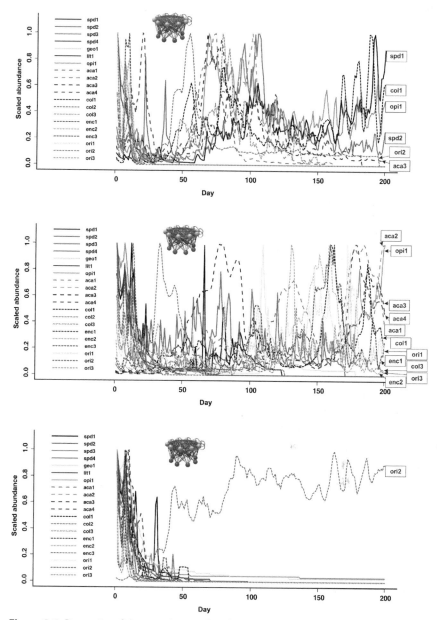

Figure 3.6 Dynamics of the population of each species (starting with 20 species) during 200 days when the connectance of the food web was very high (0.55), but islands were at different distances from each other (0 cells—superior panel; 10 cells—middle; or 40 cells—inferior panel). Figures depict one replicate out of the five ran per simulation. Abundances of each species (*Y*-axis) are shown scaled at 0–1. Dotted lines correspond to fungivores (oribatid mites, springtails and potworms), dashed lines correspond to small predators (predatory mites) and solid lines correspond to large predators (spiders, opilionids and centipedes). The codes on the right facilitate the identification of extant species in B/W printings. Names in the legend correspond to those in Table 3.1.

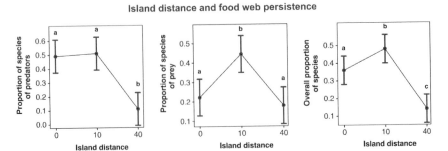

Figure 3.7 Proportion of predator (left panel), prey (central panel) and overall (right panel) species remaining at the end of the simulations ($N=5$) when islands were contiguous to each other (distance $=0$) or when they were separated by a minimum distance of either 10 or 40 cells. Values are least-squares means $\pm$ SE. Letters on top of bars denote significant differences among groups (<0.05).

results in the extinction of two species of potworms (upper panel) that would have made it to the end in a competitive environment (lower panel). Conversely, in the presence of predators, a species of springtail (col1) and another of oribatid mite (ori2) persisted a pattern that never occurred in the absence of predators (not shown). In the islands located at the shortest distances (i.e. contiguous to each other), predators had a strong negative effect on prey species, diminishing the proportion of species that remained at the end of the simulation by 40% (Fig. 3.9, 0 distance), especially after longer periods, when only two species of enchytraeids persisted in the presence of predators (lower panel of Fig. 3.8). In addition, comparing the effect according to island distance reveals that, at intermediate distances, the proportion of prey species that is maintained does not differ between environments with or without predators (GLM, distance*predator presence, $\chi^2_2=21.6$; $P<0.0001$; Fig. 3.9 see also Fig. 3.7 Tukey test on predator effect at intermediate distances, $P=1$). Therefore, predators had a strong effect on the extinction rate of prey species, but this was contingent upon the spatial composition of the food web. In particular, predators showed a strong stabilizing effect at intermediate island distances, in which half the prey species remained until the end of the simulation. Moreover, more importantly, predators affected the identity of the species that remained.

3.5. Multi-trophic spatio-temporal dynamics during a 500-day simulation

We ran one of the most stable scenarios (highly connected web, high genetic variability islands at intermediate distances) for a longer period of time

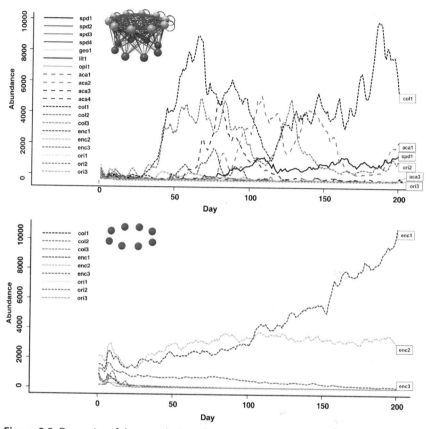

Figure 3.8 Dynamics of the population of each species during 200 days when the connectance of the food web was very high (0.55), in the presence (upper panel) or absence (lower panel) of predators. Figures depict one replicate out of the five ran per simulation. Dotted lines correspond to fungivores (oribatid mites, springtails and potworms), dashed lines correspond to small predators (predatory mites), and solid lines correspond to large predators (spiders, opilionids and centipedes). The codes on the right facilitate the identification of extant species in B/W printings. Names in the legend correspond to those in Table 3.1.

(500 days) and found that out of 20 species, 5 persisted (4 species of predators and 1 prey; Fig. 3.10), and there was a steady increase in abundance of the top predator (the centipede lit1). The spatial dynamics for these five species were highly complex and showed some emergent patterns. First, large predators (centipedes and opilionids) were highly mobile relative to fungivores and small predators. Second, the first snapshot (day 58) showed an emerging spatial segregation among the two species of predatory mites, which occupied different islands, and also for the shared prey (springtail), which was

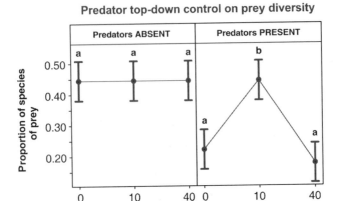

Figure 3.9 Proportion of prey species remaining at the end of the simulations ($N=5$) at three different distances between islands, and in the absence (left panel) or presence (right panel) of predators. Values are least-squares means ± SE. Letters on top of bars denote significant differences among groups (<0.05).

present mostly in one island only. Third, at day 256, prey populations had gone extinct in three of the islands, and almost all individuals (predators and prey) were concentrated in the one island in which numbers of shared prey were still large. In subsequent days, the peak of prey populations occurred in different islands and then went extinct in the originally highly populated island: thus spatial dynamics had a very strong influence on the patterns of extinction and persistence.

During the above simulation, we also recorded all the foraging interactions and built the persistent subweb with interaction strengths, defined in two ways: (a) the proportion of individuals of each species eaten by each predator species (predator perspective, red arrows) and (b) the proportion of individuals of each species that is eaten by each predator species (prey perspective, blue arrows). This subweb (Fig. 3.11) was fairly independent of the remaining web (i.e. that including all of the extinct species) as the interaction strengths both within (cannibalism) and among these five species were much higher. The centipede was the top predator, as it interacted more strongly by feeding on all other species than vice versa. Most predators fed heavily on the shared prey. Cannibalism among predator populations was in general very strong and the smaller predator species (mites) interacted much less strongly among each other than with themselves: in general, intraguild predation (IGP) was weaker than cannibalism.

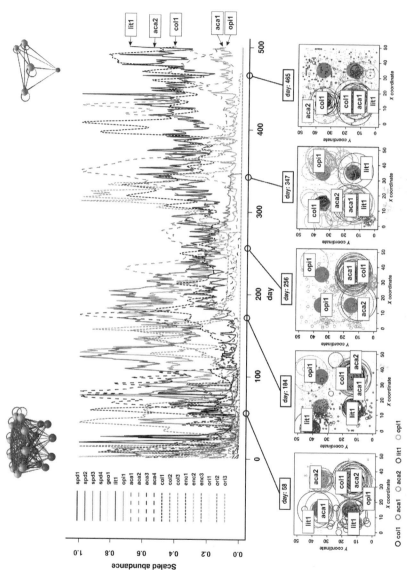

Figure 3.10 Dynamics of the most stable scenario (i.e. highly connected web, with high genetic variability and with islands at intermediate distances, solid blue (grey in the print version) circles in the bottom panels) over 500 days, when five species (four species of predators and 1 prey) persisted until the end of the simulation (superior panel). The inferior panel shows the snapshots of the spatial dynamics of these five species occurring for these 500 days. The size of the circles of equal colour correspond to the abundance of one species in that particular patch, relative to the abundance of that same species in other patches. In total, this simulation experiment included 8,891,887 individuals. The codes on the right facilitate the identification of extant species in B/W printings. Similarly, the codes within spatial snapshots facilitate the interpretation of animal mobility in B/W printings. Names in the legend correspond to those in Table 3.1.

⟶ Proportion of prey individuals of each species in a predator's diet

⟵ Proportion of individuals eaten by each of the species pointed by the arrow

Figure 3.11 Interaction strengths of a persistent web after 500 days of simulation (population and spatial dynamics in Fig. 3.10). The red (grey in the print version) arrows departing from a species denote the proportion of individuals of each species killed and consumed by that particular species. The blue (black in the print version) arrows departing from a species denote the proportion of individuals of that particular species that have been killed and consumed by the species to which the arrow is pointing. Therefore, red (grey in the print version) arrows denote the direction of predation (who eats whom) and blue (black in the print version) arrows the direction of the energy flux.

3.6. Relatively long-term micro-evolution (500 days) in a persistent web

Figure 3.12 shows the evolutionary dynamics for 14 traits associated to the 5-species food web, which persisted for 500 days (Fig. 3.10). Unless their biological relationships were established otherwise, the traits displayed in Fig. 3.12 will be discussed in turn from top to bottom. Evolution was widespread across species and traits, with some of the latter showing clear oscillation through the course of the ecological dynamics. The first apparent outcome is that the evolution of offspring mass is largely driven by the evolution of mass allocated to maintenance, growth and reproduction and not by the fixed (structural) body size of offspring (compare three top panels in Fig. 3.12). Evolution of offspring mass also differed across taxa, with some evolving larger offspring, others smaller offspring and the mesostigmata mite (aca1) showing significant oscillations but without a significant change at the endpoint. In addition, fixed body size evolved to a smaller size in the smallest animals (mites and springtails) and showed either no changes (opi1) or oscillation of trait values (lit1) without evolution.

The growth rates of the two small predatory mites showed opposite patterns, with one (mesostigmata) increasing and the other (prostigmata) decreasing. Springtails (the only fungivore) decreased in growth ratio. Within large predators, one did not evolve in growth ratio, whereas the other increased (lit1). Phenology, or egg developmental time, showed the opposite pattern in a mite species (aca1) and the shared prey (the springtail col1), and either oscillated or did not evolve at all in the remaining three predators. As expected, assimilation efficiency generally increased over time.

Some traits have both a purely additive effect and an epistatic component (phenotypic plasticity; Appendix), the latter consisting of genes of additive effect that tune the additive component of the trait value according to environmental temperature (Q_{10}; Moya-Laraño et al., 2012). Since both the purely additive and the epistatic component have effects on the final trait, we discuss them together. Voracity, which determines animal foraging activity within a patch, increased consistently (both the additive and the epistatic component) in only the mesostigmatid mite (aca1). In the other three species, either the two components evolved in opposite directions or only one of the components evolved. Sprint speed evolved to a higher value in all animals and for both the additive and the epistatic component. However, the evolutionary response was lowest for the shared prey.

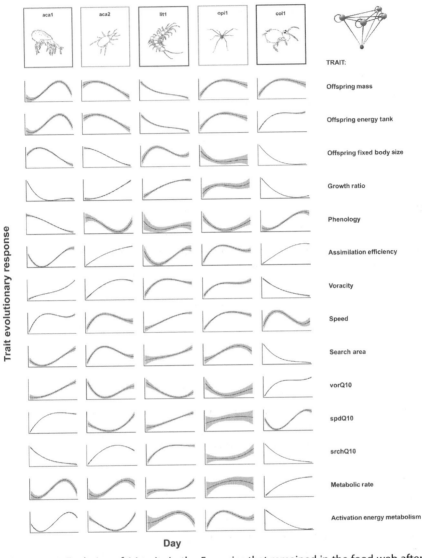

Figure 3.12 Evolution of 14 traits in the 5 species that remained in the food web after 500 days. From left to right: a mesostigmatid mite (aca1), a prostigmatid mite (aca2), a lithobiomorph centipede (lit1), an opilionid (opi1) and a collembolan (col1). The diagram in the upper right corner depicts the structure of the food web (but see Fig. 3.11). In all panels, the *X*-axis represents time, whereas the *Y*-axis represents trait values. Areas in grey correspond to 95% confidence bands, calculated across individuals. Trait definitions can be found in Table 3.1, the Appendix and in Moya-Laraño et al. (2012). The maximum number of generations attained by each population was aca1 = 58; aca2 = 39; col1 = 52; lit1 = 12; and opi1 = 14.

Search area, which determines animal foraging activity among patches, evolved consistently in both components (additive and epistatic) only for the shared prey (decrease) and for the top predator (increase). The other three species of predators showed opposite patterns of evolution for the additive and epistatic effects (aca1) or very weak oscillatory evolution in one of them (aca2) or no evolution at all (opi1). Finally, although there was significant evolution and oscillations on metabolic rates, the only strong response was for the shared prey in which the two components of metabolic rate, the scaling coefficient and the activation energy, evolved in opposite directions.

3.7. Evolutionary dynamics of potworms in the presence and absence of predators

In the simulations with high connectance (i.e. 0.55), the potworm enc2 was the only species that consistently persisted until day 200, both with and without predators. Moreover, in a purely competitive environment, it became the co-dominant species together with another potworm (Fig. 3.13). We took advantage of this persistence in both environments to test for differences in the response to selection in enc2 between the two ecological scenarios. In a purely competitive environment, selection favoured the investment in off-spring with higher energy budgets (energy tanks) and smaller fixed size, whereas these traits did not significantly evolve in the presence of predators (Fig. 3.13). Surprisingly, in the absence of predators voracity evolved to a lower value and metabolic rate to a higher value. Also, temperature-dependent plasticity of sprint speed, which is only functional under the threat of predation, evolved to a higher value. Finally, activation energy for metabolic rate evolved to a lower value. In contrast, only traits directly related to predatory avoidance (i.e. speed and temperature-dependent plasticity for speed) clearly evolved to a higher value in the presence of predators. The other traits showed either no significant response (e.g. activation energy for metabolic rate) or an oscillating response, ending up in a trait value that did not differ from the initial one (e.g. voracity and search area).

3.8. Results summary

Our simulations showed some relevant patterns that will hopefully entail a step change in our understanding of eco-evolutionary dynamics in complex systems. We found that highly connected webs, with widespread trophic omnivory, suffered far less from extinction than less-connected webs.

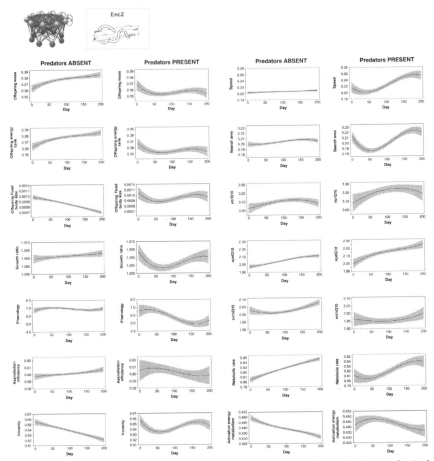

Figure 3.13 Evolution of traits in a fungivorous potworm species under two ecological scenarios: "predators absent", in which only competitive interactions are present and "predators present" in which species engage in both competition and predation. In all panels, the X-axis represents time, whereas the Y-axis represents trait values. Areas in grey correspond to 95% confidence bands, calculated across individuals.

Nevertheless, this effect differed between predators and prey, with fewer prey and more predators remaining in connected webs. Moreover, decreasing genetic variability reduced the proportion of surviving species by 72%. However, while the effect on prey was a reduction to 50%, predators suffered a much dramatic reduction, decreasing by 87% when genetic variation was reduced by half.

Furthermore, we found relevant patterns of metacommunity dynamics as the distance among four islands of identical basal productivity also explained species persistence in a highly connected food web. At intermediate distances, the proportion of prey species that persisted until the end were 2–2.5 × higher than when islands were either close to each other or farther apart. For predators, both when islands were close to each other and at intermediate distance, the proportion of species that remained was 4.5 × higher than when the islands were further apart. Overall, we found the highest diversity at intermediate distances, slightly less when islands were close to each other and much lower diversity (higher rate of extinction) when islands were further apart.

We also found that in highly connected webs with widespread omnivory and high trait genetic variation, predators had a strong top-down effect on prey diversity, tending to increase the extinction rates of prey when islands were either close to each other or at the longest distances, and greatly increasing prey diversity at intermediate distances. In addition, when the amount of prey that remained in a purely competitive environment was similar to that remaining in an environment in which both competition and predation were at play, the identity of the species tended to be different, indicating that predators modified community composition.

The 500-day simulations revealed strong spatial patterns, with clear migration–extinction dynamics typical of metacommunities and spatio-temporal segregation of species, all of which could have contributed to the overall stability of this five-species food web.

Finally, we found evidence for widespread trait evolution when high genetic variation was involved. Trait evolution differed between top predators and the rest of other predators, and also between predators and fungivores. In addition, trait evolution also differed for fungivores in a purely competitive environment when compared with the same species in an environment involving predation plus competition. Remarkably, depending on the trophic level, some traits were selected up, others down and others oscillated.

4. DISCUSSION

4.1. Ecological dynamics

Our findings that (a) highly connected webs with widespread omnivory were less prone to extinction and (b) the simulation in the long run (500 days) ended up with five species that were only weakly dependent on the extinct species, agree with the view that higher connectance and intermediately strong omnivory is stabilizing (McCann, 2000; McCann and Hastings,

1997; Solé and Montoya, 2001). It is intriguing, however, that interactions were fairly strong, while stable food webs usually have many weak and a few strong interactions (McCann et al., 1998). This apparent paradox could be explained by the high cannibalistic rates, as like in some natural food webs (e.g. Woodward and Hildrew, 2002), our persistent web was maintained with a strong degree of cannibalism (Fig. 3.11). Recent findings show that both cannibalism and anti-predatory behaviour directed towards conspecifics contribute to stabilize food webs (Rudolf, 2007a,b): the highly cannibalistic web depicted in Fig. 3.11 is in line with these predictions.

A previous very simple simulation exercise (Moya-Larano, 2011) suggested that higher trait variability would lead to higher connectance and higher variability in interaction strengths, with many weak and a few strong interactions, all of which could lead to food web stability (see above). However, the latter study did not consider the fact that higher genetic trait variability could also promote rapid or contemporary evolution with ecological time (i.e. during population dynamics). Here, genetic variation on 13 traits strongly promoted food web persistence, as the most persistent webs were those in which populations had high genetic variation. Moreover, we found widespread rapid trait evolution in the five persistent species embedded in the food web. Therefore, we can tentatively conclude that genetic variation, through its effects on contemporary evolution, is also an important factor contributing to the food web stability debate.

Genetic variability may promote stability by (a) enhancing contemporary evolution as shown here or (b) leading to a high diversity of interactions in the species embedded in the food webs, therefore increasing connectance, omnivory and a variable distribution of interaction strengths (Moya-Larano, 2011; see also Steiner and Masse, 2013). Distinguishing among these two drivers of stability is important and we propose that this is feasible in a platform such as Weaver.

Our finding that at intermediate micro-island distances, food webs are more persistent can be explained by a strong top-down extinction effect of IGP predators on prey and on each other at the closest distances (i.e. spatially compressed food webs; McCann et al., 2005), predator coupling of the four islands at intermediate distances and lack of coupling of the global dynamics at longer distances, leading to strong extinction rates in each of the four islands. These results are consistent with the predictions of recent models (McCann et al., 2005; Rooney et al., 2006, 2008) in which adaptive search and rapid responses arising from highly mobile (top) predators moving towards patches where prey are available help to stabilize the system. The high mobility displayed by top predator populations during the long-term 500-day dynamics supports this view (Fig. 3.10). Our findings confirm these models and add to other models of spatial dynamics in metacommunities involving more than one trophic level (Abrams, 2007; Amarasekare, 2008a, 2010; Koelle and Vandermeer, 2005). Note however that the effect found was only detected in highly connected webs with high genetic variation, meaning that other previously unexplored relevant ecological variables may interact with predator coupling to stabilize metacommunities.

Predators changed the identity of the dominant prey species, driving the strongest competitors to extinction, and in islands located at intermediate distances, they actually enhanced prey diversity. Therefore, our heuristic approach adds novel information for the link between keystone predation and metacommunity dynamics (Amarasekare, 2008b; Paine, 1966), in more complex systems.

4.2. Evolutionary dynamics: Relatively long-term micro-evolution (500 days) in a persistent web

Our simulations revealed that allocation of energy into offspring (offspring energy storage) had a stronger impact on the evolution of offspring mass than did structural body size at birth. This is very relevant because, given a fundamental egg–number/egg–size trade-off (Fox and Czesak, 2000), female energy allocation in offspring with materials providing starvation resistance (e.g. energy tanks) may be more important than allocation in structural body size, even though the latter provides an advantage during predator–prey interactions. This could be translated into an egg–number/starvation resistance trade-off. We found that evolution resulted in lighter offspring in two predator species and in heavier offspring in another predator species and in the shared prey (springtail). However, in another predator species, no evolutionary changes occurred in this trait. Hence, very different strategies evolved, with no clear taxonomic or trophic-level pattern. Given the egg–number/egg–size trade-off, the evolution of heavier offspring was necessarily accompanied by the concomitant evolution of lower fecundities. Therefore, selection could be targeting fecundity or egg mass in each case: further simulations and experiments could be conducted to distinguish between these two possibilities.

Despite their disadvantage in predator–prey encounters with most predators, the smaller animals (springtails and mites) evolved towards smaller offspring sizes. Possibly, in structurally smaller offspring, the overall amount of energy necessary to reach adulthood is lower and faster to accrue, and therefore viability selection (selection to reach the adult stage and reproduce) may favour smaller fixed body sizes. However, this explanation was not supported by our results, as for each of the three small species, the individuals with structurally smaller offspring matured later: no explanation for this pattern is immediately obvious, however, and further simulations and output analyses are needed.

Much variation was also found in the evolution of growth ratio. Clearly, the evolution of shorter growth ratios can be favoured again by faster egg-to-adult developmental times, increasing viability selection, as it occurred in the fungivore (springtail) and in the mesostigmatid mite. The prostigmatid mite and the top predator (a lithobiomorph centipede) evolved higher growth ratios, which should provide an advantage during predator–prey encounters. Moreover, larger adult sizes are associated with higher reproductive investments, and therefore increased fecundity. This is consistent with the fact that these two species evolved lighter offspring, suggesting that, for these two predator species, investing in offspring number rather than survival is a better strategy.

Phenology (or egg developmental time) evolved in the opposite directions in the shared prey (longer) and in one of its predators (shorter), which likely uncoupled their phenologies, allowing prey to diminish predation from at least this species. As no constraint was imposed on assimilation efficiency, all animals evolved towards higher values for this trait.

Remarkably, traits associated with animal "personalities" (Carter et al., 2013; Wolf et al., 2007), e.g., voracity (related to aggression and within patch activity) and search area (related to boldness and activity among patches) showed patterns consistent with the balance between the need to find food and the level of predation risk affecting each species. The shared prey (springtails) was the only species that most consistently evolved lower searching areas and voracities with only one of the four activity-related traits (Q_{10} on voracity) evolving to higher values. Given the amount of predation threat upon this species (Fig. 3.11), the evolution of a cautious personality might be expected. On the other hand, predators tended to evolve at least two activity-related traits to a higher value, consistent with their lower predation risk.

Searching area is also a dispersal trait, and our results are consistent with those of Pillai et al. (2012) who found that in a metacommunity under strong predation extinction pressure during predator–prey dynamics, predators evolved higher dispersal rates, and their prey evolved to lower dispersal rates. This similarity of results, stemming from two very different modelling approaches, provides some degree of robustness and increases confidence in our simulation approach. The intermediate degrees in the evolution of dispersal of the remaining predators can thus be understood by the differing degrees of predation risk and extinction.

The evolution of higher sprint speed was most consistent in the four predator species than in the shared prey, for which only the Q_{10} for speed evolved. In predators engaged in IGP, sprint speed is subject to two selection pressures: catching prey and escaping from predators. In contrast, in the shared prey, selection on this trait stems from predation avoidance only. Hence, selection pressure for sprint speed may be lower in prey than in IG prey. This hypothesis can be tested in real food webs. The prediction would be that after controlling for phylogenetic effects (Harvey and Pagel, 1991), the degree to which predators are involved in interactions with other predators (e.g. cannibalism or IGP either as predators or as prey) should correlate positively with their sprint speeds relatively to their body sizes and to prey of similar size.

The evolution of traits related to metabolic rates (scaling mass coefficient and activation energies) showed either oscillations or no consistent pattern between the scaling coefficient and activation energies. In other words, if higher energy expenditures would have evolved, both the scaling coefficient and activation energies should have been higher (see Appendix for an explanation). However, in none of the species did both coefficients evolve consistently. Therefore, deriving any mechanistic explanations for these patterns is not possible at this moment. These results, however, are consistent with previous findings that showed that activation energies for metabolic rates can evolve (Moya–Laraño et al., 2012).

The fact that a few traits oscillated during the time period of our simulations is consistent with the idea that during population dynamics, selective pressures change, probably due to changes in population numbers of the different species embedded in the food web, and to patterns of space use, which also show their own emerging dynamics (Fig. 3.10).

4.3. Evolutionary dynamics of potworms in the presence and absence of predators

In general, responses to selection were more linear in the competitive environment than in the predation environment, reflecting the higher complexity of selective pressures in the latter environment. In a purely competitive environment, potworms (enc2) evolved to produce smaller offspring (structural size) with a high energy storage (energy tank). The latter may provide a competitive advantage by preventing death from starvation, and smaller structural offspring sizes will result in smaller adult sizes (given a fixed growth ratio and amount of energy) and shorter developmental times. Therefore, the overall effect arising from the opposing evolution of the two size traits could be to increase offspring viability. However, when predation is present, investing in a higher number of larger (e.g. larger fixed size) offspring may decrease predation risk (a size refugia—Paine, 1976; Wilson, 1975). Also, given the fundamental trade-off between offspring mass and number (Fox and Czesak, 2000), which is implicit in the model (as offspring mass and number are allocated according to the energy available for reproduction in the mother), investing in more offspring also means investing in lower energy budgets. Therefore, the counterbalancing selective pressures of competition and predation might explain why neither fixed offspring size nor the offspring energy budget evolved in the environment in which predators were present.

The fact that a competitive environment selected for lower voracity is puzzling. One possible explanation is that genetic drift, rather than selection, produced these evolutionary patterns. We have not yet incorporated non-functional traits in the simulations, which will allow controlling for random genetic drift. However, it is unlikely that the robust evolutionary patterns that we found, such as that for metabolic rates, arose from drift alone in so few (<7) generations and in a pattern that was consistent across replicated simulations (not shown). Thus, these results must be seen as emerging patterns of responses to selection coming from a complex system, which are difficult to grasp from a reductionist viewpoint. For instance, we know that in Weaver—and likely in the real world—voracity and basal metabolic rates feedback on one another: animals that genetically spend more energy (higher basal metabolic rates) will as a consequence have lower energy storages, which will prompt them to be more active and voracious regardless of their constitutive (i.e. genetically determined) activity and voracity traits, leading to higher energy intake which may or may not translate into higher competitive ability depending on other traits, such as assimilation efficiencies. Similarly, animals with lower voracities will maintain lower energy tanks, which will make them more active searching for suitable patches.

Curiously, temperature-dependent sprint speeds evolved to higher values in the competitive environment, in the absence of predation. The evolution of body mass (at all stages) largely depends on how body mass constrains other traits (e.g. voracity, search area, metabolic rates or speed). Hence, selection targeting one of these traits can indirectly affect others, via the covarying effect on body mass, even in the absence of genetic correlations, as was the case here. Nevertheless, the only two traits that showed a positive response to selection were fixed body size at birth and activation energy for metabolic rate, which cannot directly explain the evolution on sprint speed.

In general, when predators were present, trait evolution was weaker and oscillating, likely explained by the higher complexity of selective pressures in the presence of predators. These oscillatory evolutionary trait dynamics in more complex environments could be the result of fluctuating selection due to fluctuating selective pressures during ecological dynamics (e.g. numbers of the different species of predators changing through time) or even co-evolution (e.g. rapid evolution in predator traits acting as dynamical selective pressures on prey traits). Future simulation experiments including different food web structures and diversity of predators should reveal if there is a positive association between trait oscillations and the diversity of selective pressures. Moreover, more simulations are needed to identify consistent evolutionary patterns in webs with or without predators. One way to test this is by playing with the dimensionality of selective agents, that is, the O matrix (MacColl, 2011; Moya-Laraño, 2012).

4.4. Future directions

We found exceptionally fast dynamics, with all animals growing and reproducing at rates that are much higher than in natural soil food webs (Coleman et al., 2004). *Lithobiomorpha* centipedes, for instance, usually take more than 1 year to mature (Lewis, 1981), which would have yielded only 1.5 generations instead of the documented 12 generations. Our models are therefore still very far from reproducing the pace of soil interactions. There are various ways that the Weaver platform could be altered in subsequent incarnations to make its constituent animals more realistic. First, we need to add periods of inactivity before moulting or during egg guarding (e.g. Geophilomorpha; Lewis, 1981). Second, the standard metabolic rates as documented by Ehnes et al. (2011) need to be accompanied by the possibility of decreasing metabolic rates to survive starvation periods, a pattern that is widespread, for instance, in spiders (Anderson, 1970, 1974; Schmalhofer, 2011), and could be linked to discontinuous gas exchange occurring during resource shortage (Chown, 2011). Allowing for a decrease in metabolic rates during starvation will allow having more realistic foraging patterns and slow down the dynamics.

Although we have simulated food webs across space, we have not included many of the typical drivers of coexistence in metacommunities, such as colonization–competition and competition–predation trade-offs, which can be parameterized by manipulating trait range values (e.g. assimilation efficiencies and searching areas, for instance). In addition, Weaver allows the inclusion of multiple genetic correlations and genetic trade-offs (antagonistic pleiotropy), which we have not exploited in this chapter (but see Moya-Laraño et al., 2012, 2013). Exploring food web persistence across space in the presence of multiple trade-offs, both ecological and evolutionary, will open new avenues of research to help us understand the eco-evolutionary stability of these complex systems. However, the problem remains of determining how realistic are the results produced *in silico*. Next, we describe how to validate this framework by continuous feedback with real systems.

4.4.1 The quest for eco-evolutionary patterns: An FRP using global optimization algorithms and approximate Bayesian computation

Ideally, to test the reliability of models one has to always confront the model results with real data (e.g., Boit et al., 2012; Hudson and Reuman, 2014). In order to ensure that Weaver is simulating realistic scenarios, our outputs should include eco-evolutionary patterns that reproduce the empirical observations of real food webs, including patterns of stability. For example, given ecological and evolutionary empirical data, what can we tell about the mechanisms that best reproduce those data? In the previous sections and in the Appendix, we have described the different components and parameters of the simulations, as well as some relevant results. Here, we briefly describe some advanced methods to compare our model predictions with the empirical data using approximate Bayesian computation (ABC), and also global optimization algorithms to produce digitally-stable webs that can be then compared with real data.

Our preliminary simulations showed that some species persisted more than others and that some traits changed monotonically (evolved) through the entire simulation, while others oscillated. Indeed, stability of the entire system (i.e. 20 species interacting with each other following a niche model food web structure) was not reached in any of the simulations: neither ecological (population dynamics) nor evolutionary (trait dynamics) dynamics showed signs of stability. Instead, a variable number of species went to extinction before the end of each simulation. Stability could be reached given the large number of parameters involved, but here we just used a small combination of them. Actually, each simulation included 20 animal species * 13 parameters + 20 animal species * 13 trait ranges + 1 World configuration * 4 parameters + 1 fungus species * 3 parameters, which gives 268 parameters and 260 trait ranges one could work with in order to search for more realistic eco-evolutionary food webs. Exploring in full all these parameter combinations would be extremely labour intensive, and many of them would be likely to be meaningless anyway (e.g. too far apart a combination of values from the real World to begin with): so, we need a more efficient way to refine the approach.

With the aim to address several main research questions, e.g., what drives stability, distinguish neutral from non-neutral scenarios, how does the evolution of adaptive traits affects food web structure, how does evolution affect food web-associated ecosystem process, etc., we propose a working protocol, which consists in a three-step iterative strategy or FRP:

1. Identify the data used to parameterize the model ecosystem as either internal (e.g. data coming for the same or very similar ecosystems) or external to the system (data or parameters extrapolated from other systems). In the absence of internal and external data to parametrize the model for some specific parameter values, in the Bayesian framework, our *a priori* data would be drawn from a uniform distribution (Beaumont, 2010; Grelaud et al., 2009).

2. Search for eco-evolutionary patterns by iteratively running models with different combinations of parameters. In this iterative process, set the internal parameters to the real (estimated) values and manipulate the external parameters. The predictions can then be compared with empirical observations. This is done by computing the distance between the empirical and simulated data for each parameter combination (Grelaud et al., 2009; Sunnaker et al., 2013).

3. Once a stable eco-evolutionary web has been reached, pick some of the parameter values and outputs found by the system and check if they apply to your system (e.g. check if animal fecundities in the outputs are within those measured in some of the animals of your food web). If this is not the case, either feed the newly measured parameters to Weaver setting them as internal or implement new algorithms (e.g. criteria for reproducing or different coefficients in the deterministic encounter probabilities—Eq. A6) and repeat step (2). In summary, this last step requires selecting the parameter combination that best matches the empirical observations. This can be done by setting a threshold to generate *a posterior* distribution, with all the parameter combinations being sufficiently similar to the empirical observations (i.e. those that are far from reality are discarded; Melian et al., 2011, 2014).

As an example of what type of data would be considered internal or external, animal body lengths that came from accurate measures in field experiments would be clearly internal. However, as these length data were then transformed to body masses using equations from other systems and assuming 70% of water body content as estimated in a different system (see Appendix), the latter would then be considered external parameters. As another example, basal metabolic rates came from published data on animals coming from similar soil systems (Ehnes et al., 2011) and were therefore considered as internal. Given the body lengths and metabolic rates as internal and external data, respectively, additional assumptions to include in the model are, for instance, the fitness consequences of body length and metabolic rate for each individual. In the neutral scenario, each individual would have the same fitness irrespective of its body length and metabolic rate. But, how can one meet the neutrality assumptions when these traits are actually functional and lead to different fitnesses across individuals? We propose that neutral eco-evolutionary dynamics, which could be used as null models to contrast the occurrence of niche-oriented dynamics and adaptive evolutionary dynamics, can be reproduced by randomizing reproductive events not only across individuals but also across species. This would require running two simulations in parallel, starting with the exact same individuals, one randomizing reproductive events (neutral) and another letting the adaptive evolution of traits drive the dynamics (functional).

Importantly, the above protocol can be applied to model assumptions and parameters. For instance, Moya-Laraño et al. (2013) simulated situations in which prey were able to assess predator threat before entering a patch and others in which predator threat could not be assessed by prey. They found that system stability involving sit-and-wait predators could only be achieved if prey were not aware about predator presence when entering a new patch. Many other biological assumptions could be tested by generating predictions and comparing them with empirical observations.

Combining parameters and algorithms in an iterative process would not be done completely at random, as this would be also almost unfeasible even for highly optimized parallel code running in high-performance computers. Instead, global optimization algorithms (Floudas and Pardalos, 2009) should be used. Next, we review these algorithms and discuss which ones would be more appropriate to accomplish step (2).

Biological systems may contain several parameters that underlie biological, evolutionary and ecological processes. Given such complexity, finding the minimal number of mechanisms (or parameters) that best predict the empirical data is challenging. Global optimization algorithms, the most popular of which in ecology and evolution are "genetic algorithms" (Hamblin, 2013), fit this purpose. The first problem we face is to explore efficiently the highly dimensional parameter space in our model. In general terms, we have to optimize the search by giving a nonempty closed (searching domain) set S and a function f, to find the minimal value f^* and all the points $x^* \in S$ such that $f^* = f(x^*) \leq f(x)$ for all $x \in S$, or show that no such a point exists. The problem to solve can be written as

$$\begin{aligned} &\min f(x) \\ &\text{s.t.} \quad x \in S \end{aligned} \quad\quad (3.1)$$

The purpose of global optimization is to find the global minimum value f^* and the set of global minimizer points $X^* = \{x^* \in S | f(x^*) = f^*\}$. The conversion of a maximization problem to a minimization one is straightforward $(\max\{f(x) | x \in S\} = -\min\{-f(x) | x \in S\})$.

Problem (1) can be classified depending on the dimension of x (e.g. number of parameters and rules), its type (discrete, continuous or mixed), the type of f (linear, quadratic or non-linear) and how the search domain S is defined (constrained or unconstrained).

When any local minimum of problem (1) is also a global minimum (e.g. in linear—Linear programming FAQ—or convex programming—Boyd and Vandenberghe, 2009), local optimization methods can find the optimum easily. However, in many problems, this is not the case (Nonlinear programming FAQ, 2013). The field of global optimization and ABC are devoted to the latter, i.e. those problems of type (1) which can have several local optima apart from the global optimum (Beaumont, 2010).

The algorithms to solve these problems can be classified into different ways. A typical classification distinguishes between deterministic or stochastic methods. Stochastic methods (for instance, tabu search, genetic algorithms or simulated annealing; Talbi, 2009) apply some random factors in the local phase to converge to a solution and in the global phase to avoid getting trapped in local optima. Although these methods can find the global optimum, they can only guarantee it when the number of iterations tends to infinity. In deterministic methods (such as outer approximation, Lipschitzian optimization or branch and bound—Scholz, 2011), no random factors are included. Some of them converge to the global optimum under certain conditions, but when the algorithm is stopped after a finite number of iterations, the accuracy of the solution may not be known with exactness.

A better classification of the algorithms, based on the degree of rigour with which they find the global optimum, is the following (Markót et al., 2006): *incomplete methods*, which may become stuck in a local optimum; *asymptotically complete methods*, which reach a global optimizer with probability one if allowed to run infinitely long, but have no means of knowing when a global minimizer has been found; *complete methods*, which reach a global optimizer with certainty (assuming exact computations and infinitely long run), but knowing after a finite time that an approximate global optimizer has been found; and finally *rigorous methods*, which reach the global optimizers with certainty within a given tolerance even in the presence of rounding errors.

According to the previous classification, rigorous methods, like the *interval branch-and-bound* algorithm (Markót et al., 2006), are desirable. However, although they reach the global solution with the desired precision, execution time increases exponentially with the dimension of the problem. Additionally, they usually need the formulation of the objective function. In summary, regardless of the methods used, the global optimum may be extremely costly to find for highly dimensional ecosystems. Thus, obtaining a Bayesian posterior distribution with the parameter values that best approximate to the empirical observations may be a useful way to decrease the number of parameter candidates to infer the mechanisms explaining the observed patterns (step 3). For example, we may obtain the number of parameters and their values to characterize the distribution of trait values or phenotypes and predation coefficients that best approximate to the empirical observations.

The Weaver algorithm can be considered a model with a large set of *a priori* parameter values and biological mechanisms. One of the desirable goals of Weaver would be to determine the feasible input parameter values to obtain the output values that best approximate to the empirical observations. Depending on the research question, output values to optimize could be the distribution of traits or phenotypes, species richness and diversity or ecosystem productivity through time.

Given the high dimensionality involved in Weaver, the optimization approach, together with ABC methods, is a candidate in this FRP protocol for inferring with higher accuracy the mechanisms that shape diversity patterns in ecological and evolutionary systems.

4.4.2 Engineering food webs for pest control
The present eco-evolutionary framework could be used successfully to help engineering food webs for effective pest control (Moya-Laraño et al., 2012).

The increasing social demand for pesticide residue-free agricultural products and recent transgovernmental policies that explicitly demand sustainability in agriculture is pushing the productive sector worldwide to put agricultural methods for improving pest control while avoiding environmental impacts (Bale et al., 2008; OJEU, 2009). As a consequence, over the last 10 years, the percentage of the World's cultivated land using biological control to fight pests has increased, and the identity of the species of natural enemies involved has suffered an ecologically rational switch, as more indigenous than exotic species are being used as biocontrol agents (van Lenteren, 2012). This tendency is expected to continue in the future because biological control will be a key feature of sustainable crop production.

When untreated with chemicals, agricultural systems typically hold multiple plant-inhabiting species, which interact with each other forming complex food webs. For example, in a typical Mediterranean greenhouse, different herbivore species listed as pest organisms (aphids, spider mites, trips, caterpillars) and several species of natural enemies (phytoseiid mites, predatory bugs, lady beetles, parasitoid wasps) can coexist, all interacting directly (i.e. predation) and indirectly (competition, apparent competition) mechanisms (Messelink et al., 2012). Furthermore, the landscape in which agriculture takes place is commonly surrounded by margins and natural and semi-natural habitats that can provide source populations of natural enemies, as well as act as a reservoir for pest species (Winkler, 2005), expanding the network of agricultural interacting species beyond the actual space occupied by crops. Finally, biotic (predation, competition) and abiotic (e.g. heat stress) selective pressures acting simultaneously on whole agricultural communities may induce rapid evolutionary responses that can be key drivers in the dynamics of populations (i.e. eco-evolutionary dynamics; Fussmann et al., 2007; Pelletier et al., 2009). Hence, to develop a sustainable agriculture with robust predictive power, pest management will need to take into account both the effects of complex interactive networks and complex architectural landscapes (Bohan et al., 2013), and rapid evolutionary responses (Loeuille et al., 2013), on the dynamics and structure of agricultural communities.

Moya-Laraño et al. (2012) coined the concept of Food Web Engineering (FWE, hereafter) as an extension of biological pest control that integrates community ecology and evolutionary biology into the management of agricultural systems. The idea is to unravel how specific community modules in agricultural systems can be manipulated so that top–down pest control is maximized. The approach implicitly requires prior knowledge on which species form the community, how they interact and which traits are relevant for interactions within and among species and with the environment, and the IBM platform presented here (Weaver) is a suitable tool to address this. Through simulations of the dynamics of specific communities by using different ranges (variability) and assumptions on the genetic determination of traits (number of loci, allele values, pleiotropic effects), different scenarios of pest control, as well as optimal solutions and conditions for stability, can be determined. Using a systems biology approach to connect the IBM and the real world could help to uncover those traits that should be selected in natural enemies to maximize trophic cascades.

The potential of Weaver in FWE is illustrated in the results presented here, which provide useful general guidelines of potential relevance to the development of more sustainable agriculture. On the one hand, simulations show that the higher the connectance in the food web, the higher the persistence of predators, hence herbivore top-down control, will be. Moreover, generalist predators are those that persist more often in our food web, thereby ensuring long-term prey control. These results support the idea that the presence of multiple generalist predators can effectively control pest populations (Faria et al., 2008; Messelink et al., 2012), contradicting the traditional view that they are less effective than specialist predators (Symondson et al., 2002; van Lenteren and Woets, 1988). However, our results also show that persistence of generalist predators in intermediate-highly connected food webs is strongly dependent on genetic trait variability. This implies that increasing predator diversity by releasing several species that are commercially available might not be appropriate. Mass-reared populations are expected to have low genetic variability because they originally started from a relatively small number of individuals, and because populations may have experienced several bottlenecks caused by the rearing methods. Instead, conservation strategies aimed at promoting naturally occurring populations of natural enemies (e.g. manipulation and recreation of the habitats they live in, adding alternative food sources to the habitat, etc.) may be a more appropriate approach.

The IBM platform Weaver also offers the possibility to mimic dynamics at a metacommunity level by increasing the distance between patches (islands). A key result of the simulations is that increasing such distance reduces the persistence of predators and decreases the stability of the systems. Therefore, persistence of generalist predators could be increased by introducing islands of diversity, i.e., non-crop plant species harbouring alternative food/prey, amidst the cropping area. There is evidence that densities of herbivorous pests and natural enemies within agricultural fields are influenced by the features of the surrounding landscape (Nicholls and

Altieri, 2007), enlightening the importance of considering the spatial layout when designing future agro-ecosystems.

In conclusion, we have shown how our heuristic IBM approach is useful to test eco-evolutionary hypotheses *in silico*. We found that connectance, rapid evolution occurring contemporarily to the ecological dynamics, as well as island distance, affected food web persistence. Moreover, depending on which of these scenarios is at play, we show that predators can exert a strong effect on prey diversity. In addition, we documented diverse evolutionary responses across the different trophic levels of the food web and for up to 14 functional traits. Although these *in silico* systems may still be far from fully realistic, we propose an FRP protocol by which we hope to fill the *in silico–in vivo* gap. We also envisage that this research protocol can also be successfully incorporated in FWE to improve biological pest control in a changing and increasingly spatially fragmented world.

ACKNOWLEDGEMENTS

Two anonymous referees provided helpful comments that greatly improved the manuscript. We thank U. Brose for helpful discussions on the use of equations to estimate temperature-dependent metabolic rates. T. Schoener provided helpful references. R. J. Williams provided a copy of NETWORK3D to produce food web drawings. E. De Mas, J. Pato, G. Jiménez-Navarro and N. Melguizo-Ruiz helped measuring some of the parameters included in this publication. J. R. B.-C. is a recipient of a Ramón y Cajal fellowship awarded by the Spanish Ministry of Economy and Competitiveness (MINECO). This work was funded by the Spanish Ministry Grants CGL2010-18602 to J. M.-L and by Grants TIN2008-01117 and TIN2012-37483 and Junta de Andalucía Grant P11-TIC-7176 to L. G. C. All grants have been funded in part by the European Regional Development Fund (ERDF).

APPENDIX

A.1. Weaver—An IBM platform to simulate eco-evolutionary dynamics in food webs

Weaver is the C++ porting of the former R code mini-Akira (Moya-Laraño et al., 2012), which has been additionally extended to 2D and 3D space. This porting means a great improvement in computer performance, in part due to algorithm optimization and in part due to the higher performance of C++ relative to R. For instance, each simulation set in Moya-Laraño et al. (2013) took about 48 h of execution in mini-Akira, which became 10 s for Weaver run in the exact same machine (17,280 times faster!). This improvement in speed has allowed us to increase the number of species and to work with more complex food webs (20 species and 20,000 individuals at initialization). Access to the executable program as well as an explanation of the input and output files can be found here (www.eeza.csic.es/foodweb). In addition, some other changes in the former R code, as well as a different application, can be found in the material of Moya-Laraño et al. (2013), also available in the above site. Updates on the code and potential encountered bugs will also be reported there.

A.2. Space and basal resources—2D and 3D and a chemostat

Weaver's spatio-temporal domain is currently conceived as a portion of soil where animal interactions will occur and in which many parameters can be manipulated. Currently this includes soil moisture and temperature. A future version will include carbon, nitrogen and phosphorous, mimicking decomposing leaf litter. Computationally, space has been modelled as a set of discrete, squared, contiguous, spatial cells or voxels. Cell size can be arbitrarily chosen by the user which yields high flexibility and adaptability for different animal sizes and movement characteristics. Reducing voxel size while increasing their number will lead to much higher spatio-temporal resolution while reducing computational performance as the amount of processing needs increases consistently. Each voxel has its own values for different parameters like moisture and fungus content and based on macro- and micro-climatic conditions and resource consumption, these values will fluctuate as the simulation proceeds.

Weaver initializes the cells (voxels) that will form the simulation domain as a 3D array with width, depth and height dimensions. This creates a data structure that will hold the living individuals and basal resources. Each individual, whether a fungus oran animal (and even plants in future releases), is initialized based on user-provided parameters: genetic trait variability, genetic correlation among traits, species identity and a series of parameters that aim at both providing realism and allowing simulation functionality. As such parameters are commonly defined as a range of possible values, each individual is initialized following a random initialization process and can be considered unique. Upon initialization, each individual is randomly located into one of the existing cells. Once soil, fungi and animals are initialized, simulation starts until one or more of the stopping conditions are satisfied (e.g. number of days).

Spatial patchiness is of extreme importance to mimic metapopulation and metacommunity dynamics, as well as divergent evolution within distant islands. Related to this, the user has major control over some aspects of soil initialization such as moisture and fungus geometric distribution. For instance, one can set a homogeneous moisture value that will be the same for all cells in the world or set fungus availability at random among cells. More complex initialization can also be carried out. The user can instruct the program to create as many patches (spheric or gaussian shaped) as needed and to select or to randomize the different parameters that define them, such as relative humidity, fungus biomass, radius of the patch (spheric patches) and sigma of the gaussian distribution (gaussian patches). Finally, these structures can be mixed, to create very complex basal resource spatial distributions. In the case that many structures affect the same voxel, the highest value (e.g. moisture) is retained for that particular voxel.

In the present version, fungi grow within the cell based on the level of relative humidity present in that cell (C–N–P will be implemented in future releases). In the absence of fungivores, fungus growth is governed by the same equations as in the former version of the framework (equations A1 and A2 in Moya-Laraño et al., 2012). Water availability affects only fungi thus far. In future versions of Weaver, animals will also have water tanks as state variables and water availability will change dynamically through space and time. All simulations presented here were run at constant 87.5% relative humidity and with a single basal resource (fungus species) that grew optimally at this humidity with a species-specific rate (i.e. r_T in equation A1 of Moya-Laraño et al., 2012 is maximum at 87.5% RH; $r_T = 0.017$, growth range, RH $= 85$–90%, $r_T = 0$–0.017). In the future, other fungus species differing in optimal RHs will be incorporated. However, in the current version, this possibility was not included because water availability does not change with time yet.

A.2.1 Chemostat

Preliminary behaviour of a set of simulations (11 species of predators, 9 species of prey and 4 of fungi, see main text) showed that basal resources would always go to extinction within less than 40 days. Since we were primarily interested in investigating eco-evolutionary dynamics of predator–prey interactions, and since pulses of basal resources have been shown to strongly contribute to community stability (Roelke et al., 2003), we incorporated in the simulator a chemostat that provided pulses of basal resources either at a certain (constant) frequency (days^{-1}) or when resources were close to depletion within a pre-established threshold. The former type of pulse mimics, for instance, rainy days allowing bursts of fungi growth, being by far the most realistic case. The second type can be used to allow predator–prey dynamics to go on for longer periods of time when one is interested in exploring the evolution of certain traits. In this chapter, we only present results using the most realistic frequency-based pulsing chemostat. However, to enhance the duration of the simulation, we set the chemostat at a refilling rate of 100%/day.

A.3. Phenotypic ranges with quantitative genetic variation

The following is a simplified description of how genetics are implemented in our approach (further details in Moya-Laraño et al., 2012). Each trait is determined by a couple of vectors which include an arrangement of loci with values varying between 0 and 1. To establish the trait value for each individual, the values across loci are added and the result interpolated between the phenotypic ranges of the trait (Table 3.1). For each trait, we established evolutionary limits, beyond which the population cannot evolve (assuming physical and physiological constraints), and therefore, the genetically based trait variability is determined within these limits. Thus, for each trait X, we describe the limits and the phenotypic range used as follows:

$$l_X = L_X + \varphi\left(\frac{U_X - L_X}{2}\right) \tag{A1}$$

$$u_X = U_X - \varphi\left(\frac{U_X - L_X}{2}\right) \tag{A2}$$

where l_X and u_X define, respectively, the lower and upper limits of the range used for trait X in the simulation, L_X and U_X define standard lower and upper limits for the trait (Table 3.1) and φ is a coefficient (range 0–1) which determines what proportion of the distance from the standard limits to the midpoint between them is used to calculate the final trait range (l_X, u_X). Thus, a higher φ involves lower trait variability. We forced $U_X < K_X$ and $L_X > \Pi_X$, where K_X and Π_X are the uppermost and lowermost evolutionary limits for trait X, respectively. The above criteria ensure that variability was sufficiently large for new phenotypes to evolve (determined by standing genetic variation), but with thresholds far enough (L_X and U_X) from the evolutionary limits (Π_X and K_X). We used $\varphi = 0.01$ or 0.99 for simulations with high or low genetic variability, respectively.

A.4. Animal traits

We refer to the reader to our repository (www.eeza.csic.es/foodweb) to learn about all the input parameters and evolutionary limits used for each trait and species. The 13 traits (or 14 if consider total body size, e.g., B

$+\varepsilon_o$ below) included in the present simulations were the following (see ranges in Table 3.1):

body size at birth (size_ini, B_0): structural body mass at birth.

energy tank at birth and after moulting (tank_ini, ε_o): percentage of mass devoted to maintenance and future growth and reproduction. Individual body mass (M) is thus the sum of body size and the energy tank, both of which are also state variables.

voracity (v): maximal consumption rate per day (implemented as a scaling coefficient v which makes voracity to scale with body mass as $0.1M^v$). Source: Yodzis and Innes (1992) and A. M. DeRoos (unpublished notes), which provide a fixed maximal consumption rate coefficient of 0.75; see also Englund et al. (2012) for variation around this value. This trait does not just constrain how much food an animal can consume per day but also affects predation risk, as the more voracious animals expend more time foraging. Thus, the probability of encounter with predators (P_p in Eq. A10) depends, among other things, on this trait. Hence, animals that are genetically highly voracious are more exposed to predation risk, which is consistent with what we know from animal personalities and behavioural syndromes (Sih et al., 2004) and does therefore include them explicitly in the framework.

speed (s): sprint speed (cm/s) when a predator (in the case of prey) or a prey (in the case of predators) is encountered and the prey tries to escape from the predator and the predator tries to catch the prey. Implemented as a scaling coefficient s which makes speed to scale with body mass as $\propto M^s$. This coefficient has been documented to vary across studies: 0.17–0.25 (Peters, 1983; Schmidt-Nielsen, 1984). Taking 4 as the normalization constant ($4M^s$), we obtain sprint speeds which fall within the observed ranges from the tiniest mites to the largest wandering spiders, covering a mass range of 0.03–465 mg.

metabolic rate (met_rate, a): Energy losses from metabolism follow the metabolic theory of ecology (Brown et al., 2004) and recent estimates in soil fauna for the separate effects on metabolic rate of temperature, activation energy and body mass (Ehnes et al., 2011):

$$\ln I = \ln I_0 + a \ln M - E\left(\frac{1}{kT}\right) \tag{A3}$$

where I is metabolic rate (J/h), I_0 is a normalization constant, a is a coefficient which relates body size to metabolic rate, E is the activation energy (in eV), k is the Boltzmann's constant (8.62×10^{-5} eV/K) and

T is the environmental temperature in Kelvin. All parameters are included as reported for each animal soil group (Ehnes et al., 2011). Genetic variability was included and modelled around the coefficient "a".

In addition, we also included field metabolic rates, which were calculated in an algorithm that includes environmental stress from encounters with predators (Hawlena and Schmitz, 2010) as well as on the state of voracity and amount of movement of each individual (Eq. A7).

growth (g): Growth is a trait that determines how much an individual grows in each moulting event. Note that we are simulating invertebrates which grow by moulting. Thus, this trait is not truly growth rate but growth ratio at moulting independently of the rate (t^{-1}) at which moulting occurs. Therefore, this trait determines how much of the available energy storage is allocated to fixed body parts in the next developmental stage (instar). Since a fraction of the energy tank at moulting should also be allocated to the post-moulting energy tank, these two traits basically decide when an individual will moult. Growth is merely included as a ratio of the linear dimension of fixed (structural) body parts of the new (target) instar relative to the previous instar. In Section A.7, we describe how the algorithm for growth has been improved in Weaver.

search area (search_area, m): Importantly, we distinguish between speed and mobility. Speed reflects sprint speed when trying to escape from a predator or trying to catch a prey. However, we consider mobility (search area) as how much one individual is able to move to search for resources or for safe patches. Lacking better information, the entire area covered in 1 day (*m*), scales with body size in a similar way as sprint speed: M^m. Differently than in mini–Akira (Moya-Laraño et al., 2012), for translating mobility into actual search area in the simulation, we have derived scaling constants for Weaver (see Section A6).

assimilation efficiency (assim): Assimilation efficiency is merely the amount of ingested food which is converted in own body mass. Following previous work on soil fauna, we can assume to be around 0.85 (85%) (Rall et al., 2010 and references therein).

phenology (pheno): Day of birth since either the beginning of the season (simulation) or since the date of oviposition. This trait can be also called egg developmental time, as the date of birth will depend on how fast eggs develop beyond what is dictated by temperature and other environmental constraints. Thus, for calculating the final phenological date, which will vary depending on temperature, we further included temperature-dependent developmental rates by using published equations (Gillooly et al., 2002) and calculating the average Q_{10} values across the range of body masses for our propagule sizes in the simulation, which gave $Q_{10} = 2.84$.

activation energy for metabolic rate (E_{met}, E in Eq. A3): To further control the effect of temperature on eco-evolutionary dynamics, we also included, in addition to simulations at different temperatures, variability around E, which will serve to study adaptive evolution around thermal sensitivity of metabolic rate, a form of thermal adaptation. Ranges in Table 3.1 were set around published coefficients in Ehnes et al. (2011).

We further included three additional traits that represented variability in plasticity to temperature (Q_{10}) for three activity traits: voracity, speed and search area (vorQ_{10}, spdQ_{10} and srchQ_{10}, respectively). We used recent published accounts from a thorough review on temperature-dependent ecological traits in predator–prey interactions (Dell et al., 2011). For activity traits, we used Q_{10} (i.e. how many times a given trait increases for a 10 °C increase in temperature) instead of E because we lacked information for how E and M combine to determine trait values, as it is the case for metabolic rate (I) in Eq. (A3) (Ehnes et al., 2011). In addition, Q_{10} values are more easily interpretable and converted to reaction norms. Together, these three traits represent thermal plastic adaptation for mobility. For simplicity, we ran all simulations at 18 °C.

Q_{10} *on voracity (vorQ_{10})*: Based on data on consumption rates (Dell et al., 2011).

Q_{10} *on speed (spdQ_{10})*: Based on data on escaping speeds (Dell et al., 2011).

Q_{10} *on search area (srchQ_{10})*: Based on data on voluntary body speed (Dell et al., 2011).

To estimate the effect of Q_{10} values in the simulation for all traits that involved temperature sensitivity, we used linear interpolation between the minimum and maximum temperatures used for all simulations (15–25 °C). Thus, real Q_{10} would be used if a simulation was performed at 25 °C, and for simulations at intermediate temperatures, we estimated the value of Q (e.g. Q_3 at 18 °C) by interpolation between the two temperatures, which assumes linearity of Q across temperatures. Since Q_{10} have a quantitative genetic basis and modify other genetically driven traits, Q_{10} genes are epistatic in nature (i.e. the action of one gene on the phenotype is affected by the expression of Q_{10} genes). This is an epistatic view of phenotypic plasticity (Roff, 1997; Scheiner, 1993), as the phenotypic effect of Q_{10} genes as the environment changes (i.e. increase in temperature) is to modify the expression of other genes. Thus, this fourth module includes genes for trait plasticity to temperature variation.

A.5. Predator and prey quantitative genetics with more realistic recombination rates

In the former version of this IBM framework (Moya-Laraño et al., 2012), the authors successfully induced genetic correlations from pleiotropic effects by including all loci that affected the same traits in arrays which were called chromosomes. This was unrealistic because genetic correlations occur by both pleiotropic effects of quantitative genes and from linkage disequilibrium (Roff, 1997), and we only considered the former. In linkage disequilibrium, loci that are close to each other in the chromosome tend to stay together for several generations (linkage), the number of which depends on their relative distance in the chromosome and on the recombination rate. Since linkage disequilibrium has its own evolutionary importance (e.g. in genetic drift) and can be driven by many mechanisms, such as selection and non-random mating (Falconer and MacKay, 2006), we decided to improve our modelling of quantitative genetics by better mimicking true recombination. The formerly described loci vectors (Moya-Laraño et al., 2012), which are useful to induce the desired degrees of genetic correlations among traits, we now term "correlosomes". In the current version of the algorithm, we include more realistic recombination rates by randomly permuting the position of each locus across the genome before cross-over. These random positions are established at the beginning of each simulation and are kept constant throughout for all animals. After cross-over, and before the new egg is built, the position of the loci is returned back to the correlosome position, which will allow assigning phenotypes to the newborns by keeping the original degree of genetic correlation (ρ parameter in Moya-Laraño et al., 2012). Note that the previous claim (Moya-Laraño, 2012) that non-expressing alleles in correlosomes could be taken as microsatellite markers depending on the distance to expressing alleles still holds. However, now these alleles, as functional alleles do, will recombine in random positions and their position to functional alleles will be also random, a much more realistic situation.

A.6. New adjustments in all mass-dependent equations and plastic traits

A.6.1 Water body content

In the former version of this framework, the authors did not explicitly model water body content (Moya-Laraño et al., 2012). Although water content is still static in the present version, as a prelude of a future version in which water loss will drive individual behaviour as well (e.g. Verdeny-Vilalta and Moya-Laraño, 2014), here we modelled all animals as having 70% of their body mass in water form (following Sabo et al., 2002).

A.6.2 Adjustment of scaling constants in mass- and temperature-dependent equations

In Moya-Laraño et al. (2012), the authors modelled animals that were smaller than 1 mg at maturation, which allowed all non-linear equations to behave consistently (e.g. larger animals had always higher speeds than smaller animals). In order to accommodate larger animals and to be sure that scaling was realistic, we used here linear interpolation to predict scaling constants (intercepts) from scaling coefficients (slopes). To estimate voracity, speed and search area of each animal (V, A and S in equations A6, A7 and A8 of Moya-Laraño et al., 2012, respectively), we estimated the intercepts (note that former equation A7—Eq. A5—now includes a scaling constant) by linear interpolation in which each constitutive trait assigned to each animal (the scaling coefficients v, m and s to obtain V, A and S, respectively) was linearly interpolated from the scale of the evolutionary limits (Π_X and K_X) to a final scale. This ensured that, for the smallest and largest animals, V, A and S were always higher for larger animals. We observed a high rate of starvation and a low rate of mobility depending on the coefficients of body condition (c) and rates of encounters with predators (e) in the former equations A6 and A7 (Moya-Laraño et al., 2012). To enhance survival and mobility, in what are now Eqs. (A4) and (A5), the term $c \cdot e$ has been changed to $(c + e)/2$. This gives equal weight to condition and anti-predator behaviour in determining animal voracity and mobility and does not penalize voracity and mobility as strongly when animals are well fed and have encountered many predators (e.g. low c and e values). Future data across a few animal taxa should determine how the internal state of the animal (c) and anti-predatory behaviour (e) interact to determine the levels of animal activity. Therefore, the new equations are as follows:

$$V = f_V M^v [(c + e)/2] Q_{VT} \tag{A4}$$

$$A = f_A M^m [(c+e)/2] Q_{AT} \qquad \text{(A5)}$$
$$S = f_S M^s c Q_{ST} \qquad \text{(A6)}$$

where the fs are the interpolated scaling constants (ranges f_V: 0.05–0.15, f_A: 1–12.85, f_S: 1–7). All other terms discussed in Moya-Laraño et al. (2012). To improve basal metabolic rate estimates, we took advantage of the strong correlation occurring between $\ln I_0$ and E across nine taxa of soil organisms in the data of Ehnes et al. (2011; equation A3) and used linear regression ($\ln I_0 = -7.29 + 43.97E$; $R^2 = 0.99$; $P < 0.0001$; $N = 9$) to predict basal metabolic losses for individuals genetically differing in the scaling coefficient (a) and activation energies for metabolic rates (E). This approach ensured that animals with higher E expended more energy for a given environmental temperature (more energy was needed to activate their metabolism) and greatly improved survival from starvation in the simulations.

A.6.3 Field metabolic rates

Field metabolic rates can be obtained by multiplying basal metabolic rates by a value of 2.5–3 (Brose et al., 2008; Yodzis and Innes, 1992). In order to make sense of individual variation, we used a more dynamic approach of assigning the multiplying coefficient of field metabolic rates (3) only to the fraction of time (P_t) that animals were really active (Moya-Laraño, 2012):

$$P_t = w_A \left(\frac{W}{A_{\max}} \right) + (1 - w_A) \left(\frac{V}{V_{\max}} \right) \qquad \text{(A7)}$$

where w_A determines how much weight is given to activity among patches (W) or within patches (V), and $A_{\max}$ and $V_{\max}$ are maximum searching area and voracities in the community (i.e. for the largest animals with lowest condition, lowest previous encounter rates with predators and at the highest temperature $T = 25\ °C$), respectively. In the current version, $A_{\max}$ and $V_{\max}$ values have been calculated for each species. This implies that the offspring of the largest species has a survival bonus, as they will spend relatively much lower energy from moving around than the offspring of the smallest species, which will be closer to their maximum. We decided to pursue this strategy for now because, due to the initialization following allometric mass–abundance scaling, too few of the large predators were included in the simulations, which were therefore highly subjected to stochasticity and went rapidly extinct. This procedure ensured that large predators persisted more. In a future parallelized code, which will allow initializations with >100.000

individuals, each individual will have its own A_{max} and V_{max}, which will in turn depend on its own traits, a much more realistic situation.

A.7. Moulting algorithm

An improvement relative to the former moulting algorithm is that now the shape of the animals, and thus the length–mass allometric relationship, is taken into account to decide when the next instar should be achieved:

$$M = aL^b \qquad\qquad (A8)$$

We obtained most of the a and b coefficients from published accounts (Edwards and Gabriel, 1998; Gruner, 2003; Hodar, 1996), and for enchytraeids (Clitellata), we used unpublished estimates ($a=0.0039$, $b=2.53$; O. Verdeny-Vilalta et al., unpublished data). In all equations, body length is measured in mm and mass in mg. Now, the mass available for moulting (i.e. 90% of the state variable energy tank, ε) is transformed into the length of the animal (by rearranging Eq. A5) in order to allow for length growth to be governed by the trait growth ratio. Once an individual moults, the original equation (A5) is applied to estimate the body mass of the instar. The mass allocated to ε is a fixed proportion of the fixed body mass (i.e. the B trait) which is the same proportion as that given at birth (i.e. the ε_o trait).

In the current version, the number of instars was species specific and was calculated by first estimating egg mass for each species by fitting the following adult–offspring allometric relationship: offspring mass $=0.03*$ adult mass$^{0.5}$, which was calculated by visually fitting the intercept for the invertebrate relationship presented in figure 3.3 of Hendriks and Mulder (2008). We then obtained adult masses from a unpublished data set on individual lengths across four beech forests (J. Moya-Laraño et al., unpublished data) to which we applied Eq. (A5) and the same coefficients as in our simulations. Finally, we applied the moulting algorithm described above to the offspring of each species, for which we included the same growth ratio (mid-point of the range) used in our simulations. By iteration of this algorithm, we calculated the number of instars which matched the targeted adult body size and this result was used as the total number of instars for each simulated species.

Since growth curves are not ruled by a single ratio parameter, as they usually have a sigmoid shape, we allowed animals to also plastically moult by including time as an alternative for moulting. We included a couple of state variables which were time elapsed in the prior instar and time elapsed in the present instar. The ratio between time in present and time in past instars was established as a second rule of thumb to moult with whatever amount of resources the animal could have accumulated to that point, forcing a moulting event which does not follow the growth ratio. This provides some realism to growth curves as arthropods are generally phenotypically plastic in developmental time (Nylin and Gotthard, 1998). If an individual has not moulted due to its growth ratio trait by the time determined by this ratio, the animal proceeds to moult. Ratios used in simulations were 0.8, 0.9 and 1.5 for large predators, small predators and fungivores, respectively. This difference was necessary to enhance maturation in predators, which were larger on average than prey.

A.8. Reproductive algorithm

In addition to the changes in recombination explained above, we simplified the criterion for reproduction. Now we calculated the biomass available for reproduction as 90% of the energy tank (ε) and when this value was twice as large as the weight of the fixed body mass of the adult individual, the animal was able to lay an egg batch. In future versions, this criterion should be changed to a species-specific mode of reproduction criterion, allowing for continuous egg laying (as many mites do) and also laying large batches at once (as spiders). Note that this is not merely a distinction between iteroparity or semelparity, as iteroparous animals can still lay either one egg at each reproductive event or an entire clutch. What is relevant for the dynamics is how much energy an individual needs to accrue before a reproductive event occurs. Continuing with this simple criterion for reproduction, and as in the former version, each prey individual was allowed to lay up to two batches before dying and predators up to 5. The number of batches and the number of eggs per batch, which are closely associated to the mode for reproduction, should be adjusted in the future in a species-specific basis.

In the former version of this IBM mate search was not necessary because we assumed that animals would encounter a mate with 100% probability as long as there was one present in the population. Since space is now relatively much larger (albeit still in arbitrary units, i.e. 4000 vs. 100 cells), we have implemented a momentary solution in which once an individual enters the reproductive state it starts searching for patches with mates (moving preferentially to patches with more reproductive individuals of their own species) and will be allowed to move up to 1000 steps per day. These individuals are both invulnerable to predation and do not expend energy during searching. This simplification had to be added to enhance mating encounters and improve ecological dynamics. However, once mate attractants are successfully implemented, future versions should also include the cost of mate search (e.g. predation risk and energy expenditure).

A.9. Restricting and controlling attack rates

Within a given patch (or cell), the probability that a predator finds a prey and vice versa as well as the probability that predation occurs upon an encounter depends on probabilities drawn at random from a uniform distribution which are contrasted with deterministic probabilities that come from the animal traits involved (equations A13 and A14 in Moya-Laraño et al., 2012). The logistic relationship between the traits of the two animals involved in a potential encounter or predation event and the deterministic probability dependent on a set of coefficients ($\alpha \ldots \eta$ for encounters and $\alpha \ldots \delta$ for predation...δ) which determined the relative importance of each predator and prey trait would take in either encounters or predation. These coefficients were called naïve coefficients because they are not based on real data. In the future, they should either be estimated from real data or be the subject of sensitivity analysis, in particular if we want to make strong inferences about the evolution of certain traits. To facilitate the task of simulating the effects of different values for these coefficients, we estimated by simulation means and standard deviations for all traits and trait combinations (e.g. predator by prey mass interaction) which allowed standardizing these traits and their combinations as if following a standard normal distribution. This ensured that all traits and combinations had identical units in the equations, which allowed the naïve coefficients to have identical weight on the final probabilities regardless of the units of measurement of the traits involved. The final equations are thus:

$$P_e = \frac{1}{\left[1 + e^{-\left(\alpha + \beta V'_{P_d} + \gamma V'_{P_y} + \delta\left(V_{P_d} V_{P_y}\right)' + \varepsilon B'_{P_d} + \zeta B'_{P_y} + \eta\left(B_{P_d} B_{P_y}\right)'\right)}\right]} \tag{A9}$$

$$P_p = \frac{1}{\left[1 + e^{-\left(\alpha + \beta V'_{P_d} + \gamma R'_B + \delta R'_S\right)}\right]} \tag{A10}$$

where the $'$s refer to the fact that the variables (or their products) were standardized. Traits are voracities (V) of the predator or attacking individual in IGP and cannibalism (P_d) or prey or attacked individual (P_y), B is structural body mass and R_B and R_S are structural body size and speed ratios, respectively. More details can be found in Moya-Laraño et al. (2012). The naïve coefficients used in the present simulation were for encounter probabilities $(\alpha \ldots \eta)$: 1, 0.1, 0.1, 0.1, 0.1, 0.1, 0.1; and for predation probabilities given an encounter $(\alpha \ldots \delta)$: 1, 0.1, 0.1, 0.1. Therefore, we assumed that all variables (and thus all traits) had equal weight in deciding an interaction outcome.

One of the problems of the above equations is that predator–prey interactions are ruled by traits and naïve coefficients and therefore attack rates could be overlay too high or excessively low to ensure any ecological dynamics. We added an additional control, which could be used for controlling attack rates upon each species. This was accomplished by restricting the range of those P-values randomly obtained from a uniform distribution which were to be contrasted against P_e and P_p to decide the outcome of interactions (i.e. an encounter occurs if $P_e >$ random P-value). We first obtained the distribution of values by running a simulation and then established the following ranges of random P-values for large and small predators, and prey, respectively (webs with connectance 0.1 or 0.55): *encounters*, 0.337–0.342, 0.337–0.340, 0.337–0.3394; *predation*, 0.606–0.623, 0.606–0.617, 0.606–0.6149. For the web with connectance 0.3 (which had different species), the values were, respectively: *encounters*, 0.337–0.3375, 0.337–0.338, 0.337–0.339; *predation*, 0.607–0.612, 0.607–0.614, 0.607–0.617. This ensured that large predators (and especially the juvenile stages) suffered less from IGP. Again, this simulation strategy allowed large predators to stay longer in the simulations, as otherwise extinction occurred too rapidly. For the same reason (improving large predator survival), and since small predators already had a predation pressure coming from higher trophic levels, all background mortalities (Moya-Laraño et al., 2012) were set to zero. Although we do not know what defensive traits allow the offspring of large predators to reach maturation, it is obvious that without these improvements, we were not able to mimic the survival of juveniles in large predatory species. This in itself opens an interesting question to investigate (a) whether the offspring of large predators have improved survival over other predators of the same size and (b) which traits may be involved in this differential survival.

To further control attack rates and make them more realistic, we allowed encounters and predation events only within the following range of log body mass ratios: −1.21 to 6.68, as data on 800 induced predator–prey encounters among individuals of a beech forest food web have shown that predation never occurs beyond these thresholds (J. Moya-Laraño et al., unpublished data).

In addition, we relaxed the 1-day one-prey criterion to allow larger predators to feed in a more realistic way. Now all predators can search and feed on several prey per day as long as the ingested body mass does not surpass that of their voracity (V) or as long as the number of visited patches in 1 day does not surpass that of the individual trait area searched (A). Since now food ingestion is at its maximum (V), we assumed that all food could be handled and digested in 1 day and set all digestion times (regardless of temperature) to zero. Therefore, all predators that rested alive were ready to hunt again the following day regardless of how much they had ingested the day before. Once we improve the capabilities of our framework, we will be able to establish simulations with time steps of hours or minutes, when real handling and digestion times will be appropriately included.

A.10. Adaptive animal movement

Animals move adaptively in 3D space similarly as they did in 1D mini-Akira (Moya-Laraño et al., 2012) with the plus that they now remember the cells they visited within the same day as not to repeat the same cell in a given day. This was implemented as to make predators to move more naturally, as once interactions (encounters plus predation) have been unsuccessfully attempted with all individuals in a patch, the predator should assess the patch as unsuitable for hunting prey. Although Weaver now implements the possibility of animals assessing cells farther away from neighbouring cells, all simulations in this paper assume that animals can only assess the 26 cells that immediate surround the cell in which the animal is located in a given time. At each move, fungivores will move to cells with the lowest ratio between predation threat and fungus biomass with the additional improvement that fungivores assess only the species to which they are linked according to the food web structure established at the beginning of the simulation (see main text) and within the predatory threshold imposed by the limits of log mass ratios (-1.21 to 6.68). Predators behave in a similar way: they assess predator threat and prey availability (both fungivores and IGP prey) considering only those species with which they are linked according to food web structure and following the log mass ratio criterion.

In addition, as in mini-Akira, both prey and predators can perform jumps so as to clear out from the areas in which edible items have been depleted. To accommodate these jumps to a 3D environment, when food availability within the 27 neighbouring cells (26 surrounding cells plus the cell where the animal is currently located) is zero, the animal will perform a jump at a distance which is established by drawing a random number from a uniform distribution which ranges between 2 and its search area trait (A). This animal then evaluates the new area and can either perform another jump or just stay put if food is available. At each jump, the animal evaluates how many cells are within its jump value (in the 3D directions) and if an edge (wall) of the world is at a shorter distance than the projected jump, the direction pointing towards that edge is immediately discarded from the universe of possible directions towards which a jump can be performed. This same edge procedure takes place at normal moves (e.g. when only one step is performed at a time).

To prevent animals from getting stuck in an area, they track the cells they have visited each day and they do not go back to those cells within the same day. Therefore, each cell is assessed by each predator only once per day and if after having attempted to catch prey in a cell, additional prey need to be caught that day to meet the voracity demands (see above), these will be sought in a different cell. This process applies also when the animals jump, so they do not jump back to an already visited cell in the very same day, encouraging them to move through new areas. As in Moya-Laraño et al. (2012), the number of maximum steps an animal can attempt each day equals its search area trait (A). However, now each displacement (either a simple step or a jump) adds only a single step unit to W for estimating field metabolic rates in Eq. (A4) (A10 in Moya-Laraño et al., 2012).

A.11. Computational demand and requirements

Computationally, Weaver can be highly demanding in terms of both memory and CPU needs. Memory will be impacted by the number of individuals living in the world at a given time and cells comprising the soil structure. This should be taken into account by the user when running potentially large simulations. For example, a simulation running for 200 days with 9 species or fungivores and 11 species of predators (community size $= 20,000$ individuals), such as that presented here, may take only about 1 h but more than 18 GB of memory and store more than 100 GB of information, including the genetics of all organisms, the predator–prey interactions at the species level, the daily reports on trait and state variable values for each alive animal (including spatial location), the constitutive traits to document evolutionary dynamics and the number of animals for each state and instar for each day. Therefore, the program outputs will require vast amounts of hard disk to store all the information needed (which is also customizable), especially if binary snapshots are stored at each simulation step to graphically visualize water and fungus levels, animal population densities, etc.

REFERENCES

Abrams, P.A., 2007. Habitat choice in predator-prey systems: spatial instability due to inter-acting adaptive movements. Am. Nat. 169 (5), 581–594.

Agrawal, A.A., Lau, J.A., Hamback, P.A., 2006. Community heterogeneity and the evolution of interactions between plants and insect herbivores. Q. Rev. Biol. 81, 349–376.

Amarasekare, P., 2008a. Spatial dynamics of foodwebs. Annu. Rev. Ecol. Evol. Syst. 39, 479–500.

Amarasekare, P., 2008b. Spatial dynamics of keystone predation. J. Anim. Ecol. 77, 1306–1315.

Amarasekare, P., 2010. Effect of non-random dispersal strategies on spatial coexistence mechanisms. J. Anim. Ecol. 79, 282–293.

Anderson, J.F., 1970. Metabolic rates of spiders. Comp. Biochem. Physiol. 33, 51–72.

Anderson, J.F., 1974. Responses to starvation in spiders lycosa-lenta hentz and filistata-hibernalis (hentz). Ecology 55, 576–585.

André, H.M., Ducarme, X., Lebrun, P., 2002. Soil biodiversity: myth, reality or conning? Oikos 96, 3–24.

Bale, J., van Lenteren, J., Bigler, F., 2008. Biological control and sustainable food production. Philos. Trans. R. Soc. Lond. B Biol. Sci. 363, 761–776.

Bassar, R.D., Ferriere, R., Lopez-Sepulcre, A., Marshall, M.C., Travis, J., Pringle, C.M., Reznick, C.N., 2012. Direct and indirect ecosystem effects of evolutionary adaptation in the trinidadian guppy (*Poecilia reticulata*). Am. Nat. 180, 167–185.

Beaumont, M., 2010. Approximate Bayesian computation in evolution and ecology. Annu. Rev. Ecol. Evol. Syst. 41, 379–406.

Becks, L., Ellner, S.P., Jones, L.E., Hairston, N.G., 2010. Reduction of adaptive genetic diversity radically alters eco-evolutionary community dynamics. Ecol. Lett. 13, 989–997.

Benkman, C., 2013. Biotic interaction strength and the intensity of selection. Ecol. Lett. 16, 1054–1060.

Bohan, D.A., Woodward, G., 2013. Editorial commentary: the potential for network approaches to improve knowledge, understanding and prediction of the structure and functioning of agricultural systems. Adv. Ecol. Res. 49, xiii–xviii.

Bohan, D., Raybould, A., Mulder, C., Woodward, G., Tamaddoni-Nezhad, A., Bluthgen, N., Pocock, M., Muggleton, S., Evans, D., Astegiano, J., Massol, F., Loeuille, N., Petit, S., Macfadyen, S., 2013. Networking agroecology: integrating the diversity of agroecosystem interactions. Adv. Ecol. Res. 49, 1–67.

Boit, A., Martinez, N.D., Williams, R.J., Gaedke, U., 2012. Mechanistic theory and modelling of complex food-web dynamics in lake Constance. Ecol. Lett. 15, 594–602.

Bolnick, D.I., Amarasekare, P., Araújo, M.S., Burger, R., Levine, J.M., Novak, M., Rudolf, V.H.W., Schreiber, S.J., Urban, M.C., Vasseur, D.A., 2011. Why intraspecific trait variation matters in community ecology. Trends Ecol. Evol. 26, 183–192.

Boyd, S., Vandenberghe, L., 2009. Convex Optimization. Cambridge University Press, Cambridge.

Brose, U., Ehnes, R.B., Rall, B.C., Vucic-Pestic, O., Berlow, E.L., Scheu, S., 2008. Foraging theory predicts predator-prey energy fluxes. J. Anim. Ecol. 77, 1072–1078.

Brown, J.H., Gillooly, J.F., Allen, A.P., Savage, V.M., West, G.B., 2004. Toward a metabolic theory of ecology. Ecology 85, 1771–1789.

Carroll, S.P., Hendry, A.P., Reznick, D.N., Fox, C.V., 2007. Evolution on ecological timescales. Funct. Ecol. 21, 387–393.

Carter, A.J., Feeney, W.E., Marshall, H.H., Cowlishaw, G., Heinsohn, R., 2013. Animal personality: what are behavioural ecologists measuring? Biol. Rev. 88, 465–475.

Chen, B.R., Wise, D.H., 1999. Bottom-up limitation of predaceous arthropods in a detritus-based terrestrial food web. Ecology 80, 761–772.

Chislock, M.F., Sarnelle, O., Olsen, B.K., Doster, E., Wilson, A.E., 2013. Large effects of consumer offense on ecosystem structure and function. Ecology 94, 2375–2380.

Chown, S.L., 2011. Discontinuous gas exchange: new perspectives on evolutionary origins and ecological implications. Funct. Ecol. 25, 1163–1168.

Coleman, D.C., Crossley, D.A., Hendrix, P.F., 2004. Fundamentals of Soil Ecology. Elsevier, Burlington, MA.

Darwin, C.R., 1959. On the Origin of Species by Means of Natural Selection. John Murray, London.

Decaëns, T., 2010. Macroecological patterns in soil communities. Glob. Ecol. Biogeogr. 19, 287–302.

Dell, A.I., Pawar, S., Savage, V.M., 2011. Systematic variation in the temperature dependence of physiological and ecological traits. Proc. Natl. Acad. Sci. U.S.A. 108, 10591–10596.

Directive 2009/128/EC of the European Parliament and of the Council. Establishing a framework for Community action to achieve the sustainable use of pesticides, 2009/128/EC. Official J. Eur. Union.

Edwards, R.L., Gabriel, W.L., 1998. Dry weight of fresh and preserved spiders (Araneida: Labidognatha). Entomol. News 109, 66–74.

Ehnes, R.B., Rall, B.C., Brose, U., 2011. Phylogenetic grouping, curvature and metabolic scaling in terrestrial invertebrates. Ecol. Lett. 14, 993–1000.

Elton, C., 1927. Animal Ecology. Sidgwick and Jackson, London.

Englund, G., Öhlund, G., Hein, C.L., Diehl, S., 2012. Temperature dependence of the functional response. Ecol. Lett. 14, 914–921.

Falconer, S.F., MacKay, T.F.C., 2006. Introduction to Quantitative Genetics, fourth ed. Prentice and Hall, Harlow.

Faria, L.D.B., Umbanhowar, J., McCann, K.S., 2008. The long-term and transient implications of multiple predators in biocontrol. Theor. Ecol. 1, 45–53.

Farkas, T.E., Mononen, T., Comeault, A.A., Hanski, I., Nosil, P., 2013. Evolution of camouflage drives rapid ecological change in an insect community. Curr. Biol. 23 (19), 1835–1843.

Floudas, C.A., Pardalos, P.M., 2009. Encyclopedia of Optimization. Springer, New York.

Fontaine, C., Guimarães, P.R., KEfi, S., Leouille, N., Memmott, J., van der Putten, W.H., van Veen, F.J.F., Thebault, E., 2011. The ecological and evolutionary implications of merging different types of networks. Ecol. Lett. 14, 1170–1181.

Fox, J., 2003. Effect displays in R for generalised linear models. J. Stat. Software 8, 1–27.

Fox, C.W., Czesak, M.E., 2000. Evolutionary ecology of progeny size in arthropods. Annu. Rev. Entomol. 45, 341–369.

Fussmann, G., Loreau, M., Abrams, P., 2007. Eco-evolutionary dynamics of communities and ecosystems. Funct. Ecol. 21, 465–477.

Gillooly, J.F., Charnov, E.L., West, G.B., Savage, V.M., Brown, J.H., 2002. Effects of size and temperature on developmental time. Nature 417, 70–73.

Grelaud, A., Robert, C.P., Marin, J.M., Rodolphe, F., Taly, J., 2009. ABC likelihood-free methods for model choice in Gibbs random fields. Bayesian Anal. 4, 317–336.

Gruner, D.S., 2003. Regressions of length and width to predict arthropod biomass in the Hawaiian Islands. Pac. Sci. 57, 325–336.

Haegeman, B., Loureau, M., 2014. General relationships between consumer dispersal, resource dispersal and metacommunity diversity. Ecol. Lett. 17, 175–184.

Hamblin, S., 2013. On the practical usage of genetic algorithms in ecology and evolution. Methods Ecol. Evol. 4, 184–194.

Hanski, I., 2011. Eco-evolutionary spatial dynamics in the Glanville fritillary butterfly. Proc. Natl. Acad. Sci. U.S.A. 108, 14397–14404.

Harmon, J.P., Moran, N.A., Ives, A.R., 2009. Species response to environmental change: impacts of food web interactions and evolution. Science 323, 1347–1350.

Harvey, P.H., Pagel, M.D., 1991. The Comparative Method in Evolutionary Biology: The Comparative Method in Evolutionary Biology. Oxford University Press, New York.

Hättenschwiler, S., Tiunov, A.V., Scheu, S., 2005. Biodiversity and litter decomposition in terrestrial ecosystems. Annu. Rev. Ecol. Evol. Syst. 36, 191–218.

Hawlena, D., Schmitz, O.J., 2010. Physiological stress as a fundamental mechanism linking predation to ecosystem functioning. Am. Nat. 176, 537–556.

Hendriks, A.J., Mulder, C., 2008. Scaling of offspring number and mass to plant and animal size: model and meta-analysis. Oecologia 155, 705–716.

Hodar, J.A., 1996. The use of regression equations for estimation of arthropod biomass in ecological studies. Acta Oecol. 17, 421–433.

Holyoak, M., Leibold, M.A., Mouquet, N., Holt, R.D., Hoopes, M.F., 2005. Metacommunities: a framework for large scale community ecology. In: Holyoak, M., Leibold, M.A., Holt, R.D. (Eds.), Metacommunities: Spatial Dynamics and Ecological Communities. University of Chicago Press, Chicago, pp. 1–31.

Hudson, L.N., Reuman, D.C., 2014. A cure for the plague of parameters: constraining models of complex population dynamics with allometries. Proc. R. Soc. Lond. B 264, 1249–1254.

Ingram, T., Svanback, R., Kraft, N.J.B., Kratina, P., Southcott, L., Schluter, D., 2012. Intraguild predation drives evolutionary niche shifts in threespine stickleback. Evolution 66, 1819–1832.

Johnson, M.T.C., Vellend, M., Stinchcombe, J.R., 2009. Evolution in plant populations as a driver of ecological changes in arthropod communities. Philos. Trans. R. Soc. Lond. B Biol. Sci. 364, 1593–1605.

Kerr, B., Neuhauser, C., Bohannan, B.J., Dean, A.M., 2006. Local migration promotes competitive restraint in a host-pathogen 'tragedy of the commons'. Nature 442, 75–78.

Koelle, K., Vandermeer, J., 2005. Dispersal-induced desynchronization: from metapopulations to metacommunities. Ecol. Lett. 8, 167–175.

Kraaijeveld, A.R., Godfray, H.C.J., 1997. Trade-off between parasitoid resistance and larval competitive ability in Drosophila melanogaster. Nature 389, 278–280.

Lau, J., 2012. Evolutionary indirect effects of biological invasions. Oecologia 170, 171–181.

Lennon, J.T., Martiny, J.B.H., 2008. Rapid evolution buffers ecosystem impacts of viruses in a microbial food web. Ecol. Lett. 11, 1178–1188.

Lensing, J.R., Wise, D.H., 2006. Predicted climate change alters the indirect effect of predators on an ecosystem process. Proc. Natl. Acad. Sci. U.S.A. 103, 15502–15505.

Levene, H., 1953. Genetic equilibrium when more than one ecological niche is available. Am. Nat. 87, 331–333.

Levins, R., 1968. Evolution in changing environments. Princeton University Press, Princeton, NY.

Lewis, J.G.E., 1981. The Biology of Centipedes. Cambridge University Press, Cambridge.

Lindberg, N., Engtsson, J.B., Persson, T., 2002. Effects of experimental irrigation and drought on the composition and diversity of soil fauna in a coniferous stand. J. Appl. Ecol. 39, 924–936.

Linear programming FAQ. 2013. http://www.neos-guide.org/content/lp-faq.

Loeuille, N., Barot, S., Georgelin, E., Kylafis, G., Lavigne, C., 2013. Eco-evolutionary dynamics of agricultural networks: implications for sustainable management. Adv. Ecol. Res. 49, 339–435.

Losos, J.B., Schoener, T., Spiller, D.A., 2004. Predator-induced behaviour shifts and natural selection in field-experimental lizard populations. Nature 432, 505–508.

MacColl, A.D.C., 2011. The ecological causes of evolution. Trends Ecol. Evol. 26, 514–522.

Maraldo, K., Schmidt, I.K., Beier, C., Holmstrup, M., 2008. Can field populations of the enchytraeid, *Cognettia sphagnetorum*, adapt to increased drought stress? Soil Biol. Biochem. 40, 1765–1771.

Markót, M., Fernández, J., Casado, L., Csendes, T., 2006. New interval methods for constrained global optimization. Math. Program. Ser. A 106, 287–318.

Martins, N.E., Faria, V.G., Teixeira, L., Magalhães, S., Sucena, E., 2013. Host adaptation is contingent upon the infection route taken by pathogens. PLoS Pathog. 9, e1003601.

McCann, K., 2000. The diversity–stability debate. Nature 405, 228–233.

McCann, K., Hastings, A., 1997. Re-evaluating the omnivory-stability relationship in food webs. Proc. R. Soc. Lond. B 264, 1249–1254.

McCann, K.S., Hastings, A., Huxel, G., 1998. Weak trophic interactions and the balance of nature. Nature 395, 794–798.

McCann, K.S., Rasmussen, J.B., Umbanhowar, J., 2005. The dynamics of spatially coupled food webs. Ecol. Lett. 8, 513–523.

McLaughlin, O.B., Jonsson, T., Emmerson, M.C., 2010. Temporal variability in predator–prey relationships of a forest floor food web. In: Woodward, G. (Ed.), Advances in Ecological Research: Ecological Networks, vol. 42. Academic Press, Amsterdam, pp. 171–264.

Melguizo-Ruiz, N., Verdeny-Vilalta, O., Arnedo, M.A., Moya-Larano, J., 2012. Potential drivers of spatial structure of leaf-litter food webs in south-western European beech forests. Pedobiologia 55, 311–319.

Melián, C., Vilas, C., Baldó, F., González-Ortegón, E., Drake, P., Williams, R., 2011. Eco-evolutionary dynamics of individual-based food webs. Adv. Ecol. Res. 45, 225–268.

Melian, C.J., Krivan, V., Altermatt, F., Stary, P., Pellisier, L., De Laender, F., 2014. Multi-trophic metacommunities and the biogeography of ecological networks. Am. Nat. in press.

Messelink, G., Sabelis, M., Janssen, A., 2012. Generalist predators, food web complexities and biological pest control in greenhouse crops. In: Larramendy, M.L., Soloneski, S. (Eds.), Integrated Pest Management and Pest Control—Current and Future Tactics. InTech, Rijeka, pp. 191–214.

Moore, J.C., Berlow, E.L., Coleman, D.C., de Ruiter, P.C., Dong, Q., Hastings, A., Johnson, N.C., McCann, K.S., Melville, K., Morin, P.J., Nadelhoffer, K., Rosemond, A.D., Post, D.M., Sabo, J.L., Scow, K.M., Vanni, M.J., Wall, D.H., 2004. Detritus, trophic dynamics and biodiversity. Ecol. Lett. 7, 584–600.

Moya-Larano, J., 2011. Genetic variation, predator-prey interactions and food web structure. Philos. Trans. R. Soc. Lond. B Biol. Sci. 366, 1425–1437.

Moya-Laraño, J., 2012. O matrices and eco-evolutionary dynamics. Trends Ecol. Evol. 27, 139–140.

Moya-Larano, J., Wise, D.H., 2007. Direct and indirect effects of ants on a forest-floor food web. Ecology 88, 1454–1465.

Moya-Laraño, J., Verdeny-Vilalta, O., Rowntree, J., Melguizo-Ruiz, N., Montserrat, M., Laiolo, P., 2012. Climate change and eco-evolutionary dynamics in food webs. In: Woodward, G., Jacob, U., Ogorman, E.J. (Eds.), Global Change in Multispecies Systems, Pt. 2. In: Advances in Ecological Research, vol. 47, pp. 1–80.

Moya-Laraño, J., Foellmer, M.W., Pekár, S., Arneda, M.A., Bilde, T., Lubin, Y., 2013. Evolutionary ecology: linking traits, selective pressures and ecological functions. In: Penney, D. (Ed.), Spider Research in the 21st Century: Trends and Perspectives. Siri Scientific Press, Manchester, pp. 122–153.

Nicholls, C., Altieri, M., 2007. Agroecology: Contributions Towards a Renewed Ecological Foundation for Pest Management. Cambridge University Press, Cambridge, MA.

Nonlinear programming FAQ. 2013. http://www.neos-guide.org/content/nlp-faq.

Nylin, S., Gotthard, K., 1998. Plasticity in life-history traits. Annu. Rev. Entomol. 43, 63–83.

Olesen, J.M., Dupont, Y.L., O'Gorman, E.J., Ings, T.C., Layer, K., Melian, C.J., Trojelsgaard, K., Pichler, D.E., Rasmussen, C., Woodward, G., 2010. From broadstone to zackenberg: space, time and hierarchies in ecological networks. In: Woodward, G. (Ed.), Advances in Ecological Research: Ecological Networks, vol. 42. Academic Press, Amsterdam, pp. 1–69.

Orsini, L., Spanier, K.I., De Meester, L., 2012. Genomic signature of natural and anthropogenic stress in wild populations of the waterflea Daphnia magna: validation in space, time and experimental evolution. Mol. Ecol. 21, 2160–2175.

Paine, R.T., 1966. Food web complexity and species diversity. Am. Nat. 100, 65–75.

Paine, R.T., 1976. Size-limited predation—observational and experimental approach with Mytilus-Pisaster interaction. Ecology 57, 858–873.

Palkovacs, E.P., Post, D.M., 2009. Experimental evidence that phenotypic divergence in predators drives community divergence in prey. Ecology 90, 300–305.

Pelletier, F., Garant, D., Hendry, A., 2009. Eco-evolutionary dynamics. Philos. Trans. R. Soc. Lond. B Biol. Sci. 364, 1483–1489.

Peters, R.H., 1983. The Ecological Implications of Body Size. Cambridge University Press, Cambridge.

Pillai, P., Gonzalez, A., Loreau, M., 2011. Metacommunity theory explains the emergence of food web complexity. Proc. Natl. Acad. Sci. U.S.A. 108, 19293–19298.

Pillai, P., Gonzalez, A., Loreau, M., 2012. Evolution of dispersal in a predator-prey metacommunity. Am. Nat. 179, 204–216.

Pimentel, D., 1968. Population regulation and genetic feedback. Science 159.

Pólya, G., 1957. How to Solve It. Doubleday, Garden City, NY.

Rall, B.C., Vucic-Pestic, O., Ehnes, R.B., Emmerson, M., Brose, U., 2010. Temperature, predator-prey interaction strength and population stability. Glob. Chang. Biol. 16, 2145–2157.

Reuman, D.C., Mulder, C., Banasek-Richter, C., Blandenier, M.F.C., Breure, A.M., Den Hollander, H., Kneitel, J.M., Raffaelli, D., Woodward, G., Cohen, J.E., 2009. Allometry of body size and abundance in 166 food webs. In: Caswell, H. (Ed.), Advances in Ecological Research, vol. 41. Academic Press, Amsterdam, pp. 1–44.

Reznick, D.A., Bryga, H., Endler, J.A., 1990. Experimentally-induced life-history evolution in a natural population. Nature 346, 357–359.

Roelke, D., Augustine, S., Buyukates, Y., 2003. Fundamental predictability in multispecies competition: the influence of large disturbance. Am. Nat. 162, 615–623.

Roff, D.A., 1997. Evolutionary Quantitative Genetics. Chapman and Hall, New York.

Rooney, N., McCann, K., Gellner, G., Moore, J.C., 2006. Structural asymmetry and the stability of diverse food webs. Nature 442, 265–269.

Rooney, N., McCann, K.S., Moore, J.C., 2008. A landscape theory for food web architecture. Ecol. Lett. 11, 867–881.

Rowntree, J.K., Shuker, D.M., Preziosi, R.F., 2011. Forward from the crossroads of ecology and evolution introduction. Philos. Trans. R. Soc. Lond. B Biol. Sci. 366, 1322–1328.

Rudolf, V.H., 2007a. Consequences of stage-structured predators: cannibalism, behavioral effects, and trophic cascades. Ecology 88, 2991–3003.

Rudolf, V.H., 2007b. The interaction of cannibalism and omnivory: consequences for community dynamics. Ecology 88, 2697–2705.

Sabo, J.L., Bastow, J.L., Power, M.E., 2002. Length–mass relationships for adult aquatic and terrestrial invertebrates in a California watershed. J. North Am. Bentholog. Soc. 21, 336–343.

Savill, N.J., Rohani, P., Hogeweg, P., 1997. Self-reinforcing spatial patterns enslave evolution in a host-parasitoid system. J. Theor. Biol. 188, 11–20.

Scheiner, S.M., 1993. Genetics and evolution of phenotypic plasticity. Annu. Rev. Ecol. Syst. 24, 35–68.

Scheu, S., Schaefer, M., 1998. Bottom-up control of the soil macrofauna community in a beechwood on limestone: manipulation of food resources. Ecology 79, 1573–1585.

Schmalhofer, V.R., 2011. Impacts of temperature, hunger and reproductive condition on metabolic rates of flower-dwelling crab spiders (Araneae: Thomisidae). J. Arachnol. 39, 41–52.

Schmidt-Nielsen, K., 1984. Scaling, Why is Animal Size so Important?. Cambridge University, Press, Cambridge.

Schneider, F.D., Scheu, S., Brose, U., 2012. Body mass constraints on feeding rates determine the consequences of predator loss. Ecol. Lett. 15, 436–443.

Schoener, T., 2011. The newest synthesis: understanding the interplay of evolutionary and ecological dynamics. Science 331, 426–429.

Scholz, D., 2011. Deterministic Global Optimization: Geometric Branch-and-Bound Methods and Their Applications. Springer, New York.

Schume, H., Jost, G., Katzensteiner, K., 2003. Spatio-temporal analysis of the soil water content in a mixed Norway spruce (Picea abies (L.) Karst.)-European beech (Fagus sylvatica L.) stand. Geoderma 112, 273–287.

Sih, A., Bell, A., Johnson, C., 2004. Behavioral syndromes: an ecological and evolutionary overview. Trends Ecol. Evol. 19, 372–378.

Singer, M.C., McBride, C.S., 2012. Geographic mosaics of species' association: a definition and an example driven by plant-insect phenological synchrony. Ecology 93, 2658–2673.

Singer, M.C., Thomas, C.D., 1996. Evolutionary responses of a butterfly metapopulation to human- and climate-caused environmental variation. Am. Nat. 148, S9–S39.

Solé, R.V., Montoya, J.M., 2001. Complexity and fragility in ecological networks. Proc. Biol. Sci. 268, 2039–2045.

Steiner, C.F., Masse, J., 2013. The stabilizing effects of genetic diversity on predator-prey dynamics ([v1; ref status: approved 2, approved with reservations 1, http://f1000r.es/9u) F1000Research 2013, 2:43.

Stewart, R.I.A., Dossena, M., Bohan, D.A., Jeppesen, E., Kordas, R.L., Ledger, M.E., Meerhoff, M., Moss, B., Mulder, C., Shurin, J.B., Suttle, B., Thompson, R., Trimmer, M., Woodward, G., 2013. Mesocosm experiments as a tool for ecological climate-change research. In: Woodward, G., Ogorman, E.J. (Eds.), Advances in Ecological Research: Global Change in Multispecies Systems. Part 3, vol. 48. Academic Press, Amsterdam, pp. 71–181.

Sunnaker, M., Busetto, A.G., Numminen, E., Corander, J., Foll, M., Dessimoz, C., 2013. Approximate Bayesian computation. PLoS Comput. Biol. 9, e1002803.

Swift, M.J., Heal, O.W., Anderson, J.M., 1979. Decomposition in Terrestrial Ecosystems. University of California Press, Berkeley and Los Angeles, CA.

Symondson, W.O.C., Sunderland, K.D., Greenstone, M.H., 2002. Can generalist predators be effective biocontrol agents? Annu. Rev. Entomol. 47, 561–594.

Talbi, E.G., 2009. Metaheuristics: From Design to Implementation. Wiley, Hoboken, NJ.

Tamadoni-Nezhad, A., Milani, G.A., Raybould, A., Muggleton, S., Bohan, D.A., 2013. Construction and validation of food webs using logic-based machine learning and text mining. Adv. Ecol. Res. 49, 225–289.

Terhorst, C.P., Miller, T.E., Levitan, D.R., 2010. Evolution of prey in ecological time reduces the effect size of predators in experimental microcosms. Ecology 91, 629–636.

Thomas, C.D., Singer, M.C., Bougthon, D.A., 1996. Catastrophic extinction of population sources in a butterfly metapopulation. Am. Nat. 148, 957–975.

Thompson, J., 1998. Rapid evolution as an ecological process. Trends Ecol. Evol. 13, 329–333.

Thuiller, W., Münkemüller, T., Lavergne, S., Moiullot, D., Mouquet, N., Schiffers, K., Gravel, D., 2013. A road map for integrating eco-evolutionary processes into biodiversity models. Ecol. Lett. 16, 94–105.

Travis, J., Reznick, D., Bassar, R.D., López-Sepulcre, A., Ferriere, R., Coulson, T., 2014. Do eco-evolutionary feedbacks help us understand nature? Answers from studies of the Trinidadian guppy. In: Moya-Laraño, J., Rowntree, J., Woodward, G. (Eds.), Advances in Ecological Research, Vol. 50: Eco-evolutionary Dynamics. Academic Press, Amsterdam, pp. 1–40.

Urban, M., 2013. Evolution mediates the effects of apex predation on aquatic food webs. Proc. Biol. Sci. 280, 1–9.

Urban, M.C., Leibold, M.A., Amarasekare, P., De Meester, L., Gomulkiewicz, R., Hochberg, M.E., Klausmeier, C.A., Loeuille, N., de Mazancourt, C., Norberg, J., Pantel, J.H., Strauss, S.Y., Vellend, M., Wade, M.J., 2008. The evolutionary ecology of metacommunities. Trends Ecol. Evol. 23, 311–317.

van Lenteren, J., 2012. The state of commercial augmentative biological control: plenty of natural enemies but a frustrating lack of uptake. BioControl 57, 1–20.

van Lenteren, J., Woets, J., 1988. Biological and integrated pest control in greenhouses. Annu. Rev. Entomol. 33, 239–269.

Verdeny-Vilalta, O., Moya-Laraño, J., 2014. Seeking water while avoiding predators: moisture gradients can affect predator-prey interactions. Anim. Behav. 90, 101–108.

Walsh, M.R., Delong, J.P., Hanley, T.C., Post, D.M., 2012. A cascade of evolutionary change alters consumer-resource dynamics and ecosystem function. Proc. Biol. Sci. 279, 3184–3192.

Wardle, D.A., 2006. The influence of biotic interactions on soil biodiversity. Ecol. Lett. 9, 870–886.

Williams, R.J., Martinez, N.D., 2000. Simple rules yield complex food webs. Nature 404, 180–183.

Wilson, D.S., 1975. Adequacy of body size as a niche difference. Am. Nat. 109, 769–784.

Wilson, D.S., 1992. Complex interactions in metacommunities, with implications for biodiversity and higher levels of selection. Ecology 73, 1984–2000.

Winemiller, K., Polis, G., 1996. Food Webs: What Can They Tell Us About the World? In: Polis, G., Winemiller, K. (Eds.), Food Webs. Chapman and Hall, New York, pp. 1–22.

Winkler, K., 2005. Assessing the risks and benefits of flowering field edges: strategic use of nectar sources to boost biological control. Ph.D. Wageningen University.

Wise, D.H., Chen, B.R., 1999. Vertebrate predation does not limit density of a common forest-floor wolf spider: evidence from a field experiment. Oikos 84, 209–214.

Wolf, M., van Doorn, G.S., Leimar, O., Weissing, F.J., 2007. Life-history trade-offs favour the evolution of animal personalities. Nature 447, 581–584.

Woodward, G., Hildrew, A.G., 2002. Body-size determinants of niche overlap and intraguild predation within a complex food web. J. Anim. Ecol. 71, 1063–1074.

Yodzis, P., Innes, S., 1992. Body size and consumer-resource dynamics. Am. Nat. 139, 1151–1175.

Yoshida, T., Jones, L.E., Ellner, S.P., Fussmann, G.F., Hairston Jr., N.G., 2003. Rapid evolution drives ecological dynamics in a predator-prey system. Nature 424, 303–306.

CHAPTER FOUR

Eco-Evolutionary Interactions as a Consequence of Selection on a Secondary Sexual Trait

Isabel M. Smallegange*,1, Jacques A. Deere†
*Institute for Biodiversity and Ecosystem Dynamics (IBED), University of Amsterdam, Amsterdam, The Netherlands
†Department of Zoology, University of Oxford, Oxford, United Kingdom
1Corresponding author: e-mail address: i.smallegange@uva.nl

Contents

Abstract

Ecological and evolutionary population changes are often interlinked, complicating the understanding of how each is affected by environmental change. Using a male dimorphic mite as a model system, we studied concurrent changes in the expression of a conditional strategy and in the population in response to harvesting over 15 generations. We found evolutionary divergence in the expression of alternative male reproductive morphs—fighters and defenceless scramblers (sneakers)—caused by the selective harvesting of each male morph. Regardless of which morph was targeted, the direction of evolution of male morph expression in response to harvesting was always towards scramblers, which, in case of the harvesting of scramblers, we attributed to strong ecological feedback (reduced cannibalism opportunities for fighters) within the closed populations. Current evolutionary theory, however, predicts that the frequency of a morph always decreases when selected against: to understand phenotypic trait

Advances in Ecological Research, Volume 50
ISSN 0065-2504
http://dx.doi.org/10.1016/B978-0-12-801374-8.00004-9

145

evolution fully, evolutionary theory would benefit from including ecological interactions, especially if traits have ecological consequences that in turn feedback to their evolutionary trajectory.

1. INTRODUCTION

Assessing the likely population dynamic consequences of environmental change and their consequent effects on community structure and ecosystem functioning requires in-depth knowledge of the complex interactions between ecological and evolutionary population dynamics (Schoener, 2011). Conceptually, we already have a solid understanding of the structure of this interaction (Carroll et al., 2007; Chevin et al., 2010; Kokko and López-Sepulcre, 2007; Post and Palkovacs, 2009; Smallegange and Coulson, 2013). For example, any process that changes ecological characteristics, such as population density or structure, can also elicit concurrent phenotypic responses, including change in the conditional expression of alternative phenotypes (e.g. Buzatto et al., 2012; Tomkins et al., 2011), which underpins life-history decisions and many dichotomous traits (Gross, 1996; Hazel et al., 2004; Roff, 1996; Smallegange, 2011a,b; West-Eberhard, 1989, 2003). Evolved change in the conditional expression of alternative phenotypes in turn affects population structure and dynamics (Myers, 1984; Piou and Prévost, 2013), creating an eco-evolutionary feedback with population and community-level consequences. However, few studies have disentangled ecological from evolutionary change in either natural or experimental settings. Studies that have separated observed life-history change (mainly change in population growth rate) into both compartments have rarely tested whether phenotypic change came about through environmental or through genetic, and therefore, evolutionary processes (Coulson et al., 2011; Ezard et al., 2009; Hairston et al., 2005; Losos et al., 1997; but see Ozgul et al., 2009). In contrast, experimental studies that explicitly focused on disentangling environmental from genetic drivers of phenotypic change rarely recorded concurrent ecological responses to treatments (Dodd, 1989; Garland and Rose, 2009; Kawecki et al., 2012; Moya et al., 1995; but see Cameron et al., 2013; Reznick et al., 2001; van Doorslaer et al., 2009). This is largely because such experimental evolution studies employ discrete generation methods, in which a sample of individuals of one generation is used to start the next (e.g. Kolss et al., 2009; Santos et al., 1997; Tomkins et al., 2011). This alters the structure of populations at the start of each new generation, interrupting ongoing ecological change, so the population dynamical response to experimental treatments cannot be tracked. Thus, there is an important caveat in our understanding of the mechanisms and the population consequences of concurrent ecological and evolutionary change and their interaction.

Conditional strategies, whereby phenotype expression depends on a threshold response to an environmental cue (Roff, 1996), allow individuals to respond rapidly to environmental change (Gross, 1991). Because this response depends on system productivity, understanding it is also of economic significance to the commercial exploitation of populations (Myers, 1984). For example, the conditional expression of alternative reproductive phenotypes (ARPs) in males of many salmon species can vary in response to selective harvesting (Gross, 1991), as well as climate change (Piou and Prévost, 2013). ARPs in males are commonplace and usually comprise fighters, which compete over access to females, and 'sneakers' that avoid direct competition (Oliveira et al., 2008). Although we have a thorough understanding of how drivers such as food quality and habitat complexity influence the evolution and expression of ARPs (e.g. Łukasik et al., 2006; Smallegange, 2011a,b; Tomkins et al., 2011), how a change in the environment (e.g. by selective harvesting) affects the interplay between ecological and evolutionary dynamics within ARP populations is still unclear (Johnstone et al., 2013; Piou and Prévost, 2013; Smallegange and Johansson, 2014).

The environmental threshold (ET) model is currently at the forefront to explain the evolution of conditional strategies (Buoro et al., 2012; Tomkins and Hazel, 2007): this model was first conceptualized over 30 years ago (Hazel, 1977; Hazel and West, 1982) and subsequently formalized mathematically over the last couple of decades (Hazel et al., 1990, 2004). It can take into account frequency-dependent selection (but this is not a requirement) and can model both the trajectory and the outcome of selection on any conditional strategy (Hazel et al., 2004). Its results are almost identical to those of an earlier model on the conditional strategy (Lively, 1986), but unlike the latter, the ET model uses quantitative genetic theory (Hazel et al., 1990, 2004). The ET model assumes that the two male morphs have different, crossing fitness functions (i.e. the slopes of fitness plotted against status (e.g. size) for each morph differ) and that male morph expression depends on whether or not an individual reaches a critical threshold during ontogeny. This threshold, in turn, is assumed to be under polygenic control and influenced by a cue such as juvenile body size. If a juvenile reaches the size threshold, it develops into a fighter; otherwise, it develops into a sneaker.

In this study, we use the ET model to derive predictions on the evolutionary response to selection against male morph in the bulb mite *Rhizoglyphus robini*. Adult fighter bulb mites have a greatly thickened third pair of legs with sharp ends that can be used to kill rival males by grabbing them and puncturing their skin. Fighters also use their fighter legs to kill other adults as well as larger juveniles to cannibalize them if other food is scarce (Łukasik, 2010). This third pair of legs is unmodified in defenceless sneakers, which are called scramblers (Radwan et al., 2000). Whether or not a male bulb mite develops into a fighter predominantly depends on whether a male reaches a critical size threshold during its final instar stage (quiescent tritonymph stage): males larger than this quiescent tritonymph size threshold (henceforth, referred to as tritonymph size threshold) are most likely to develop into a fighter and males smaller than this size into a scrambler (Smallegange, 2011a). The development of most male bulb mites follows this conditional expression of male morphology (Buzatto et al., 2012; Smallegange, 2011a,b), but, sometimes, small individuals develop into a fighter and large individuals into a scrambler (Leigh and Smallegange, 2014; Smallegange, 2011a). This can occur because there is large genetic variation in the size threshold (Buzatto et al., 2011), or because some males have lost the ability to express the fighter morph (Buzatto et al., 2012). Male morph expression in *R. robini* is also heritable. Smallegange and Coulson (2011) assumed, based on Radwan (1995), that the male dimorphism in the bulb mite is a genetic polymorphism, but we now know that environmental conditions play a large role in male morph expression in the bulb mite (Smallegange, 2011a). The estimated heritability for scrambler morph is 0.41 and for the fighter morph 0.30 (Smallegange and Coulson, 2011); hence, a conditional strategy, which assumes male morph expression to be a polygenic trait (Tomkins and Hazel, 2007), best matches the underlying genetics of the bulb mite male dimorphism. Unlike in other soil and bulb mites (e.g. *R. echinopus* and *Sancassania berlesei*), male morph expression in *R. robini* is unaffected by signals of population density such as airborne substances (Radwan, 1995) or by male morph frequency (Deere and Smallegange, 2014). Further differences between the two morphs are that scramblers live longer than fighters (Radwan and Bogacz, 2000). The reproductive success of each morph, in turn, depends on a number of variables. In the absence of male–male competition, the reproductive success of fighters is independent of their adult size but the reproductive success of scramblers increases with adult size (Smallegange et al., 2012). When in competition with other males, scrambler mating duration, and thereby the number of offspring produced by its mate, was unaffected by the size of its mate, whereas fighter mating duration decreased with increasing size of its mate (Smallegange et al., 2012). Which morph has the highest fitness as determined by its survival rate and reproductive success, is, therefore, context dependent.

Predictions from the ET model are derived from functions that relate male morph fitness to a cue such as body size (Tomkins and Hazel, 2007). Selection against a morph lowers its fitness, shifting the intersection point of the fitness functions and hence the mean size threshold from its position prior to selection to one associated with a reduced expression of this morph. According to the ET model, we therefore expect that if fighter fitness changes due to, e.g., selective harvesting, then this should result in evolutionary divergence of the mean final instar size threshold: if fighter (scrambler) expression is suppressed, the mean size threshold should increase (decrease) and fighter frequency should decrease (increase) as fewer (more) males will reach the higher (lower) size threshold and develop into fighters (scramblers) (cf. Tomkins et al., 2011). Results from a previous experiment, where we created male morph selection lines using the aforementioned discrete generation method, are in line with this expectation as we found that selection against fighters reduced fighter expression (i.e. the proportion of males that are fighters) and selection against scramblers reduced scrambler expression (Smallegange and Coulson, 2011). To test our hypothesis within a population setting with overlapping generations and uninterrupted ecological change, we established replicate experimental populations and applied proportional harvesting regimes to fighters and scramblers. The experimental populations were created by collecting mites from source populations and culturing them in enclosed populations with a constant food supply. To record the ecological response of experimental populations, we regularly performed a census of the total size of each population and its structure, i.e., the number of individuals of each life stage, sex and male morph. To score the evolutionary response of populations, we measured the size of final instars (as final instar size is the cue for fighter expression) and the size of adult fighters, scramblers and females (females were included to compare the response in male body size with that of females) throughout the experiment. After approximately nine generations (at the end of the experiment), we performed a common garden environment life–history assay to assess whether phenotypic change in body size and morph expression observed during the experiment was plastic or genetic. Using these data, we reveal complex links between ecological and evolutionary change and an unexpected response to selection in populations of a sexually reproducing organism.

2. METHODS
2.1. Predictions on evolution of fighter expression in response to harvesting

The ET model uses fitness functions of the two male morphs to predict the direction of evolution (Hazel et al., 1990, 2004; Tomkins and Hazel, 2007; Tomkins et al., 2011). Fitness functions of fighters and scramblers in relation to adult body size intersect (fitness was measured as reproductive success) (Smallegange et al., 2012). From this, we predict that selection against fighters (scramblers) through selective harvesting shifts the intersection point and mean tritonymph size threshold below and above which males most likely develop into scramblers and fighters, respectively, reducing the frequency of fighters (scramblers) in the male population (Appendix). We did not incorporate frequency-dependence into the ET model as this is non-significant in bulb mites (Deere and Smallegange, 2014; Radwan and Klimas, 2001).

2.2. Source population and bulb mite life cycle

Mites were collected from storage rooms of flower bulbs in North Holland (the Netherlands) in December 2010 and kept on yeast as described in Smallegange (2011a). The life cycle of the bulb mite consists of six stages: egg, larva, protonymph, deutonymph, tritonymph and adult. Mites moult to grow from one life stage to the next, and during this quiescent stage, they are immobile and do not feed. The deutonymph is a facultative dispersal stage to escape unfavourable environmental conditions and its development is induced by low food quality and quantity (Díaz et al., 2000). Few were observed in this study (on average three on each census day, which is less than 1% of the total population) and were therefore not included in the analyses.

2.3. Experimental procedure

The experiment comprised three treatments: (i) harvesting of fighters (FH), (ii) harvesting of scramblers (SH) and (iii) no harvesting (control treatment) (C) and was conducted from July 2011 to July 2012 (365 days). Each treatment was replicated three times, resulting in nine populations. Each population was initiated with 50 randomly selected adult mites from the source population. To reduce founder effects, we removed these founder mites after 15 days, by which time the next generation had matured (founder mites were much larger than mites from subsequent generations and could therefore easily be distinguished). Each population was fed eight rods of yeast per day (rods range in size from 750 to 1000 μm and eight rods per day amounts to ~0.50 mg yeast per day). After initialization, for the first 230 days, on each Tuesday and Friday, mites of all life stages in each population were counted using a hand counter at 15 × magnification, and photos were taken at 15 × magnification of up to 10 individual adult females, fighters, scramblers and preimaginal (final instar) quiescent tritonymphs using a Lumenera Infinity 3.1 camera connected to a Meiji EMZ-8TRD (10–45 ×) stereomicroscope. The body length of mites (with mouthparts) on each photo was measured to the nearest 0.1 μm using Infinity Analyze Imaging Software (Lumenera Corporation, Ottawa, Ontario, Canada). For the next 135 days of the experiment, mites were counted every week on Tuesday and in addition their sizes were measured every other Tuesday. Under a constant feeding regime, lab-conditioned acarid mite populations' stabilize at about 40 days after initialization (Cameron and Benton, 2004). We started our harvesting treatment 60 days after populations were initialized. Harvesting was carried out once per week (on Tuesday), after counting and before feeding, by removing 50% of fighters (FH treatment) or 50% of scramblers (SH treatment). Generation time of mites varies with food supply, but under good conditions the minimum egg-to-egg time is 11 days (Smallegange, 2011b). Assuming a generation time of 35 days under strong, density-dependent conditions (Cameron et al., 2013), this means that harvesting lasted for $305/35 \approx 9$ generations, which is sufficiently long for evolutionary shifts in fighter expression to occur (Smallegange and Coulson, 2011). Populations were kept in 20 mm diameter, flat-bottomed glass tubes with a plaster of Paris and powdered charcoal base, which was kept moist to avoid desiccation of the mites. Tubes were sealed by a circle of very fine mesh (allowing gaseous diffusion), which was held in place by the tubes' standard plastic caps with ventilation holes cut into them. Populations were kept in an unlit incubator at 25 °C and >70% relative humidity.

After completion of the experiment, we performed a common garden environment life-history assay to assess whether differences in male morph expression between treatments were due to evolutionary shifts in the mean tritonymph size threshold, and whether differences in body size between treatments were plastic or genetic. Ten adult females were taken from each population, individually isolated, given *ad lib* access to yeast, and allowed to lay eggs. From these ten females, three were selected (several females died or did not lay eggs) and their offspring followed until they reached the quiescent tritonymph stage. No adults, except the mother, were ever present among the offspring. Once quiescent tritonymphs were present, they were collected, their size measured as before, after which we individually isolated each quiescent tritonymph and scored its sex, morph and adult size 2–5 days after maturation. Up to 20 individuals from each female were measured this way, resulting in a total of 418 observations. Nutrition during ontogeny is the strongest environmental determinant of male morph development. Paternal morph only has an effect on offspring male morph expression under poor food quality conditions (here, mites were kept under rich food conditions) and effects of maternal nutritional conditions are negligible in this species (the effect size of offspring environment is 15 times larger than that of maternal environment; Smallegange, 2011a). Therefore, rearing mites in a common garden environment for one generation is sufficient to eliminate any maternal effects. We also isolated 100 quiescent tritonymphs from the source population to assess their mean tritonymph size threshold above and below which males are most likely to develop into scramblers and fighter, respectively. Females and their offspring, and all isolated quiescent tritonymphs were given *ad lib* access to yeast and kept in 25 mm diameter (females and offspring) and 10 mm diameter tubes (quiescent tritonymphs) with a plaster of Paris and powdered charcoal base, which were kept in an unlit incubator at 25 °C and >70% relative humidity.

2.4. Statistical analyses

To analyse the effects of harvesting on the number and size of individuals, we used data from day 100 onwards, i.e., 40 days after the start of the harvesting regimes, to reduce the impact of transients (cf. Cameron and Benton, 2004) (see Appendix for an overview of the observed population dynamics). This resulted in 783 observation days on life stage numbers and 466 observation days on life stage body sizes. We used bootstrap resampling to estimate 95% confidence intervals (CIs) for the mean number and the mean size of individuals of relevant life stages per treatment group (cf. Benton et al., 2004). We furthermore divided the experimental period into three periods, day 100–190, day 191–280 and day 281–365, and calculated 95% CIs per period to assess long-term (instead of transient) temporal changes in life stage number and mean body size over time within each treatment group. Bootstrap resampling was done by taking 1000 resamples, which were stratified by population tube within each treatment group to ensure that there were no biases due to tube effects. These samples were used to estimate the bias–corrected and adjusted (BCa) 95% CI. If a statistic's 95% CI does not overlap between treatment groups or between time periods within a treatment group then, by definition, the statistic differs across groups at $\alpha = 0.05$.

To analyse the results from the life–history assay, we used a generalized linear mixed model (GLMM) with Gaussian errors to analyse the effects of harvesting treatment, morph (including fighters, scramblers and females) and their interaction on quiescent tritonymph size and adult size (μm). We also used a GLMM with binomial errors to analyse the effects of harvesting treatment and quiescent tritonymph size on male morph expression (0 if scrambler; 1 if fighter) (cf. Tomkins et al., 2011) and used a GLMM with binomial errors to analyse the effects of harvesting treatment on the probability of female expression (0 if male; 1 if female). In each GLMM, maternal identity and population tube were included as random terms. To assess significance of treatment effects on quiescent tritonymph size and adult size, a model simplification procedure was used whereby the full model was fitted after which the least significant term was removed (starting with the highest order interaction) if the deletion caused an insignificant increase in deviance. Model simplification using likelihood ratio tests is not recommended for GLMMs (Bolker et al., 2009) and instead we used a Markov Chain Monte Carlo (MCMC) approach and calculated 95% CIs of the model parameters using the underlying Gaussian distribution of the model residuals. First, an MCMC sample was generated from the posterior distribution of each parameter estimate using the function *mcmcsamp* in the R package *lme4* (Bates and Sarkar, 2007). The Bayesian highest posterior density (HPD) 95% CIs of the MCMC sample for each parameter estimate was then computed using the function *HPD interval* in the R package *coda* (Plummer et al., 2006). If the HPD interval of a parameter estimate overlaps with zero, then the associated factor has no significant effect on the response variable. If the full model contained non–significant parameter estimates, the parameter that was the least significant was removed and new HPD intervals calculated. This process was repeated until the model only contained significant terms. In Section 3, we report the parameter estimates ($\hat{e}$) and associated statistics of (non)significant terms. Model assumptions on Gaussian errors and homoscedacity were confirmed by inspection of probability plots and error structures. Models were fitted by maximum likelihood in R.

3. RESULTS

We first present the results of the life–history assay conducted at the end of the experiment to show that evolutionary change in fighter expression had occurred, followed by the ecological changes in population number and structure associated with changes in fighter expression.

3.1. Evolution of fighter expression

The life-history assay conducted at the end of the experiment revealed that the FH resulted in evolutionary change in fighter expression. Fighter expression in the assay differed significantly between treatments ($\hat{e} = -2.34 \pm 1.13\text{SE}$, $z = -2.081$, $p = 0.037$) and the probability of fighter expression was significantly lower in the FH lines than in the control lines (Fig. 4.1A). The probability of fighter expression in the FH and control lines of the life-history assay was not significantly different from the proportion of fighters observed during the final period of the population experiment (day 280–365) in the FH and C treatments (Fig. 4.2A): this result is inferred from the fact that, for the FH and C treatments, the standard error bars of mean probability of fighter expression observed in the life-history assay (diamonds in Fig. 4.2A [which are the same as in Fig. 4.1A]) overlap with the CIs of mean probability of fighter expression observed for the final period in the population experiment (third triangle of each treatment in Fig. 4.2A). Quiescent tritonymph size significantly affected the probability of fighter

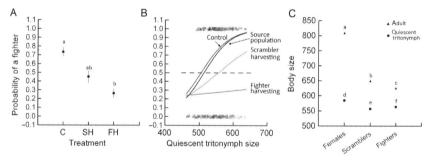

Figure 4.1 Life-history assay. (A) The probability of fighter expression differed between different treatment groups (C, control; SH, scrambler harvesting; FH, fighter harvesting). (B) The mean tritonymph size threshold (the quiescent tritonymph size at which the expected probability of fighter expression is 0.5: denoted by the horizontal dashed line) was shifted to the right for the SH treatment compared to the control treatment and for the FH treatment the predicted mean size threshold is outside of the observed range of quiescent tritonymph sizes. For comparison, the fighter expression-body size response curve of the source population is also included. (C) The body size (μm) of quiescent tritonymphs and adults differed significantly between females, scramblers and fighters. Vertical lines in (A) and (C) are standard error bars that in (C) are covered by the symbols. Letters in panels (A) and (C) denote significant differences between treatments at $\alpha < 0.05$.

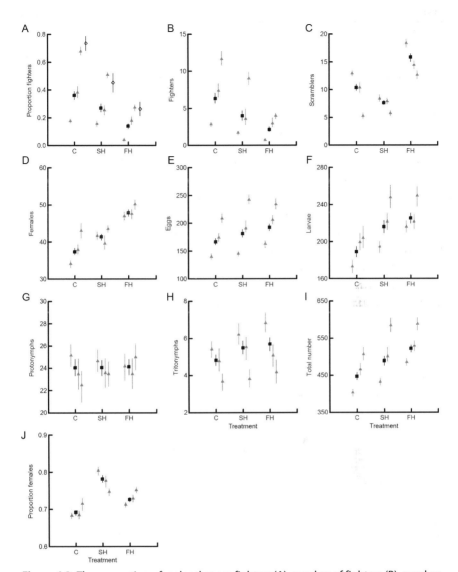

Figure 4.2 The proportion of males that are fighters (A), number of fighters (B), number of scramblers (C), number of adult females (D), number of eggs (E), number of larvae (F), number of protonymphs (G), number of tritonymphs (H), the total number of individuals in the population (I) and the proportion of the adult population that are females (J) shown per treatment group: C, control; SH, scrambler harvesting; FH, fighter harvesting. The black squares are the observed means across replicate populations from day 100 till day 365. Per treatment, the three-grey triangles denote from left to right the mean across replicate populations for day 100–190, day 190–280 and day 280–365, respectively. The probability of fighter expression values observed in the life-history assay (Fig. 4.1A) are added to panel (A) (open diamonds) to aid the visual comparison between these values and those observed at the end of the population experiment (grey triangle directly left of each diamond). Vertical lines are 95% CIs estimated by bootstrap sampling, stratified by tube, except in (A) where vertical lines associated with the open diamonds are standard error bars.

expression in the life–history assay (HPD interval: -36.61 to -23.73 μm). As expected from the selection against fighters through harvesting fighters in the FH treatment, the mean tritonymph size threshold at which the probability of fighter expression equals 0.5 was much higher for mites from the FH lines than for mites from the control lines (Fig. 4.1B: note that the predicted mean tritonymph size threshold for the FH treatment is outside of the observed quiescent tritonymph size range). An important assumption in the conditional expression of size-dependent ARPs is that, if the body size distribution changes location (e.g. mean body size becomes smaller), then the mean tritonymph size threshold for morph expression evolves such that its new location is at the same (relative) position within the body size distribution as before the change in mean body size (Tomkins et al., 2011). This means that, if there is no effect of harvesting but body size distributions differ between harvesting treatments in the life–history assay for some other, unknown reason, then the mean tritonymph size threshold will also differ between treatments as it will have tracked the change in body size distribution (all else being equal). This would then lead one to conclude incorrectly that harvesting has led to evolutionary shifts in mean tritonymph size threshold. Here, however, body size distributions (i.e. mean adult body size and mean quiescent tritonymph size) did not differ between the FH, SH and control lines in the life–history assay (HPD interval: -14.38 to 17.64 μm), so we can conclude that the differences in fighter expression and mean tritonymph size threshold observed in the life–history assay were indeed due to the different harvesting regimes. Quiescent tritonymph size furthermore differed between females and the male morphs (HPD interval: -36.61 to -23.73 μm): female quiescent tritonymphs were significantly larger than fighter quiescent tritonymphs, which were in turn significantly larger than scrambler quiescent tritonymphs (Fig. 4.1C).

Fighter expression in response to harvesting scramblers was not as we expected. The life–history assay revealed that there was evolutionary change in fighter expression in the SH treatment, but the probability of fighter expression in the SH lines was lower, instead of higher, than in the control lines (Figs. 4.1A and 4.2A). This evolutionary change was reflected in the mean tritonymph size threshold of the SH lines, which was at a larger quiescent tritonymph size than in the control lines (Fig. 4.1B). Like in the FH and control lines, fighter expression in the SH lines of the life–history assay matched the observed proportion of fighters in the SH treatment during the final period of the population experiment (standard error bars in Fig. 4.1A and CIs for the final period in Fig. 4.2A (third triangle) overlap for the SH treatment).

3.2. Effects of fighter expression on population size and structure

Overall, as a result of harvesting fighters, their mean number was signifi-cantly lower in the FH treatment than in the control treatment (squares in Fig. 4.2B: significance was inferred from the fact that the 95% CIs (vertical lines around each mean) are non-overlapping), whereas the mean number of scramblers showed the opposite relationship (squares in Fig. 4.2C). Within each treatment, the number of fighters increased significantly over the course of the population experiment and, at the same time, the number of scramblers decreased significantly over time within each treatment (trian-gles in Fig. 4.2B and C). Similarly, when the mean number of fighters was high for a particular treatment or time period (e.g. control treatment), the mean number of tritonymphs in that same treatment or time period was always low (e.g. in the control treatment) and vice versa. From this, we infer that there is a negative association between mean number of fighters and tritonymphs (squares and triangles in Fig. 4.2B and H).

Across treatments, the mean number of fighters was also negatively asso-ciated with the mean number of females, as, when the mean number of fighters was high for a particular treatment or time period, the mean number of females in that same treatment or time period was always low and vice versa (Fig. 4.2B and D). When more females were present, significantly more eggs were laid (Fig. 4.2E) and more larvae emerged (Fig. 4.2F). These knock-on effects of female numbers on eggs and larvae were no longer evident in the protonymph stage (Fig. 4.2G). In fact, protonymph numbers did not differ significantly across treatments and only showed a slight significant decrease over time in the control treatment (Fig. 4.2G). Within treatments, there was a positive association between the mean number of fighters and females as both increased in number over time (Fig. 4.2B and D). As the number of females increased significantly over time within treatments, so did the number of eggs and larvae (Fig. 4.2D–F). Finally, across treatments, the mean number of fighters showed a negative associated with the mean total number of indi-viduals in the population (Fig. 4.2B and I). Within treatments, however, the association between mean fighter number and mean total number of individ-uals was positive and all populations increased in size over time (Fig. 4.2I).

3.3. Realized sex ratio and plasticity in body size

Harvesting had a significant effect on the sex ratio within populations: across treatments, the proportion of females was highest in the SH treatment, followed by the FH and control treatment (squares in Fig. 4.2J). Within treatments, the proportion of females varied slightly: towards the end of the experiment the proportion of females increased significantly in the control treatment (triangles in Fig. 4.2J). In the SH treatment, the proportion of females decreased significantly over time, whereas in the FH treatment the proportion of females increased significantly over time (triangles in Fig. 4.2J). The life-history assay revealed that these differences in proportion of females were not due to evolutionary change as in the life-history assay there was no significant difference in the probability of female expression between the treatments ($\hat{e} = -0.088 \pm 0.264$SE, $z = -0.310$, $p = 0.757$). The probability of female expression was on average 0.52 ± 0.24 SE: lower than in the population experiment (Fig. 4.2J) and not significantly different from 0.50, as expected in a diploid species (Oliver, 1971, 1977).

Finally, harvesting significantly affected the mean body size of adults. Fighters, scramblers and adult females were on average significantly larger in the control treatment than in both harvesting treatments (squares in Fig. 4.3A–C). The life-history assay revealed that this variation was completely due to phenotypic plasticity as there was no significant difference in mean adult body size across the SH, FH and control lines (HPD interval: -36.57 to 15.46 μm). The life-history assay also showed that adult females, scramblers and fighters differed significantly in body size (HPD interval: -196.12 to 171.94 μm): females were larger than scramblers, which were larger than fighters (Fig. 4.1C). In the population experiment, females were on average also larger than the males (Fig. 4.3A–C), but scramblers were not significantly larger than fighters either across or within treatments (Fig. 4.3A and B). Within treatments, there was some variation in body size for each of the male morphs: fighter mean body size significantly increased towards the end of the experiment in the SH treatment (triangles in Fig. 4.3A), and scramblers significantly increased in size towards the end of the experiment in the FH treatment (triangles in Fig. 4.3B). Females, however, significantly increased in average size over time within each treatment (triangles in Fig. 4.3C). Quiescent tritonymphs, like the adults, were on average significantly larger in the control than in the harvesting treatments (squares in Fig. 4.3D). Within treatments, the average size of quiescent tritonymphs increased significantly over time in the harvesting treatments but not in the control (Fig. 4.3D). It is interesting to note here that, despite the fact that there was a difference in quiescent tritonymph size between mites in the life-history assay and mites in the experimental populations (all means in Fig. 4.3D are lower than means shown in Fig. 4.1C), the mean tritonymph size threshold (above and below which males are most likely to,

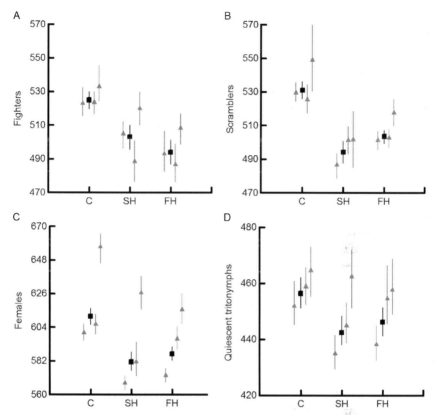

Figure 4.3 The mean size (μm) of individuals in each life stage shown per treatment group: C, control; SH, scrambler harvesting; FH, fighter harvesting. The black squares are the observed mean size across replicate populations from day 100 till day 365. Per treatment, the three-grey triangles denote from left to right the mean size across replicate populations for day 100–190, day 190–280 and day 280–365, respectively. Vertical lines are 95% CIs estimated by bootstrap sampling, stratified by tube.

respectively, develop into scramblers and fighter) of the control populations was nearly identical to that of mites in the source population (Fig. 4.1B). This suggests that mean size thresholds track plastic changes in body size.

4. DISCUSSION

To assume that selection against the expression of a particular phenotype reduces the prevalence of that phenotype in a population seems intuitive and in line with standard evolutionary theory (Falconer and MacKay, 1996). However, this effect may be very different in a setting where selection can be modified by population dynamical properties including density and structure (Kokko and López-Sepulcre, 2007). Here, we used state-of-the-art theory on the evolution of conditional strategies, the ET model, to predict the evolutionary response of male morph expression to selective harvesting. Regardless of which male morph was targeted, the mean tritonymph size threshold above and below which they are most likely to develop into scramblers and fighters, respectively, increased and fighter frequency decreased in response to the FH and scramblers. Theory predicts this will occur if fighter fitness is reduced, but not when scrambler fitness is reduced. In a previous selection experiment, where we created male morph selection lines using the discrete generation method, we *were* able to increase fighter frequency by selecting against scramblers by killing them (Smallegange and Coulson, 2011). Because, here, unlike in the previous study, we selected against male morph in closed populations, we postulate that ecological feedback from within the population plays a role in this seemingly counterintuitive response in fighter expression to scrambler harvesting. We surmise that this feedback comprises cannibalism by fighters when food is limited (Łukasik, 2010) (but has not been observed in other contexts, e.g., when competing for mates (Deere and Smallegange, 2014; Radwan and Klimas, 2001; Smallegange et al., 2012)). Although frequency-dependence could also explain this response, its strength is non-significant in this species (Deere and Smallegange, 2014; Radwan and Klimas, 2001) and a density-dependent response is more likely. We often observed fighters killing and eating scramblers (and also tritonymphs but not smaller juveniles or adult females) and fighter and scrambler numbers always showed a negative association. We can indirectly test for the occurrence of cannibalism by inspecting the sex ratio in the control treatment where no selection through harvesting occurred: if there is no cannibalism then we expect the sex ratio to be unity, but if cannibalism occurs then we expect it to be female biased, because the consumption of scramblers by fighters will reduce the frequency of males. The latter is indeed what we found, from which we infer that the reduction in scrambler numbers (i.e. density) due to harvesting reduces food availability for fighters. We therefore propose that (at least) two selective forces determine the evolution of fighter expression in these populations: harvesting scramblers selects for fighter expression (cf. Smallegange and Coulson, 2011), but reduced food availability selects against fighter expression, resulting in an eco-evolutionary response that is intermediate between that observed in the fighter-harvested and in the control populations. Fighter fitness is determined by the ability not only to survive but also to reproduce. Given that the sex ratio in all experimental populations was female biased, competition for access to females may have been low so that fighter fitness was more affected by perturbations to their survival than to their ability to gain access to females. One could furthermore argue that the removal of fighters increases opportunities for cannibalism as there are fewer competitor cannibalists. However, any positive selection for fighter expression through the removal of competitor cannibals is likely to be small as the number of fighters in the populations was always lower than the number of scramblers. The negative effect of the direct selection against fighters through harvesting, as well as the sustained removal of their food (scramblers); therefore, likely outweighs any positive effects of the indirect reduction in the strength of competition over food by the removal of a few fighters. Overall, we therefore conclude that our results confirm Kokko and López-Sepulcre's (2007) suspicion that the effect of a specific factor, such as high extrinsic mortality, may be very different in isolation than in a real-world setting where feedback loops between ecology and evolution are at play. Standard theories on the evolution of phenotypes, including the conditional strategy, can therefore not ignore eco-evolutionary interactions whenever phenotypic traits have ecological consequences and ecology affects the evolutionary trajectory of these traits.

We also observed that, across all treatments, fighter frequency gradually increased over time as the experiment progressed. Initially, fighter frequency was much lower than in the source population. This is likely a plastic response to the reduction in food availability (cf. Smallegange, 2011a) but could also have been partly genetic as a sufficiently high number of generations (approximately five since the start of the experiment) had passed for an evolutionary shift in fighter expression to occur in response to environmental change (Smallegange and Coulson, 2011: five generations; Tomkins et al., 2011: 10 generations). We suspect that the slow, temporal increase in fighter frequency occurred in response to natural selection to the experimental conditions (e.g. Cameron et al., 2013). This response was suppressed in the harvesting treatments compared to the control. Even though density dependence was high in our experiment (larval survival particularly was very low), which can have a suppressive effect on fighter expression because the development of fighter legs is costly (Radwan et al., 2002), the gradual increase in fighter frequency over time suggests that our laboratory circumstances favour a high proportion of fighter males. It is interesting to note, though, that fighter frequencies did not reach the frequency observed in the source population until many generations had passed. Had we terminated the experiment earlier, for example after a few generations, our interpretation would have been that the new, density-dependent conditions favour a reduction rather than an increase in fighter frequency.

By selectively harvesting fighters and scramblers, we removed some of the largest individuals from the populations. Such size-selective harvesting is common in both marine and terrestrial habitats and can lead to plastic as well as genetically based declines in age and size at maturity (e.g. Barot et al., 2004; Grift et al., 2003; Morita and Fukuwaka, 2006; Olsen et al., 2004; Sharpe and Hendry, 2009). If, in our study, harvesting resulted in a decrease in size at maturity then this could explain why males were on average smaller in the harvested than in the control populations. Interestingly, female body size varied in the same manner. Because there is a positive correlation between female body size and fecundity, population biomass and yield from harvested populations decrease if reduced average body size is not compensated for by an increase in population abundance (Ratner and Lande, 2001). In our study, the population response appears overcompensatory, as harvested populations contained in total more individuals than the control populations (e.g. De Roos and Persson, 2013). However, total population size was directly related to the number of females: the more females, the more eggs and larvae, and, since populations largely consist of eggs and larvae (Fig. 4.2), the larger the total size of the population. We therefore suspect that any (over)compensatory response in population size to mean body size is overshadowed by the population consequences of variation in female numbers. What drives this variation is, however, unclear. Mean female numbers showed a negative association with mean fighter numbers across treatments. However, because fighters are not very successful in killing adult females (Łukasik, 2010) (probably because adult females are of a much larger size: Fig. 4.1C) and because mean female and fighter numbers were positively and not negatively associated within treatments, it is likely that cannibalism is not at play here. But, if there is a causal link between fighter and female numbers, then the knock-on consequences of variation in fighter density for population structure and size are substantial.

Our results illustrate that it is imperative to consider the concurrent ecological and evolutionary consequences of selectively harvesting individuals of only one type for the sustainable management of harvested populations, which is not always strongly guided by an evolutionary framework (Young, 2004). For example, human activities in salmon breeding environments and in the oceans are altering the proportion of male salmon maturing as sneakers (Gross, 1991; Piou and Prévost, 2013), which can reduce population growth (Myers, 1984). Unravelling the ecological and evolutionary mechanisms at play here is essential to maintain viable salmon populations that can be harvested sustainably. Furthermore, sport hunting of large trophy males in wild game populations can result in evolutionary shifts in trophy size and body size of males (Coltman et al., 2003), which, in turn, can bias the population sex ratio towards females and lower reproductive success through the (delayed) production of fewer offspring of lower quality (Milner et al., 2007). The management of such wild game populations would therefore profit from a detailed understanding of how evolutionary processes driven by hunting feedback to affect ecological change. Rigorous monitoring and carefully conducted observations, however, are required to unravel the fine details of how interactions between evolutionary and ecological processes affect wild populations, their structure, stability and growth rates (Milner et al., 2007).

In conclusion, we have identified an eco-evolutionary interaction where phenotypic trait evolution has consequences for population size and structure, and ecology imposes selection on phenotypic traits. The importance of such interactions cannot be overstated. For example, if Costantino et al. (1995) had not replaced all individuals in their experimental populations with fresh ones from separate stock cultures during each census (to counteract possible genetic changes in life-history traits), would they still have been able to experimentally induce transitions in the dynamic behaviour of *Tribolium* populations? The existence of eco-evolutionary interactions also implies that experimental evolution studies would benefit from considering ecological responses in understanding trait evolution, whereas eco-evolutionary studies would gain from identifying the drivers of phenotypic change. Merging these fields to conduct carefully designed lab and field experiments (e.g. Cameron et al., 2013; Reznick et al., 2001; Chapter 5; this study) paves the way to a deeper understanding of the components and causal routes of the eco-evolutionary response of populations to environmental change.

ACKNOWLEDGEMENTS

We thank Izabela Lesna for collecting the mites and Maus Sabelis and Tim Coulson for discussion. This work was sponsored by an ERC advanced grant awarded to Tim Coulson.

APPENDIX. PREDICTING THE EVOLUTION OF FIGHTER EXPRESSION

Predictions from the environmental threshold (ET) model rely on fitness functions from which we predict that a reduction in fighter (scrambler) fitness through selective harvesting results in a lower frequency of fighters (scramblers) in the population (Fig. 4.A1).

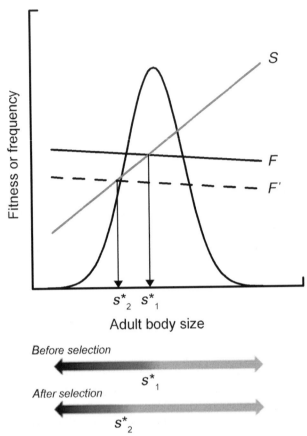

Figure 4.A1 Evolutionary predictions for selection on male morph expression using the ET model (Hazel et al., 1990, 2004). The solid straight lines are a schematic representation of the fitness functions found for scramblers (*S*; in grey) and fighters (*F*; in black) (Smallegange et al., 2012). Selection against fighters (in this case through harvesting) results in a decrease in the intercept of the fitness function of fighters (dashed line: *F'*), as a result of which the intersection point moves from its position before selection, s_1^*, to a smaller adult body size at s_2^* after selection (horizontal arrows: scramblers in grey and fighters in black), increasing scrambler expression. Note that fighter expression is negatively correlated with adult body size but positively correlated with quiescent tritonymph size (see Fig. 4.3 in Smallegange et al., 2012). This means that a reduction in fighter frequency (through reduced fighter fitness) comes about through an increase in mean tritonymph size threshold to a larger quiescent tritonymph size. Similarly, selection against scramblers increases the frequency of fighters in the population. The frequency distribution of body size (bell-shaped curve) is assumed not to be affected by selection, which was indeed the case in this study (see Section 3).

A1. Population dynamics

Time series of the number of individuals in each life stage and their mean size in each treatment is given in Fig. 4.A2.

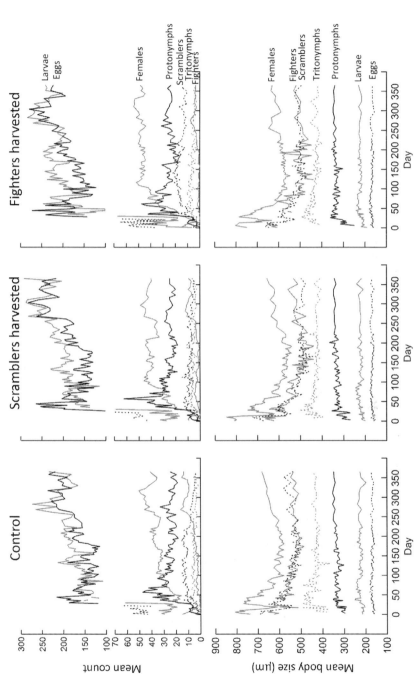

Figure 4.A2 Time series averaged per treatment. The top panels show the mean number of individuals in each life stage and the bottom panels show the mean body size of each life stage. Note the break and change of scale in the *y*-axis in the top panels. The harvesting treatment began on day 60. The dip in mean adult size of females, scramblers and fighters at the start of the experiment is caused by the removal of the large adults from the source populations at day 15.

REFERENCES

Barot, S., Heino, M., O'Brien, L., Dieckmann, U., 2004. Long-term trend in the maturation reaction norm of two cod stocks. Ecol. Appl. 14, 1257–1271.

Bates, D. and Sarkar, D., 2007. lme4: linear mixed-effects models using S4 classes. R package version 0.9975-13.

Benton, T.G., Cameron, T.C., Grant, A., 2004. Population responses to perturbations: predictions and responses from laboratory mite populations. J. Anim. Ecol. 73, 983–995.

Bolker, B.M., Brooks, M.E., Clark, C.J., Geange, S.W., Poulsen, J.R., Stevens, M.H.H., White, J.-S.S., 2009. Generalized linear mixed models: a practical guide for ecology and evolution. Trends Ecol. Evol. 24, 127–135.

Buoro, M., Gimenez, O., Prévost, E., 2012. Assessing adaptive phenotypic plasticity by means of conditional strategies from empirical data: the latent environmental threshold model. Evolution 66, 996–1009.

Buzatto, B.A., Requena, G.S., Lourenço, R.S., Munguía-Steyer, R., Machado, G., 2011. Conditional male dimorphism and alternative reproductive tactics in a Neotropical arachnid (Opiliones). Evol. Ecol. 25, 331–349.

Buzatto, B.A., Tomkins, J.L., Simmons, L.W., 2012. Maternal effects on male weaponry: female dung beetles produce major sons with longer horns when they perceive higher population density. BMC Evol. Biol. 12, 118.

Cameron, T.C., Benton, T.G., 2004. Stage-structured harvesting and its effects: an empirical investigation using soil mites. J. Anim. Ecol. 73, 966–1006.

Cameron, T.C., O'Sullivan, D., Reynolds, A., Piertney, S.B., Benton, T.G., 2013. Eco-evolutionary dynamics in response to selection on life-history. Ecol. Lett. 16, 754–763.

Carroll, S.P., Hendry, A.P., Reznick, D.N., Fox, C.W., 2007. Evolution on ecological time-scales. Funct. Ecol. 21, 387–393.

Chevin, L.-M., Lande, R., Mace, G.M., 2010. Adaptation, plasticity, and extinction in a changing environment: towards a predictive theory. PLoS Biol. 8, e1000357.

Coltman, D.W., O'Donoghue, P., Jorgenson, J.T., Hogg, J.T., Strobeck, C., Festa-Bianchet, M., 2003. Undesirable evolutionary consequences of trophy hunting. Nature 426, 655–658.

Costantino, R.F., Cushing, J.M., Dennis, B., Desharnais, R.A., 1995. Experimentally induced transitions in the dynamic behaviour of insect populations. Nature 375, 227–230.

Coulson, T., MacNulty, D.R., Stahler, D.R., Vonholdt, B., Wayne, R.K., Smith, D.W., 2011. Modeling effects of environmental change on wolf population dynamics, trait evolution, and life history. Science 334, 1275–1278.

de Roos, A.M., Persson, L., 2013. Population and Community Ecology of Ontogenetic Development. Princeton University Press, Princeton, New Jersey.

Deere, J.A., Smallegange, I.M., 2014. Does frequency-dependence determine male morph survival in the bulb mite *Rhizoglyphus robini*? Exp. Appl. Acarol. 62, 425–436.

Díaz, A., Okabe, K., Eckenrode, C.J., Villani, M.G., O'Connor, B.M., 2000. Biology, ecology, and management of the bulb mites of the genus Rhizoglyphus (Acari: Acaridae). Exp. Appl. Acarol. 24, 85–113.

Dodd, D.M.B., 1989. Reproductive isolation as a consequence of adaptive divergence in Drosophila pseudoobscura. Evolution 43, 1308–1311.

Ezard, T.H.G., Côté, S.D., Pelletier, F., 2009. Eco-evolutionary dynamics: disentangling phenotypic, environmental and population fluctuations. Philos. Trans. R. Soc. Lond. B Biol. Sci. 364, 1491–1498.

Falconer, D.S., Mackay, T.F.C., 1996. Introduction to Quantitative Genetics. Addison Wesley Longman, Harlow, Essex.

Garland, T., Rose, M.R. (Eds.), 2009. Experimental Evolution. University of California Press, Berkeley, California.

Grift, R.E., Rijnsdorp, A.D., Barot, S., Heino, M., Dieckmann, U., 2003. Fisheries-induced trends in reaction norms for maturation in North Sea plaice. Mar. Ecol. Prog. Ser. 257, 247–257.

Gross, M.R., 1991. Salmon breeding behavior and life history evolution in changing environments. Ecology 72, 1180–1186.

Gross, M.R., 1996. Alternative reproductive strategies and tactics: diversity within sexes. Trends Ecol. Evol. 11, 92–98.

Hairston, N.G., Ellner, S.P., Geber, M.A., Yoshida, T., Fox, J.A., 2005. Rapid evolution and the convergence of ecological and evolutionary time. Ecol. Lett. 8, 1114–1127.

Hazel, W., 1977. The genetic basis of pupal colour dimorphism and its maintenance by natural selection in *Papillio polyxenes* (Papilionidae: Lepidoptera). Heredity 59, 227–236.

Hazel, W., West, D.A., 1982. Pupal colour dimorphism in swallowtail butterflies as a threshold trait—selection in *Eurytides Marcellus* (Cramer). Heredity 49, 295–301.

Hazel, W.N., Smock, R., Johnson, M.D., 1990. A polygenic model for the evolution and maintenance of conditional strategies. Proc. R Soc. Lond. B Biol. Sci. 242, 181–187.

Hazel, W.N., Smock, R., Lively, C.M., 2004. The ecological genetics of conditional strategies. Am. Nat. 163, 888–900.

Johnstone, D.L., O'Connell, M.F., Palstra, F.P., Ruzzante, D.E., 2013. Mature male parr contribution to the effective size of an anadromous Atlantic salmon (*Salmo salar*) population over 30 years. Mol. Ecol. 22, 2394–2407.

Kawecki, T.J., Lenski, R.E., Ebert, D., Hollis, B., Oliveiri, I., Whitlock, M.C., 2012. Experimental evolution. Trends Ecol. Evol. 27, 547–560.

Kokko, H., López-Sepulcre, A., 2007. The ecogenetic link between demography and evolution: can we bridge the gap between theory and data? Ecol. Lett. 10, 773–782.

Kolss, M., Vijendravarma, R.K., Schwaller, G., Kawecki, T.J., 2009. Life history consequences of adaptation to larval nutritional stress in Drosophila. Evolution 63, 2389–2401.

Leigh, D.M., Smallegange, I.M., 2014. Effects of variation in nutrition on male morph development in the bulb mite *Rhizoglyphus robini*. Exp. Appl. Acarol. http://dx.doi.org/10.1007/s10493-014-9822-y.

Lively, C.M., 1986. Canalization versus developmental conversion in a spatially variable environment. Am. Nat. 128, 561–572.

Losos, J.B., Warheit, K.I., Schoener, T.W., 1997. Adaptive differentiation following experimental island colonization in *Anolis* lizards. Nature 387, 70–73.

Łukasik, P., 2010. Trophic dimorphism in alternative male reproductive morphs of the acarid mite *Sancassania berlesei*. Behav. Ecol. 21, 270–274.

Łukasik, P., Radwan, J., Tomkins, J.L., 2006. Structural complexity of the environment affects the survival of alternative male reproductive tactics. Evolution 60, 399–403.

Milner, J.M., Nilsen, E.B., Andreassen, H.P., 2007. Demographic side effects of selective hunting in ungulates and carnivores. Cons. Biol. 21, 36–47.

Morita, K., Fukuwaka, M., 2006. Does size matter most? The effect of growth history on probabilistic reaction norm for salmon maturation. Evolution 60, 1516–1521.

Moya, A., Galiana, A., Ayala, F.J., 1995. Founder effect speciation theory: failure of experimental corroboration. Proc. Natl. Acad. Sci. U.S.A. 92, 3983–3986.

Myers, R.A., 1984. Demographic consequences of precocious maturation of Atlantic salmon *(Salmo salar)*. Can. J. Fish. Aquat. Sci. 41, 1349–1353.

Oliveira, R.F., Taborsky, M., Brockmann, H.J. (Eds.), 2008. Alternative Reproductive Tactics. Cambridge University Press, Cambridge.

Oliver Jr., J.H., 1971. Parthenogenesis in mites and ticks. Am. Zool. 11, 283–299.

Oliver Jr., J.H., 1977. Cytogenetics of mites and ticks. Annu. Rev. Entomol. 22, 407–429.

Olsen, E.M., Heino, M., Lilly, G.R., Morgan, M.J., Brattey, J., Ernande, B., Dieckmann, U., 2004. Maturation trends indicative of rapid evolution preceded the collapse of northern cod. Nature 428, 932–935.

Ozgul, A., Tuljapurkar, S., Benton, T.G., Pemberton, J.M., Clutton-Brock, T.H., Coulson, T., 2009. The dynamics of phenotypic change and the shrinking sheep of St. Kilda. Science 325, 464–467.

Piou, C., Prévost, E., 2013. Contrasting effects of climate change in continental versus oceanic environments on population persistence and micro-evolution of Atlantic salmon. Glob. Chang. Biol. 19, 711–723.

Plummer, M., Best, N., Cowles, K., and Vines, K., 2006. Coda: output analysis and diagnostics for MCMC. R package version 0.10-7.

Post, D.M., Palkovacs, E.P., 2009. Eco-evolutionary feedbacks in community and ecosystem ecology: interactions between the ecological theatre and the evolutionary play. Philos. Trans. R. Soc. Lond. B Biol. Sci. 364, 1629–1640.

Radwan, J., 1995. Male morph determination in two species of acarid mites. Heredity 74, 669–673.

Radwan, J., Bogacz, I., 2000. Comparison of life-history traits of the two male morphs of the bulb mite, *Rhizoglyphus robini*. Exp. Appl. Acarol. 24, 115–121.

Radwan, J., Klimas, M., 2001. Male dimorphism in the bulb mite, *Rhizoglyphus robini*: fighters survive better. Ethol. Ecol. Evol. 13, 69–79.

Radwan, J., Czyż, M., Konior, M., Kolodziejcyk, M., 2000. Aggressiveness in two male morphs of the bulb mite *Rhizoglyphus robini*. Ethology 106, 53–62.

Radwan, J., Unrug, J., Tomkins, J.L., 2002. Status-dependence and morphological trade-offs in the expression of a sexually selected character in the mite, *Sancassania berlesei*. J. Evol. Biol. 15, 744–752.

Ratner, S., Lande, R., 2001. Demographic and evolutionary responses to selective harvesting in populations with discrete generations. Ecology 82, 3093–3104.

Reznick, D., Butler IV, M.J., Rodd, H., 2001. Life history evolution in guppies. VII. The comparative ecology of high- and low-predation environments. Am. Nat. 157, 126–140.

Roff, D.A., 1996. The evolution of threshold traits in animals. Q. Rev. Biol. 71, 3–35.

Santos, M., Borash, D.J., Joshi, A., Bounlutay, N., Mueller, L.D., 1997. Density-dependent natural selection in Drosophila: evolution of growth rate and body size. Evolution 51, 420–432.

Schoener, T.W., 2011. The newest synthesis: understanding the interplay of evolutionary and ecological dynamics. Science 331, 426–429.

Sharpe, D.M.T., Hendry, A.P., 2009. Life history change in commercially exploited fish stocks: an analysis of trends across studies. Evol. Appl. 2, 260–275.

Smallegange, I.M., 2011a. Complex environmental effects on the expression of alternative reproductive phenotypes in the bulb mite. Evol. Ecol. 25, 857–873.

Smallegange, I.M., 2011b. Effects of paternal phenotype and environmental variability on age and size at maturity in a male dimorphic mite. Naturwissenschaften 98, 339–346.

Smallegange, I.M., Coulson, T., 2011. The stochastic demography of two coexisting male morphs. Ecology 92, 755–764.

Smallegange, I.M., Coulson, T., 2013. Towards a general, population-level understanding of eco-evolutionary change. Trends Ecol. Evol. 28, 143–148.

Smallegange, I.M., Johansson, J., 2014. Life history differences favor evolution of male dimorphism in competitive games. Am. Nat. 183, 188–198.

Smallegange, I.M., Thorne, N., Charalambous, M., 2012. Fitness trade-offs and the maintenance of alternative male morphs in the bulb mite (*Rhizoglyphus robini*). J. Evol. Biol. 25, 972–980.

Tomkins, J.L., Hazel, W., 2007. The status of the conditional evolutionarily stable strategy. Trends Ecol. Evol. 22, 522–528.

Tomkins, J.L., Hazel, W.N., Penrose, M.A., Radwan, J., LeBas, N.R., 2011. Habitat complexity drives experimental evolution of a conditionally expressed secondary sexual trait. Curr. Biol. 21, 569–573.

Van Doorslaer, W., Stoks, R., Duvivier, C., Bednarska, A., De Meester, L., 2009. Population dynamics determine genetic adaptation to temperature in *Daphnia*. Evolution 63, 1867–1878.

West-Eberhard, M.J., 1989. Phenotypic plasticity and the origins of diversity. Annu. Rev. Ecol. Evol. Syst. 20, 249–278.

West-Eberhard, M.J., 2003. Developmental Plasticity and Evolution. Oxford University Press, New York.

Young, K.A., 2004. Toward evolutionary management: lessons from salmonids. In: Hendry, A.P., Stearns, S.C. (Eds.), Evolution Illuminated. Oxford University Press, New York, pp. 358–376.

CHAPTER FIVE

Eco-Evolutionary Dynamics: Experiments in a Model System

Tom C. Cameron*,[1], Stewart Plaistow[†], Marianne Mugabo[‡], Stuart B. Piertney[§], Tim G. Benton[‡]

*Environmental & Plant Sciences, School of Biological Sciences, University of Essex, Colchester, United Kingdom
[†]Institute of Integrative Biology, University of Liverpool, Liverpool, United Kingdom
[‡]School of Biology, Faculty of Biological Sciences, University of Leeds, Leeds, United Kingdom
[§]Institute of Biological and Environmental Sciences, University of Aberdeen, Aberdeen, United Kingdom
[1]Corresponding author: e-mail address: tcameron@essex.ac.uk

Contents

Abstract

Understanding the consequences of environmental change on both long- and short-term ecological and evolutionary dynamics is a basic pre-requisite for any effective conservation or management programme but inherently problematic because of the complex interplay between ecological and evolutionary processes. Components of such complexity have been described in isolation or within conceptual models on numerous occasions. What remains lacking are studies that characterise effectively

Advances in Ecological Research, Volume 50
ISSN 0065-2504
http://dx.doi.org/10.1016/B978-0-12-801374-8.00005-0

171

the coupled ecological and evolutionary dynamics, to demonstrate feedback mechanisms that influence both phenotypic change, and its effects on population demography, in organisms with complex life histories. We present a systems-based approach that brings together multiple effects that 'shape' an organism's life history (e.g. direct and delayed life-history consequences of environmental variation) and the resulting eco-evolutionary population dynamics. Using soil mites in microcosms, we characterise ecological, phenotypic and evolutionary dynamics in replicated populations in response to experimental manipulations of environment (e.g. the competitive environment, female age, male quality). Our results demonstrate that population dynamics are complex and are affected by both plastic and evolved responses to past and present environments, and that the emergent population dynamic itself shaped the landscape for natural selection to act on in subsequent generations. Evolutionary and ecological effects on dynamics can therefore be almost impossible to partition, which needs to be considered and appreciated in research, management and conservation.

1. INTRODUCTION

A fundamental goal in evolutionary ecology is to understand the mechanisms responsible for generating the phenotypic variation upon which selection acts. Similarly, a fundamental goal in population ecology is to understand the role that individual phenotypic variation, created by density–independent and/or density-dependent processes, plays in shaping population dynamic patterns. Thus, understanding between–individual phenotypic variation is key to understanding both ecological and evolutionary dynamics (Benton et al., 2006). Traditionally, an individual's phenotype has been considered a consequence of interaction between its genes and the environment in which they are expressed. Phenotypic variation has thus been envisaged as the sum of direct environmental and genetic effects, plus their interactions. Despite this recognition, for most of the history of ecology it has been assumed that the ways in which genes and environments interact are relatively unimportant for population dynamics (i.e. the trait changes from life-history evolution are either small or take too long to influence short-term dynamics). Two major conceptual advances have recently occurred that casts doubt on this traditional view. First, we now recognise that the environment experienced in previous generations can have consequences for contemporary phenotypes (Beckerman et al., 2002), reflecting the importance of non–genetic modes of inheritance that relate parental and offspring life histories (Bonduriansky and Day, 2009; Qvarnstrom and Price, 2001; Rasanen and Kruuk, 2007). Second, there is a growing realisation that evolutionary change can occur over ecological timescales, which has highlighted the need to better understand how ecological and evolutionary processes interact to drive population dynamics and demographic change (Bassar et al., 2010; Carroll et al., 2007; Coulson et al., 2006, 2010; Ellner et al., 2011; Ezard et al., 2009; Hairston et al., 2005; Olsen et al., 2004; Pelletier et al., 2007, 2009; Schoener, 2011; Stockwell et al., 2003).

Teasing apart parental, plastic, ecological and reversible responses from evolved and irreversible responses of life histories to environmental change is inherently problematic, as it is rarely possible to study parental environment effects, genetics, life histories and population dynamics simultaneously and in sufficient detail (Andersen and Brander, 2009a,b; Becks et al., 2012; Bonenfant et al., 2009; Coulson and Tuljapurkar, 2008; Coulson et al., 2010; Darimont et al., 2009; Morrissey et al., 2012; Ozgul et al., 2009, 2012; Uller, 2008). However, this is exactly what is required to understand how, or even if, populations will be able to respond to rapid anthropogenic environmental stressors such as selective harvesting (Andersen and Brander, 2009a,b; Browman et al., 2008; Coltman et al., 2003; Ezard et al., 2009; Kinnison et al., 2009; Law, 2007), the potential for species to respond to environmental change through evolution (Bell and Gonzalez, 2009; Ezard et al., 2009; Stockwell et al., 2003) and the role that parental effects have in those adaptive responses to environmental change (Uller, 2008).

Our research with an invertebrate model system has gone some way towards understanding the role of parental environments, and the significance of plastic responses and rapid evolution in delimiting individual phenotypic variation. Here, we describe how we have approached these challenging questions by presenting our conceptual framework of eco-evolutionary population dynamics (Fig. 5.1) and reporting on what progress we have made in determining each process within this framework. To this end, we review previously published material and report new results from ongoing empirical studies. We use our findings to identify new avenues for research necessary to properly understand how contemporary, historical and evolutionary determinants of individual life histories interact to shape population-level responses.

2. AIMS AND SCOPE

The aim of this chapter is to introduce the mite model system, a soil invertebrate microcosm-based experimental system, and show how it has been used to test and develop our understanding of individual phenotypes,

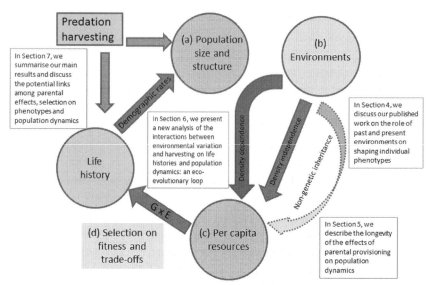

Figure 5.1 A diagrammatic representation of eco-evolutionary dynamics based on the results of mite model system experiments. The eco-evolutionary loop is moving between the three circled states: from (a) population structure is dependent on life-history transition rates, and interacts with the environment (b) via an interaction between density-dependent and -independent mechanisms and parental effects to determine per capita resources (c). Per capita resources interact with genetic and environmental determinants of individual life histories (d), which leads to a closure of the eco-evolutionary loop by creating population structure. We consider here the effects of predation and harvesting as external to the loop (bordered and shaded box), affecting the loop directly by selecting against life histories or changing population size and structure.

how they form and how they scale up to population dynamics (i.e. Fig. 5.1). We will begin by introducing our study organism, its general biology and the various experimental methods we have used to explore individual and population biology (Section 3). In Section 4, we will review our previously published work on the development of individual phenotypes as a function of resource availability. This has been a key empirical proof-of-principle of the L-shaped reaction norms predicted to arise when developmental thresholds determine age- and size-at-maturity (Day and Rowe, 2002). Again referring to our published works, using this L-shaped age- and size-at-maturity reaction norm as a background measurement, we will describe our current understanding of when and how parental environments shape offspring phenotypes. The role of non-genetic inheritance of parental traits is important in the development of our later arguments that describe how current and historical environmental effects interact with natural selection to create eco-evolutionary population dynamics. If, and how, parental effects manifest themselves beyond effects on individual, offspring will be presented in Section 5. Here, we will present our published work on the magnitude and longitude of detectable effects of ancestral environments on soil mite population dynamics.

In Section 6, we will present a new analysis of how selection on individual phenotypes, caused by feedbacks from population dynamics in the form of strong density-dependent competition, leads to the evolution of population dynamics. This extends the analysis of soil mite populations living in periodically fluctuating resource environments and subject to experimental harvesting (Cameron et al., 2013). Here, we are able to present data across constant, randomly variable and periodically variable resource environments. Crucially, it is the imposition of experimental harvesting that reveals that the environmental variation is important in the evolutionary responses of populations to environmental change. Finally, in Section 7, we summarise what we have presented in the form of previously published and new analyses and discuss how the different routes we have found to influence population dynamics through changes in individual phenotypes might interact. The overall scope of this contribution therefore is to stress that it is by understanding how the different routes that lead to phenotypic variation interact that we will come to a more than conceptual understanding on eco-evolutionary population ecology.

3. MODEL SYSTEM AND METHODS

The soil mite *Sancassania berlesei* (Michael) is common in soil, poultry litter and stored food products. Populations of *S. berlesei* have been collected from a variety of sources in different years since 1996 and have been kept in separate stock lines ever since (stock cultures kept in 10-cm diameter containers maintained at $24\,^\circ\text{C}$ in unlit incubators, number $c1\text{--}2.5\times10^5$ individuals).

3.1. The mite model system and generic methods

The life cycle consists of five stages, beginning with eggs (length: $0.16\pm\text{SD}$ 0.01 mm), continuing through a six-legged larvae (length: 0.22 ± 0.01 mm), a protonymph, tritonymph and then to adulthood (female length at maturity: 0.79 ± 0.17 mm, range 0.47 (low food) to 1.17 (high food), $n=64$; males: 0.72 ± 0.11 mm, range 0.55 (low food) to 1.02 (high food),

$n = 39$). As indicated by the standard deviations of the adult lengths, there is considerable variation in the life history and much of it is governed by intake rates of food (Plaistow et al., 2004). An individual's intake rate is a function of a number of factors: population density, stage structure and the amount of food supplied and its spatial configuration; together these factors create the individual's competitive environment (Benton and Beckerman, 2005).

Eggs hatch 2–5 days after being laid. Juveniles can mature from as little as 4–50 + days after hatching (Beckerman et al., 2003), depending on food and density. The longevity of the adults can also vary from ca 10 to ca 50 days. Thus, total longevity varies from 3 weeks (high food, low density) to 7 + weeks (low food, high density). Fecundity is related to resources, and so to body size, and to survival. The relationship between fecundity and the growth-survival trade-off is in itself dependent on resources (Plaistow et al., 2006, 2007).

3.2. General experimental procedures

Generally, mite cultures are supplied with food in the form of powdered or granulated yeast. Different feeding regimes were used in different experiments and consisted of controlled feeding of balls or rods of dried baking yeast, filtered to minimise variation in their size (diameter of 1.25–1.40 mm for standard size balls). Experimental vessels are either glass tubes (20 mm in diameter and 50 mm in height) or small non-static plastic vials (3-7 ml). These are half-filled with plaster of Paris, which, when kept moist, maintains humidity in the tubes. The tops of the tubes are sealed with a circle of filter paper held in place by the tubes' cap with ventilation holes cut into it. For some shorter experiments (24 h), the plastic vials were sealed with cling film. For population experiments, the mites are censused using a Leica MZ8 binocular microscope and a hand counter. In each tube, a sampling grid is etched into the plaster surface to facilitate more accurate counting and observation. All adults are counted in the tube, but juveniles and eggs are counted in a randomly chosen quarter.

3.2.1 Common garden environments

Common garden tubes were used to both standardise and manipulate parental and offspring environments prior to carrying out life-history assays or population dynamic experiments. A common garden was created by placing standardised numbers of eggs (from either stock culture females or experimental animals) into identical tubes with controlled food access/competitor density and rearing them until maturation. Upon maturation, these individuals are paired and either placed in a new common garden or in egg-laying tubes for the collection of eggs for life-history assays, reproduction allocation measurements or population dynamic experiments (i.e. Plaistow and Benton, 2009; Plaistow et al., 2004).

3.2.2 Life-history assays

Life-history assays are used to quantify the life history or phenotype of an individual, full-sib family or population from a given treatment. Life-history assays are conducted by placing individuals or groups of random or full-sib eggs in a small vial that is half-filled with plaster (7–20-ml plastic or glass vials). These individuals are observed daily, either with density being standardised by replacement of dead individuals or not. At maturation, individuals are photographed for later measurement and then removed from the vial. We can collect data on age- and size-at-maturity, fecundity at maturity or any other stage of development (e.g. egg size, hatching, protonymphs). Reproductive allocation is a measure of the differences between mite eggs laid by mothers from different parental environments (i.e. Plaistow et al., 2007). We have measured reproductive allocation in terms of numerical (e.g. total eggs, eggs-at-age), physical (e.g. length, volume) and biochemical properties of eggs laid (e.g. total protein). Measurements of individuals and eggs are made from digital images captured from the microscope (e.g. Leica MZ8, Nikon SMZ15) and measured using ImageJ 1.28u (http://rsb.info.nih.gov/ij) or Nikon Elements D software (v3.2 64bit).

3.2.3 Population dynamic experiments

Population dynamic experiments involve monitoring free-running populations over multiple generations. Such experiments have been started in different ways depending on the purpose of the experiment. Where the purpose was to investigate the timescale of parental effects, populations were started with controlled numbers of eggs from parents of different environmental backgrounds or ages (Pinder, 2009; Plaistow et al., 2006, 2007). To investigate the interplay between population and phenotypic dynamics, populations were initiated with a mix of sexed adults ($n = 75–150$/sex) and juveniles ($n = 500–1000$), approximately at stable stage distribution to minimise transient dynamics. To investigate the links between ecological plasticity and life-history change, populations were initiated with mites recently collected from the wild to maximise genetic diversity ($n = 150$ adult/sex and 1000 juveniles).

In the population experiments, we have often manipulated stochasticity by varying the timing and amount of food supplied, while trying to maintain other factors as close to constant as possible. Our rationale for this is that many natural environmental factors will either vary the absolute food supply (e.g. the weather), the requirement for food (e.g. temperature) or the availability of food (e.g. patchiness, territoriality, inter-specific competition). Each treatment supplied food at the same mean daily rate (equivalent to one or two balls of yeast per day), but at a variable amount on different days. The algorithms we developed were to supply balls of yeast randomly, or periodically, within each window of time, such that over repeating window lengths, the cultures received a constant number of balls of yeast. Other populations were maintained on constant food regimes either to act as contrasts to those in the variable environments, or on their own for some parental effect experiments. Effects of the different distributions of food supply on variation in population abundance are described elsewhere (Benton et al., 2002).

4. WITHIN AND BETWEEN INDIVIDUAL PHENOTYPIC VARIATIONS

In this section, we review our previously published work explaining how environment-induced changes in the growth rate and maturation decisions are responsible for generating a L-shaped age- and size-at-maturity reaction norm. We then summarise our previously published work explaining how variation in age- and size-at-maturity alters the provisioning of individual offspring and the developmental environment of those same offspring, leading to inter-generational phenotypic variation.

4.1. Age- and size-at-maturity reaction norms

Population growth rates are intrinsically linked to the trade-off between the age and size at which individuals mature because age-at-maturity determines how quickly individuals start to reproduce and because fecundity is often closely associated with age and body size (Plaistow et al., 2006, 2007; Roff, 2002). Consequently, understanding how populations respond to environmental change is likely to depend upon how individuals, within those populations, respond to environmental change. Organisms that live in variable environments, due to environmental forcing or density dependence, for example, are expected to evolve plasticity in age- and size-at-maturity because of fluctuations in resource availability (DeWitt et al.,

1998; Via et al., 1995). We demonstrated that in soil mites, the trade-off between age- and size-at-maturity is extremely plastic in response to food availability. Offspring reared on high food matured five times faster and at double the body size of offspring reared in a poor food environment. Moreover, the age- and size-at-maturity reaction norm is L-shaped (Plaistow et al., 2004) (Fig. 5.2). This pattern arises because an individual's decision to mature is controlled by a developmental threshold, which is the minimum size below which maturation cannot occur (Day and Rowe, 2002). Fast growing individuals in good food environments overshoot the minimum threshold size considerably by the time maturation is complete. In contrast, slow-growing individuals in poor food environments have to delay maturation until the minimum threshold size is reached. Consequently, in good food environments, all individuals mature at young age but individual differences in growth rates translate into variation in size at maturation. In contrast, in poor food environments, all individuals mature at the same minimum threshold size but individual differences in growth rates translate into differences in age-at-maturity (Plaistow et al., 2004).

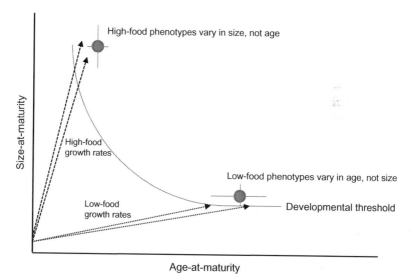

Figure 5.2 A model of the L-shaped developmental threshold model predicting growth rates to maturation along an environmental gradient of food availability (i.e. norm of reaction). This model, developed by Day and Rowe (2002), is supported by our results in the mite model system and captures the feedback caused by the interaction between population size and environmental quality on per capita resources, and the resulting density-dependent effects on individual phenotype. *Based on Beckerman et al. (2003) and Plaistow et al. (2004).*

As we will see later, this fundamental difference in how environmental variation is translated into phenotypic variation has important implications for understanding how individual plasticity influences population dynamics.

4.2. Inter-generational parental effects on individual phenotypic variation

Parental effects are defined as any effect that parents have on the development of their offspring over and above directly inherited genetic effects (Uller, 2008). Two types of mechanisms can be involved in the transmission of parental effects to offspring phenotypes. In the first mechanism, parental effects can arise from alterations of the developmental environment experienced by offspring through variation in allocation of non-genetic resources such as nutrients (e.g. Benton et al., 2005; Plaistow et al., 2007), immune factors (e.g. Hasselquist and Nilsson, 2009) and hormones (e.g. Meylan et al., 2012). Traditionally, studies of environmental parental effects have focused on maternal influences on her offspring's developmental environment because, in most species, females invest more resources in offspring than males. However, a few examples of paternal effects arising from variation in food provisioning (e.g. Isaksson et al., 2006) and transmission of immune factors (e.g. Jacquin et al., 2012; Roth et al., 2012) exist in the literature. In addition, females can alter their investment in offspring in response to males' characteristics (e.g. Gil et al., 1999; Pinder, 2009), leading to indirect paternal effects. In the second mechanism, parental effects can arise from alterations of gene expression through epigenetic modifications of regulatory regions of the genome in the germline, for instance mediated by DNA methylation and histone modifications, and without changes in DNA sequences (Bonduriansky and Day, 2009). Trans-generational inheritance of epigenetic modifications have been suspected to be involved in some parental age effects (e.g. Bonduriansky and Day, 2009; Perrin et al., 2007), in some heritable disorders (e.g. Champagne, 2008; Olsen et al., 2012), and, more generally in paternal effects transmitted through variation in allocation of non-genetic resources (e.g. Rando, 2012). In addition, there is increasing evidence that maternal and paternal effects arising from variation in offspring's provisioning or from epigenetic modifications are context-dependent (e.g. Badyaev and Uller, 2009), and can interact to shape offspring phenotype (e.g. Ducatez et al., 2012). In soil mites, we have explained how age- and size-at-maturity is critically dependent on food availability in the offspring's current environment (Plaistow et al., 2004). However, we have also demonstrated how variation in the maternal provisioning of offspring and the age of the mother can influence both off-spring growth rates (Plaistow et al., 2006) and their decision to mature (Benton et al., 2008). In this contribution, we are specifically dealing with the first mechanism described above (i.e. alterations of the developmental environment). Consequently, individual variation in developmental or somatic growth is not just a result of the environment that the individual experiences, but also the environment experienced by its ancestors (e.g. Pinder, 2009) (Fig. 5.3A). From a population dynamic perspective, these

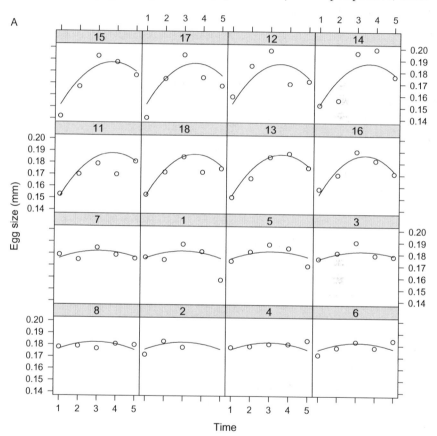

Figure 5.3 (A) Male age and condition influences female allocation patterns. Sixteen different males were mated to virgin females at each of five time-points during their lifetime (time). Males (sub-panels) were well fed (males 11–18) or poorly fed (males 1–8) and are presented in the order of the two male conditions. Graphs show egg size (mm) as a function of male age. Lines are fitted values from mixed effects' model. Time, food and male are all significant. Virgin females mating with 'prime' males (time class 3) laid larger eggs (Pinder, 2009).

(Continued)

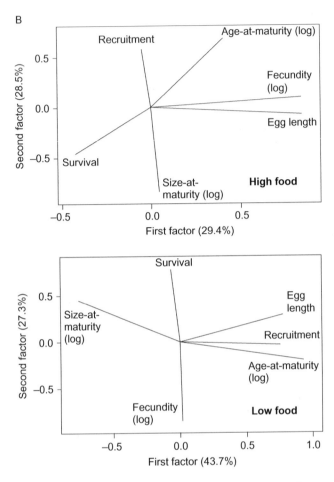

Figure 5.3—cont'd (B) Vector plots of the factor loadings from a factor analysis of parental effects (variation in egg length) between life-history traits for individuals reared in high- or low-food current environments. In high-food current environments, variation in egg length predominantly influenced a negative trade-off between fecundity and adult survival and had little effect on recruitment or age- and size-at-maturity. In contrast, in low-food environments variation in egg length translated into differences in the probability of recruiting and variation in age- and size-at-maturity. *Modified from figure 4 in Plaistow et al. (2006) with the kind permission of University of Chicago Press.*

effects are important because they mean that a population's response to environmental change may be time-lagged to some degree, with inter-generational effects operating as a source of intrinsic delayed density dependence (Beckerman et al., 2002; Rossiter, 1994).

4.3. Understanding the context dependence of parental effects

Our results have suggested that the importance of parental environments for the variation of offspring phenotypes in soil mites is trait-dependent and may be highly context-dependent (Beckerman et al., 2006; Plaistow et al., 2006). For instance, in low-food current environments, variation in egg size produced by different parental food environments altered the trade-off between age- and size-at-maturity, but had little effect on the size of eggs produced in subsequent generations. Consequently, the variation in egg size that affected inter-generational effects decreased over time. In contrast, in high-food environments, variation in egg size predominantly influenced a trade-off between fecundity and adult survival and generated increasing variation in egg size (Fig. 5.3B). As a result, maternal effects transmitted through variation in egg provisioning persisted and we have observed great grand-maternal effects on descendant's life histories (Plaistow et al., 2006). We therefore predicted that the persistence and significance of inter-generational effects for population dynamics would itself be context-dependent. However, it is important to realise that in an eco-evolutionary sense 'context' is itself something that is derived from the traits and maternal strategies that have evolved in the population.

In viscous populations with overlapping generations, mothers and offspring are forced to compete for the same resources and may, therefore, directly influence each other's probability of survival and future reproductive success. The close co-variation between the quality and number of offspring produced and maternal survival means that any change in one offspring-provisioning trait may have consequences for the others (Beckerman et al., 2006). It is necessary, therefore, to understand how females change their offspring-provisioning strategy as a whole (e.g. egg numbers, egg size, maternal survival) in order to interpret the adaptive significance of maternal responses to changes in their environment. We have shown that in soil mites, offspring-provisioning strategies are dynamic, switching from investment in many small eggs in young females to fewer, better provisioned eggs in older females (Plaistow et al., 2007). This strategy may be adaptive if it increases the survival of younger offspring that must compete with older, larger siblings that had been laid previously. This age-related dynamic shift in egg provisioning was greater in high-food environments in which females lived longer, creating a greater asymmetry in offspring competitive abilities. Such conditions are likely to be common in an opportunistic species such as soil mites that have evolved a life history that specialises in strong competition between individuals exploiting patchily distributed resources, such as carcasses and dung (Houck and Oconnor, 1991). In the following section, we examine the effects that these complex environmentally driven parental effects have on patterns of population dynamics.

5. FROM PHENOTYPIC VARIATION TO POPULATION DYNAMICS

Parental effects may be especially important from a population dynamic perspective because they generate a lag in the response of a population to an environmental change (Beckerman et al., 2002, 2006; Benton et al., 2005). This could make it harder to predict changes in population size, but may also theoretically lead to long-term deterministic population dynamic patterns, such as population cycles (Ginzburg, 1998; Ginzburg and Taneyhill, 1994; Inchausti and Ginzburg, 1998). Consequently, we have been interested in how parental effects might influence population dynamics (Benton et al., 2001). This is not easy to study in the wild, or in many laboratory systems, due to the difficulty of measuring parental effects and following population dynamics in sufficient demographic detail. However, it is possible in the soil mite system because replicated populations can first be initiated with different numbers of eggs, changing the initial environment experienced by offspring; but also initiated with eggs from different types of mothers, enabling us to experimentally manipulate parental effects (e.g. Benton et al., 2005, 2008; Plaistow and Benton, 2009).

5.1. Transient population dynamics and parental effects

In the first of these types of experiments, all replicated populations were initiated with 250 eggs. However, half the populations were set up with large eggs from mothers experiencing low food, the other half were set up with small eggs from well-provisioned mothers (see Benton et al., 2005 for details). This manipulation of the maternal effect alone was sufficient to generate differences in the transient population dynamics of the populations that were still present after three generations, even though the populations were experiencing the same constant environment with respect to the food supplied to them each day. Such deviations in population dynamics arise because differences in the hatching success, growth rate, size and fecundity and survival in the initial cohort generate differences in the competitive environment experienced by offspring produced in the second cohort.

Changes in the competitive environment creates further phenotypic variation between individuals from the two treatments that ultimately leads to large differences in the population dynamics of the populations sustained over multiple generations (Benton et al., 2005).

In a second experiment, but this time using similarly sized eggs that either came from young (3 days) or old (9 days) mothers, the effects on transient population dynamics again lasted three generations (Benton et al., 2008) (Fig. 5.4). The results clearly demonstrate that deterministic differences in eggs, which are not obviously related to their size, and so may be undetectable in a population setting, may have a significant effect on population dynamics. Comparing these two experiments, the effects of parental background or age were of a similar magnitude. However, as we discussed earlier, our individual-level studies of maternal effects in soil mites suggested that the exaggeration and the transmission of maternal effects from one generation to the next increased in high-food environments, but decreased in low-food environments (Plaistow et al., 2006). Consequently, we hypothesized that maternal effects would be more likely to persist, and have a bigger influence on population dynamics, in high-food environments compared to

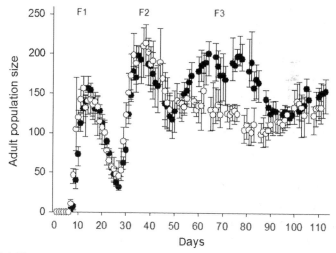

Figure 5.4 The inter-generational effects of variation in parental investment in offspring on population dynamics. The graphs show the transient dynamics of populations initiated with eggs that were laid by either younger 3-day-old (white points) or older 9-day-old mothers (black points). The error bars represent bootstrapped 95% confidence intervals. The individual cohorts are marked approximately on the figures as F1, F2 and F3 and were identified by inspection of the age-structured dynamics. *Modified from Benton et al. (2008) with permission from Wiley and the British Ecological Society.*

low-food environments. In order to test this hypothesis, we created maternal effects by initiating populations with eggs from young mothers or old mothers but we also simultaneously manipulated the initial resource environment by changing the initial density from high (500 eggs, low food) to low (50 eggs, high food) (see Plaistow and Benton, 2009 for details). The results clearly supported our hypothesis that the importance of maternal effects for population dynamics is context-dependent. An influence of maternal age treatment on both population and egg and body size dynamics was only observed in the populations initiated under low density rather than high density (Plaistow and Benton, 2009).

In summary, we have explained how an interaction between current and historical maternal states (transmitted as parental effects) interacts to shape patterns of individual phenotypic variation (e.g. size-at-hatch, growth rate to maturity, size-at-maturity, offspring's own egg-provisioning patterns) and how this phenotypic variation is then translated into fluctuations in population size. Understanding the various factors that can determine such fluctuations is crucial for predictive modelling of populations for management purposes. From an eco-evolutionary perspective, it is also critical because it is those fluctuations in the number, size and age structure of populations that determine the temporal resource heterogeneity that ultimately shape how individual traits and life-history strategies evolve (Roff, 2002). In the following section, we summarise our current understanding of how differences in temporal resource heterogeneity, created by environmental variation and harvesting, influence the evolution of mite life histories and, in turn, how this evolution influences population dynamics.

6. ECO-EVOLUTIONARY POPULATION DYNAMICS—THE FULL LOOP

Debate on the role of genetic change in ecological dynamics is not new (Lenski, 1984; Pimentel, 1961; Pimentel and Stone, 1968; Pimentel et al., 1978; Wilcox and Maccluer, 1979), and it includes predictions of cyclic consumer-resource dynamics caused by evolution (Abrams and Matsuda, 1997; Lenski, 1984). It is only more recently that the search for the role of the gene in ecology has been termed 'eco-evolutionary dynamics'.

It has largely been assumed that this emerging field of eco-evolutionary dynamics has demonstrated that evolutionary 'loops' exist in nature, where loops are defined as genetic selection pressures placed on populations from ecological interactions that have significant effects on population dynamics, additive to that of the ecological interaction itself (Kinnison and Hairston, 2007). For example, while a predator can reduce population growth by killing individuals, does it have an additional detectable effect on prey population growth rate by causing the average somatic growth rate to maturation to evolve? Such an evolutionary response of the prey life history, causing a feedback to prey population dynamics, and subsequently predator dynamics would be an evolutionary loop (Post and Palkovacs, 2009).

There is however a dearth of robust empirical evidence for such evolutionary loops. An early study by Nelson Hairston, Jr., described the pattern of rapid evolution of toxin resistance in *Daphnia galeata* in Lake Constance in response to eutrophication (Hairston et al., 1999, 2001). While not evidence of a loop *per se*, the Lake Constance study led to a series of experiments on zooplankton–phytoplankton interactions that demonstrated that rapid evolution in response to an ecological interaction can alter predator–prey cycles (Yoshida et al., 2003), that rapid evolution can mask interactions normally identified through changes in predator and prey abundance (Yoshida et al., 2007) and that rapid prey evolution can affect predator dynamics more than changes in prey abundance (Becks et al., 2012). Other studies on microcosm-based asexual communities have followed to show the generality of the importance of rapid evolution on ecological dynamics (e.g. Friman et al., 2014).

A common thread across all these aquatic predator–prey studies, with few exceptions (e.g. Fussmann et al., 2003), is the evolution of traits associated with either defence from predators or digestion of prey. This is clearly important in a community setting, but it is difficult to make the jump from proof-of-principle in these systems to studies that consider the role of environmental change (e.g. trends in mean annual temperature) or high rates of harvesting against life-history traits such as somatic growth rate in well-studied populations of fishes, birds and mammals (Darimont et al., 2009). Other differences between demonstrated eco-evolutionary dynamics in freshwater microorganisms and proposed eco-evolutionary dynamics in larger animals exist, not least of which is asexual versus sexual reproduction and *more* complex life histories based on significant growth from birth. Experimental studies have shown that rapid life-history evolution in vertebrates is possible, through response to selection caused by predation (Reznick et al., 1996) and harvesting (van Wijk et al., 2013), but trait change from selection on vertebrates in itself is not an eco-evolutionary loop.

Analyses of empirical data demonstrates that eco-evolutionary feedback from an environmental change to population dynamics could explain observed trait distributions and population sizes (Coulson et al., 2010; Ozgul et al., 2010, 2012), but this generally lacks evidence of genetic selection, but see similar studies of trait demography in birds (Charmantier et al., 2008; Nussey et al., 2005). Other studies have identified where eco-evolutionary dynamics are likely to occur, for example, by demonstrating how changes in selection have led to changes in animal behaviour and/or distribution (Strauss et al., 2008). Fewer studies, however, have been able to manipulate the eco-evolutionary loop in more complex organisms and ask what role ecological conditions have on selection on traits, and does this trait change feedback to influence population dynamics (Cameron et al., 2013; Walsh et al., 2012).

The role of predation in life-history evolution has long been recognised (Law, 1979; Michod, 1979; Reznick, 1982; Stenson, 1981), and it remains a contemporary interest (Beckerman et al., 2013). There has been a fever of interest in the role of high rates of trait-selective exploitation on shifts in the trait distributions of many harvested animal populations, in particular of body size or age and traits that would otherwise be under sexual selection, such as male ornamentation (Biro and Post, 2008; Bonenfant et al., 2009; Bunnefeld et al., 2009; Ciuti et al., 2012; Coltman et al., 2003; Darimont et al., 2009; Hamilton et al., 2007; Milner et al., 2007; Olsen et al., 2009; Pelletier et al., 2007). There has also been a concomitant interest in the role that these shifts in trait distributions may play in eco-evolutionary dynamics (Coulson et al., 2006, 2010). In those animal species that we exploit at some of the highest rates, specifically the marine and freshwater fishes, there is an ongoing debate about the mechanisms that lead to these shifts in body size distributions (Andersen and Brander, 2009a,b; Anderson et al., 2008; Browman et al., 2008; Kinnison et al., 2009; Kuparinen and Merila, 2007, 2008; Law, 2007). There are several more robust explanations for reduced mean body size-at-age in exploited fishes including body condition effects (Marshall and Browman, 2007), size-structured community interactions (Anderson et al., 2008; De Roos et al., 2003; Persson et al., 2007; Van Leeuwen et al., 2008) and fisheries-induced evolution (Jorgensen et al., 2007). Intuitively, these more prominent explanations are not mutually exclusive and have each been more plausible an explanation for responses to harvesting in different case studies. Here, we will investigate the role of evolutionary responses of phenotypes to exploitation, and in particular to stage-selective harvesting.

Stage-selective harvesting, occurring at times of the year or in places where particular life-history stages dominate the harvest (e.g. adult Barents Cod at spawning ground), or where there are other stage-based vulnerabilities in likelihood of harvest mortality (e.g. in cryptic selection of hunted birds (Bunnefeld et al., 2009), or killing only adults or juveniles of pest species), is predicted to lead to shifts in growth rate to maturity that are distinct from size-selection harvesting. Here, it is expected that life histories will evolve such that individuals who minimise their time in the most vulnerable stages will be selected for (Stearns, 1992). So, we expect that harvesting juveniles will lead to faster developmental growth to maturity, while harvesting adults will reduce developmental growth via a trade-off with increased juvenile survival and adult fecundity (Ernande et al., 2004).

Previous investigations with soil mites in seasonal environments where we exposed populations to adult or juvenile mortality resulted in statistically different growth rates to maturity in harvested populations, and compared to unharvested populations, the shifts in growth rate were exactly as predicted by theory (Cameron et al., 2013). Here, we extend this analysis to the evolved responses of growth rate to maturity when harvesting juveniles or adults across constant, random and periodic environments. Mite populations were harvested at a rate of 40% per week (proportional harvest) or as an additional threshold harvest treatment in randomly variable environments of all adults above 60% of the long-term adult population size. We estimated these rates to be close to the maximum that soil mite populations can sustain without collapsing (Benton, 2012). We report the life-history results on low-food conditions as we assume that this is most representative of the conditions in long-term experimental populations (e.g. Cameron et al., 2013).

In summary of this introduction, we present new empirical data from the mite model system where we have investigated the role that evolution plays in the contemporary responses of population dynamics to environmental change. We will summarise our main finding on the role of phenotypic evolution on population responses to highly competitive environments and building on this, we will discuss the roles of environmental variation (i.e. variation in food availability) and harvesting on the development of the eco-evolutionary feedback loop.

6.1. Methods

Soil mites were collected from several wild populations and allowed to mate for two generations in the laboratory before being placed in our standard microcosm population tubes (see Section 3) (Cameron et al., 2013). Sixty populations were started with 150 of each sex of adult and approximately 1000 juveniles in order to minimise transient dynamics. Each population received the same average access to resources of two balls of yeast per day, but it was randomly assigned to one of three experimentally induced levels of resource variability (i.e. environmental variation): constant (replicates $(n) = 18$); periodically variable $(n = 18)$ and randomly variable $(n = 24)$. The periodically variable treatment was designed to represent seasonality as best as possible by having a 28–day cycle (e.g. Cameron et al., 2013). The randomly variable treatment was designed to be entirely unpredictable with daily food provisions being chosen from a random distribution with a mean of two balls over a 56-day window, with a maximum daily provision of 12 balls (Benton et al., 2002). The mite populations were censused each week for 2 years, where a generation is approximately 5 weeks (Ozgul et al., 2012).

From week 13 to 83, the populations from each environmental variation treatment were subjected to a factorial stage-structured harvest treatment where: populations were either unharvested; juveniles were proportionally harvested (where 40% of juveniles were removed each week) or adults were proportionally harvested (where 40% of adults were removed each week). In the randomly variable treatment, there was an additional treatment of a threshold adult harvest, sometimes called a fixed-escapement harvest (Fryxell et al., 2005), where all adults above 60% of the long-term mean number of adults was removed. This number was set to 176 adults based on 60% of the long-term mean adult population size from previous studies on the same mean resources (Benton and Beckerman, 2005). Threshold harvest strategies have been said to be more conservative in affecting the variance in population size and therefore minimise extinction risks to harvested populations (Lande et al., 1997), but such claims have not been tested experimentally in variable environmental conditions.

In tandem with the population census, we conducted less frequent common garden life-history assays to measure the development to maturation of seven full-sib families for two of the six replicate populations per treatment combination. For the common garden, 100 juveniles were removed from populations and reared to the F2 generation on fixed per capita resources to standardise parental effects (e.g. Plaistow et al., 2006). Single F2 male–female pairs were allowed to mate and their eggs were collected. Twenty offsprings from each pair were each reared collectively in either high- or low-food resource availability. Only the results from the low-food life–history assay will be presented in this paper as this was found to best represent the competitive conditions in experimental populations. Age (days) and body sizes (body length in mm) at maturity were recorded for each adult individual of each sex. Daily survival rates until maturity of the cohort of 20 juveniles were calculated using standard methods (e.g. Mayfield estimates). Fecundity at maturity was estimated for each female individual using a linear regression of the age- and size-at-maturity with cumulative fecundity from day 3–7 post-eclosion from existing data (Plaistow et al., 2006, 2007). These data led to average trait values representing family and treatment phenotypes.

Twenty-four adult females per population were sampled from the common garden F3 generation in weeks (i.e. time-points) 0, 18, 37, 63 and 95 and their genotype was characterised using amplified fragment length polymorphisms (AFLP). The assay used 299 loci and the methodology has been described in detail elsewhere (Cameron et al., 2013), but here incorporated the constant, periodic and random environmental variation treatments.

6.1.1 Quantitative methods and statistical analysis

Life–history trait data on age- and size–at–maturity are presented in the text as full–sib female or treatment means with standard deviations at the beginning (week 0) and end (week 95) of the experiment (e.g. Plaistow et al., 2004). Statistical differences in daily Mayfield survival estimates between environmental and harvesting treatments were most appropriately tested using a generalised linear model with a quasi–Poisson error distribution. Significance of treatments was tested while correcting for the highly over-dispersed distribution using F tests (Crawley, 2007). The significance of environmental variation and harvesting treatments on the mean female phenotype and the age- and size–at–maturity of each family per treatment at the end of the study was assessed using MANOVA to jointly model log(age) and log(size) in Low–food conditions while controlling for population density in the life–history assay tubes by using tube covariates (weighted density, median density and total tube survival), see Cameron et al. (2013). Owing to the extra threshold harvest treatment in random variation treatments, a full model was first built without this one treatment to independently test for an environment*harvest interaction. Following this, and for predictions of treatment means, a separate MANOVA was built for each environmental variation treatment. Age- and size–at–maturity trait values were then plotted as model predicted means with associated standard errors of the model estimates.

To test for any link between low-food phenotypic change and changes in observed population growth, we estimated the mean and confidence intervals of the basic reproductive rate per treatment, R_0 ($R_0 = \exp((\ln (l_x{}^*m_x))/T_c$, where l_x is the chance of an individual surviving to age x, m_x is the number of offspring produced during age $x-1$ to x and T_c is the average generation time) (Stearns, 1992). R_0 was corrected by the average generation time due to the overlapping generations. For further details of this method, refer to supplementary material associated with Cameron et al. (2013). Average population growth rate (pgr = Nt + 1/Nt) was calculated from a smoother fitted across replicate population time series per treatment (observed population growth = change in total population size from week to week, over a 10-week window around assay time-points), and a Pearson's correlation test between the two estimates of population growth was undertaken. All analyses described above were performed in R (R3.1.0, 2014).

For each environmental variation treatment, genetic diversity in age-at-maturity in a low-food assay was apportioned using an analysis of molecular variation (AMOVA) approach into: (1) differences among individuals within replicate populations; (2) differences among replicate populations within time-points within harvesting regimes; (3) differences among time-points within harvesting treatment; and (4) differences among harvesting treatments across time-points (AMOVA, Arlequin Version 3.5, Excoffier and Lischer, 2010). The relative magnitude of differences can highlight the effects of deterministic and stochastic microevolution acting across the populations. It is expected that drift would cause significant differences to accumulate among replicates within time-points for any treatment, whereas selection would cause significant differences across time-points within a treatment or among the treatments themselves.

6.2. Results—Evolution of population dynamics in variable environments

All mite populations initially declined across all three environments and then recovered (Fig. 5.5). Before the recovery, the mean population growth rate of the populations was 0.980 (=2% decline per week), 0.978 and 0.980 at week 20 for the constant, periodic and random environments, respectively. During the recovery, the population growth had increased to 1.010 (=1% increase per week), 1.013 and 1.012, respectively, by week 60. At the start of the experiment, in low food and hence highly competitive conditions, soil

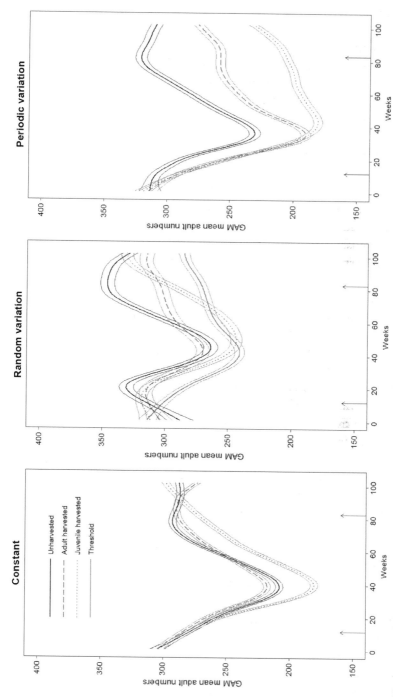

Figure 5.5 Adult population size (±95%CI) from generalized additive model (GAM) fits across a 5-week centred moving average of replicate weekly counts per treatment (6 d.f., minimum model across all environments). All other stage counts show a similar pattern of initially decreasing in abundance then increasing. Arrows at weeks 13 and 83 mark start and end of harvesting period, respectively.

mites took an average of 12.3 days to mature. By the end of the experiment, we observed a large reduction in the growth rate to maturity of the average mite family from all three environments, equating to a 35%, 76%, and 83% delay in age-at-maturity in the constant (16.6 ± 2.6 SD days), periodic (22.1 ± 3.6 SD days) and variable environments (21.6 ± 4.27 SD days), respectively. The observed increasing delays in developmental growth rate over the course of the experiment in resource poor conditions are positively correlated with increases in fecundity in adult mites (Cameron et al., 2013; Plaistow et al., 2006, 2007). This is suggestive that the delays in maturity are adaptive. There was no significant difference in daily survival rate between families from the three environments (Quasipoisson GLM:$F_{env} = 0.29_{2,123}$, $P > 0.7$). Consequently, while the earlier maturation phenotype we see in constant environments would have reduced fecundity compared to other environment phenotypes, this appears to be offset by increased overall survival to maturity. The question of interest, which separates our experiment from only demonstrating that the traits of mites change when they are placed in different laboratory environments, was to determine if the change in growth rates observed was caused by selection and if that selection led to the recovery of the populations after only eight generations.

The basic reproductive rates R_0 estimated from the common garden life-history data at weeks 0, 18, 37, 63 and 95 were highly correlated with the average of observed population growth rates estimated from replicated experimental time series (Pearson's $= 0.88$, $t_{2,13} = 4.81$, $P < 0.001$). Furthermore, there is no significant difference between the estimates of population growth from life-history data or the time series (e.g. R_0 vs, pgr, paired t-test, $P = 0.34$). Given that the phenotype data used to estimate R_0 (i.e. age- and size-at-maturity, survival to maturity, reproduction at maturity) are collected in similar competitive conditions to those in the population experiments but after three generations in a common garden environment, this is very strong evidence that we are observing evolved changes in mean life history that lead to changing population dynamics; a requirement for the demonstration of an eco-evolutionary feedback loop (Schoener, 2011). However, it does not prove that the phenotypic change observed is being caused by genetic evolution (e.g. Chevin et al., 2010). The AMOVA analysis on AFLP variation confirms that both genetic drift and selection are operating in concert to affect the levels and distribution of genetic variation in growth rates within the microcosm system (Fig. 5.6). All of the partitions explained a significant proportion of the variation observed (e.g. more than 5%) except for the difference among harvesting treatments within the

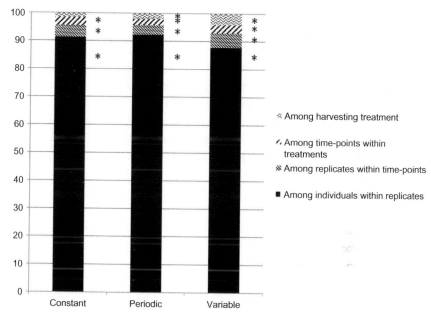

Figure 5.6 Analysis of molecular variance for 299 AFLP loci for (black) differences among individuals within replicate populations; (back hatching) differences among replicate populations within time-points; (forward hatching) differences among time-points within harvesting regimes; (waves) differences among harvesting regimes. * indicates statistical significance of treatment group at $P < 0.05$.

constant food environment. This need not reflect a lack of selection caused by harvesting acting on growth rates in constant environments, but that among individual variation is likely masking its importance in this treatment. This highlights that within each environmental variation treatment, genetic drift is acting to force populations into different evolutionary trajectories (given that replicate populations within harvesting treatments within time-points and within environments accumulated significant genetic differences). It also demonstrates that selection operates to generate differences in the growth rate to maturity across time-points, within harvesting regimes, in the different environment treatments as well as between environments across time-points.

6.3. Results—Life-history responses to harvesting in variable environments

We found a significant interaction between environmental variation and harvesting treatment on the age- and size-at-maturity (MANOVA:

age–at–maturity $F_{\text{env:har}} = 2.45_{4,123}$ $P < 0.05$; size–at–maturity $F_{\text{env:har}} = 3.15_{4,123}$ $P < 0.02$). To understand this interaction, and by controlling for stochastic differences in mite densities between life-history assay tubes, we standardised survival and density covariates to the mean values per environmental treatment and predicted the mean and variance of trait values from a MANOVA for each environment. In both constant and randomly variable environments, harvesting adults or juveniles led to a significant delay in maturation in comparison to unharvested controls (Fig. 5.7, left and centre panels). This contrasts with what was observed in periodic environments where harvesting juveniles reduced age-at-maturity in line with reducing

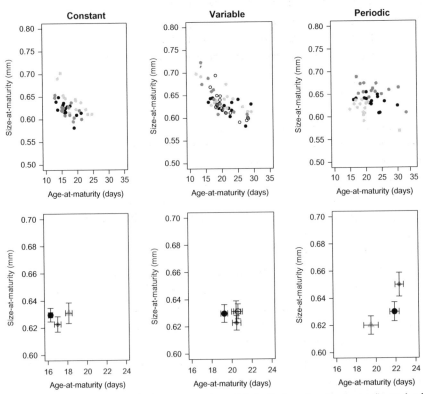

Figure 5.7 Mean age- and size-at-maturity of full-sib females (top panel), and of harvesting treatment means and twice standard error bars predicted from MANOVA when controlling for differences in tube densities (bottom panel). Panels represent constant (left panels), randomly variable (centre panels) and periodically variable resource environments (right panels). Colours represent juvenile (green (pale grey in the print version)), adult (red (dark grey in the print version)), threshold adult (orange (unfilled circle in the print version)) and un-harvested harvesting treatments (black).

risk of increased harvesting mortality (Fig. 5.7, right panel). In both constant and randomly variable environments, there was no significant effect of harvesting on size at maturation (constant: $F_{har} = 2.25_{2,28}$ $P > 0.1$; random: $F_{har} = 0.76_{3,40}$ $P > 0.5$), unlike the small but significant increase in size-at-maturity in adult harvested phenotypes from periodic environments originally described in Cameron et al. (2013). As we discussed in Sections 6.1–6.3, we detected a statistically significant effect of selection caused by harvesting on the variation in developmental growth rates in both random and periodically variable environments (Fig. 5.6). It is surprising that given the clear phenotypic differences found between unharvested and harvested constant environment populations at the end of the experiment, that the AFLP response was not more pronounced. However, selection was observed, and this assay method is a blunt tool given that we only have a snapshot of phenotype and genotype differences from a small number of individuals from two of six replicate populations at the F3 generation.

6.4. Discussion of evolution of life histories in response to environmental variation and harvesting

Life-history research increasingly focuses on understanding the links between environmental variation and population demography. Stochastic demography, which often uses a matrix-based approach, estimates optimum life histories that maximise fitness averaged over variable environments, when variable environments lead to variation in vital rates (Caswell, 2010; Haridas and Tuljapurkar, 2005; Trotter et al., 2013; Tuljapurkar et al., 2003, 2009). Not all such approaches have focussed or presented the same traits we have considered here, i.e. developmental growth. However, stochastic demographic approaches have shown that the generation time, measured variously as cohort generation time (T_c) or longevity, buffers against the negative effects of environmental variation on fitness (Morris et al., 2008; Tuljapurkar et al., 2009). Shertzer and Ellner present a dynamic energy budget approach that, while not strictly evolving *per se*, sought out optimum energy allocation strategies to growth, storage or reproduction that maximised R_0 in a genetic algorithm model of a rotifer population (Shertzer and Ellner, 2002). In the Shertzer and Ellner study, what is relevant is that environmental variation was experienced over the timescale of an individual's lifetime, as in soil mites (e.g. day-to-day variation instead of between-generation or inter-annual variation). Life-history strategies that delayed age to maturity were optimum in more variable environments and/or environments with periods of resource limitation (Shertzer and

Ellner, 2002). Tenhumberg and colleagues also focussed on stochastic variation in prey availability within a predator lifetime that led to a negative relationship between growth rate and mortality arising from the physiological constraints of 'digestion and gut capacities' in syrphids (Tenhumberg et al., 2000). The negative relationship led to increased fitness of those strategies that delayed growth rate to maturity in variable environments. Negative relationships between vital rates have been suggested to increase fitness in variable environments in other analytical approaches (Tuljapurkar et al., 2009). In *Caenorhabditis elegans*, mutants that aged slower were also found to have higher fitness in more stressful environments, including when food availability was variable. This is suggested to lead to altered allele frequencies in more heterogeneous environments in ecological time that feeds into evolutionary dynamics (Savory et al., 2014). All these predictions fit with our main result that strong competition and more variable food supply led to larger delays in maturity, which led to increased population growth rates. There is great consistency therefore, across a number of empirical and theoretical approaches that the evolution of slow life histories is likely in variable environments. However, the relative importance of the magnitude of environmental variability, its predictability or autocorrelation in the evolution of slow life histories is not yet clear and should be an interesting avenue of future research.

While our experiment was designed to investigate potential links between phenotypic change and population dynamics, it shows the potential for populations to recover from an extinction trajectory through evolution: evolutionary rescue (Bell and Gonzalez, 2009). Across all three of our environmental variation treatments, the initial trajectory of population growth is negative (i.e. an extinction trajectory), but becomes positive after evolution in response to laboratory conditions leads to delayed maturity and increased fecundity.

It is a key result that increased juvenile mortality can generate faster or slower life histories relative to controls depending on the temporal variability in the strength of resource competition. The constant and random environments produced more similar juvenile harvested mite life histories when compared to the periodic treatment. While the variation in food provision in the constant and random treatments was different (coefficient of variation (CV): 0 vs. 0.36), the resulting variation in mite abundance was more similar due to demographic noise in constant populations (Benton et al., 2002; Cameron, submitted)(CV_{adults}: 0.20 vs. 0.34; $CV_{juveniles}$: 0.46 vs. 0.50). In periodic environments, the variation of food provision, and therefore adult and juvenile mite abundance is much greater (CV = 0.86, 0.46 and 0.76, respectively). However, the greatest difference between constant, random and periodic variation is that periodicity is caused by highly autocorrelated resource provisioning. We predict that this is where the different life-history responses to harvesting arise, in the interaction between density-dependent demographic responses to mortality and evolutionary responses to more (periodic) or less (noisy-constant and random) predictable resource pulses between harvesting events. Such interactions could increase the positive relationship between age-at-maturity and fecundity if the increase in risk of harvesting mortality from delaying maturity was less than the potential gains to lifetime fitness from receiving a glut of resources just before maturation. Theoretical understanding of the interaction between intra-generation environmental noise and selective mortality at this temporal scale is currently lacking, largely due to the taxonomic bias in evolutionary demography studies towards long-lived mammals and birds.

What we have presented in Section 6 by describing ecological dynamics of a wild population adapting to a controlled laboratory environment provides a much higher level of resolution on the consequences of ecological and evolutionary interaction. We demonstrate how individuals maximise their lifetime fecundity in response to resource poor conditions, or high selective mortality and highlight how complex population dynamics can be maintained despite long-term erosion of genetic diversity caused by both stochastic and deterministic processes. The latter is difficult to reconcile with classical ideas of extinction debt in conservation population genetics (e.g. Fagan and Holmes, 2006), whereby positive feedback occurs between reduced population growth rate and loss of genetic diversity that leads to an inevitable extinction. Clearly, there is a need to address how evolutionary rescue can interrupt an ongoing extinction vortex and the limits to the recovery of populations in relation to extant and introduced genetic variation.

7. SUMMARY

The aim of this contribution was to explore the complexity of the route from individual phenotypic variation to population dynamics and back again in a model system: the eco-evolutionary loop. The mite model system has provided a rich series of experiments that have highlighted the level of information on individual life histories we require to make predictions about transient population dynamics following environmental perturbations is often considerable. The study of ecology has been described as the investigation of variation in space and time of the abundance and density of organisms (Begon et al., 2006), and while demography may be a main objective of ecology, it is clear from our work and others in this volume that the proposal that all evolutionary biologists should be demographers goes both ways (Metcalf and Pavard, 2007).

We have presented the study of three distinct pathways between environments, phenotypes and population dynamics: the role of current and historical environments on offspring phenotypes; the multigenerational effects of environmentally determined phenotypes on short-term population dynamics and finally the feedback between population abundance and resource availability to selection on phenotypes and evolution of population dynamics. In our diagram of eco-evolutionary interactions (Fig. 5.1), we have represented those pathways as independent routes. It is, however, clear from the context dependency of our results that the selection on life histories that determines population dynamics will very much depend on the interaction between historical (parental effects) and current environments (growth rate to developmental thresholds).

Through our demonstration that soil mite population trends are determined by their life histories, which evolve in response to density-dependent competition and predation (the eco-evolutionary loop), we have shown that in populations in which density-dependent competition is common, there is selection for individuals with life-history strategies that permit individuals to mature later in low-food conditions, but still retain the ability to mature early when conditions improve (Cameron et al., 2013). If this is evidence of eco-evolutionary dynamics selecting for increased phenotypic plasticity, it highlights the potential importance of the parental effects we previously found to shape reaction norms such that selection can act on novel phenotypes (e.g. Plaistow et al., 2006). Selection on more novel phenotypes would have the potential to allow more rapid feedbacks between natural selection and population dynamics. This is particularly relevant in light of the interest in rapid evolutionary responses to environmental change. Our current research in the mite model system is examining how variation in the population dynamic patterns created in different environments influences the evolution of offspring-provisioning strategies and epigenetic variation in gene expression during development and the effect that this has on later population dynamic patterns. This should lead to a less conceptual, and more mechanistic, understanding of eco-evolutionary population dynamics.

While we have identified much complexity, we have also shown when the role of environmentally determined phenotypic variation is less important in a population dynamics context (e.g. maternal effects when resources are low), but it was only through experimentation that we were able to say this. This is in some ways the most important conclusion of this review, that carefully planned experiments in well-studied systems are what is required to separate potential consequences of eco-evolutionary dynamics from those which are likely to have important consequences in natural populations.

REFERENCES

Abrams, P.A., Matsuda, H., 1997. Prey adaptation as a cause of predator-prey cycles. Evolution 51, 1742–1750.

Andersen, K.H., Brander, K., 2009a. Expected rate of fisheries-induced evolution is slow. Proc. Natl. Acad. Sci. U.S.A. 106, 11657–11660.

Andersen, K.H., Brander, K., 2009b. Reply to Kinnison et al.: effects of fishing on phenotypes. Proc. Natl. Acad. Sci. U.S.A. 106, E116.

Anderson, C.N.K., Hsieh, C.H., Sandin, S.A., Hewitt, R., Hollowed, A., Beddington, J., May, R.M., Sugihara, G., 2008. Why fishing magnifies fluctuations in fish abundance. Nature 452, 835–839.

Badyaev, A.V., Uller, T., 2009. Parental effects in ecology and evolution: mechanisms, processes and implications. Philos. Trans. R. Soc. Lond. B Biol. Sci. 364, 1169–1177.

Bassar, R.D., Marshall, M.C., Lopez-Sepulcre, A., Zandona, E., Auer, S.K., Travis, J., Pringle, C.M., Flecker, A.S., Thomas, S.A., Fraser, D.F., Reznick, D.N., 2010. Local adaptation in trinidadian guppies alters ecosystem processes. Proc. Natl. Acad. Sci. U.S.A. 107, 3616–3621.

Beckerman, A., Benton, T.G., Ranta, E., Kaitala, V., Lundberg, P., 2002. Population dynamic consequences of delayed life-history effects. Trends Ecol. Evol. 17, 263–269.

Beckerman, A.P., Benton, T.G., Lapsley, C.T., Koesters, N., 2003. Talkin' 'bout my generation: environmental variability and cohort effects. Am. Nat. 162, 754–767.

Beckerman, A.P., Benton, T.G., Lapsley, C.T., Koesters, N., 2006. How effective are maternal effects at having effects? Proc. R. Soc. B Biol. Sci. 273, 485–493.

Beckerman, A.P., De Roij, J., Dennis, S.R., Little, T.J., 2013. A shared mechanism of defense against predators and parasites: chitin regulation and its implications for life-history theory. Ecol. Evol. 3, 5119–5126.

Becks, L., Ellner, S.P., Jones, L.E., Hairston, N.G., 2012. The functional genomics of an eco-evolutionary feedback loop: linking gene expression, trait evolution, and community dynamics. Ecol. Lett. 15, 492–501.

Begon, M., Townsend, C.R., Harper, J.L., 2006. Ecology: From Individuals to Ecosystems. Blackwell Publishing, Oxford, UK.

Bell, G., Gonzalez, A., 2009. Evolutionary rescue can prevent extinction following environmental change. Ecol. Lett. 12, 942–948.

Benton, T.G., 2012. Individual variation and population dynamics: lessons from a simple system. Philos. Trans. R. Soc. B Biol. Sci. 367, 200–210.

Benton, T.G., Beckerman, A.P., 2005. Population dynamics in a noisy world: Lessons from a mite experimental system. In: Desharnais, R. (Ed.), Population Dynamics and Laboratory Ecology, In: Advances in Ecological Research, vol. 37, pp. 143–181.

Benton, T.G., Ranta, E., Kaitala, V., Beckerman, A.P., 2001. Maternal effects and the stability of population dynamics in noisy environments. J. Anim. Ecol. 70, 590–599.

Benton, T.G., Lapsley, C.T., Beckerman, A.P., 2002. The population response to environmental noise: population size, variance and correlation in an experimental system. J. Anim. Ecol. 71, 320–332.

Benton, T.G., Plaistow, S.J., Beckerman, A.P., Lapsley, C.T., Littlejohns, S., 2005. Changes in maternal investment in eggs can affect population dynamics. Proc. R. Soc. B Biol. Sci. 272, 1351–1356.

Benton, T.G., Plaistow, S.J., Coulson, T.N., 2006. Complex population dynamics and complex causation: devils, details and demography. Proc. R. Soc. B Biol. Sci. 273, 1173–1181.

Benton, T.G., St Clair, J.J.H., Plaistow, S.J., 2008. Maternal effects mediated by maternal age: from life histories to population dynamics. J. Anim. Ecol. 77, 1038–1046.

Biro, P.A., Post, J.R., 2008. Rapid depletion of genotypes with fast growth and bold personality traits from harvested fish populations. Proc. Natl. Acad. Sci. U.S.A. 105, 2919–2922.

Bonduriansky, R., Day, T., 2009. Nongenetic inheritance and its evolutionary implications. Annu. Rev. Ecol. Evol. Syst. 40, 103–125.

Bonenfant, C., Pelletier, F., Garel, M., Bergeron, P., 2009. Age-dependent relationship between horn growth and survival in wild sheep. J. Anim. Ecol. 78, 161–171.

Browman, H.I., Law, R., Marshall, C.T., 2008. The role of fisheries-induced evolution. Science 320, 47–50.

Bunnefeld, N., Baines, D., Newborn, D., Milner-Gulland, E.J., 2009. Factors affecting unintentional harvesting selectivity in a monomorphic species. J. Anim. Ecol. 78, 485–492.

Cameron, T.C., O'Sullivan, D., Reynolds, A., Piertney, S.B., Benton, T.G., 2013. Eco-evolutionary dynamics in response to selection on life-history. Ecol. Lett. 16, 754–763.

Carroll, S.P., Hendry, A.P., Reznick, D.N., Fox, C.W., 2007. Evolution on ecological timescales. Funct. Ecol. 21, 387–393.

Caswell, H., 2010. Life table response experiment analysis of the stochastic growth rate. J. Ecol. 98, 324–333.

Champagne, F.A., 2008. Epigenetic mechanisms and the transgenerational effects of maternal care. Front. Neuroendocrinol. 29, 386–397.

Charmantier, A., McCleery, R.H., Cole, L.R., Perrins, C., Kruuk, L.E.B., Sheldon, B.C., 2008. Adaptive phenotypic plasticity in response to climate change in a wild bird population. Science 320, 800–803.

Chevin, L.-M., Lande, R., Mace, G.M., 2010. Adaptation, plasticity, and extinction in a changing environment: towards a predictive theory. PLoS Biol. 8, e1000357.

Ciuti, S., Muhly, T.B., Paton, D.G., Mcdevitt, A.D., Musiani, M., Boyce, M.S., 2012. Human selection of elk behavioural traits in a landscape of fear. Proc. R. Soc. B Biol. Sci. 279, 4407–4416.

Coltman, D.W., O'Donoghue, P., Jorgenson, J.T., Hogg, J.T., Strobeck, C., Festa-Bianchet, M., 2003. Undesirable evolutionary consequences of trophy hunting. Nature 426, 655–658.

Core Team, R., 2014. R: A Language and Environment for Statistical Computing. R Foundation for Statistical Computing, Vienna, Austria. http://www.R-project.org/.

Coulson, T., Tuljapurkar, S., 2008. The dynamics of a quantitative trait in an age-structured population living in a variable environment. Am. Nat. 172, 599–612.

Coulson, T., Benton, T.G., Lundberg, P., Dall, S.R.X., Kendall, B.E., Gaillard, J.M., 2006. Estimating individual contributions to population growth: evolutionary fitness in ecological time. Proc. R. Soc. B Biol. Sci. 273, 547–555.

Coulson, T., Tuljapurkar, S., Childs, D.Z., 2010. Using evolutionary demography to link life history theory, quantitative genetics and population ecology. J. Anim. Ecol. 79, 1226–1240.

Crawley, M.J., 2007. The R Book. Wiley-Blackwell, England.

Darimont, C.T., Carlson, S.M., Kinnison, M.T., Paquet, P.C., Reimchen, T.E., Wilmers, C.C., 2009. Human predators outpace other agents of trait change in the wild. Proc. Natl. Acad. Sci. U.S.A. 106, 952–954.

Day, T., Rowe, L., 2002. Developmental thresholds and the evolution of reaction norms for age and size at life-history transitions. Am. Nat. 159, 338–350.

De Roos, A.M., Persson, L., Thieme, H.R., 2003. Emergent allee effects in top predators feeding on structured prey populations. Proc. R. Soc. B Biol. Sci. 270, 611–618.

Dewitt, T.J., Sih, A., Wilson, D.S., 1998. Costs and limits of phenotypic plasticity. Trends Ecol. Evol. 13, 77–81.

Ducatez, S., Baguette, M., Stevens, V.M., Legrand, D., Freville, H., 2012. Complex interactions between parental and maternal effects: parental experience and age at reproduction affect fecundity and offspring performance in a butterfly. Evolution 66, 3558–3569.

Ellner, S.P., Geber, M.A., Hairston, N.G., 2011. Does rapid evolution matter? Measuring the rate of contemporary evolution and its impacts on ecological dynamics. Ecol. Lett. 14, 603–614.

Ernande, B., Dieckmann, U., Heino, M., 2004. Adaptive changes in harvested populations: plasticity and evolution of age and size at maturation. Proc. R. Soc. Lond. B Biol. Sci. 271, 415–423.

Excoffier, L., Lischer, H.E.L., 2010. Arlequin suite ver 3.5: a new series of programs to perform population genetics analyses under linux and windows. Mol. Ecol. Resour. 10, 564–567.

Ezard, T.H.G., Cote, S.D., Pelletier, F., 2009. Eco-evolutionary dynamics: disentangling phenotypic, environmental and population fluctuations. Philos. Trans. R. Soc. B Biol. Sci. 364, 1491–1498.

Fagan, W.F., Holmes, E.E., 2006. Quantifying the extinction vortex. Ecol. Lett. 9, 51–60.

Friman, V.P., Jousset, A., Buckling, A., 2014. Rapid prey evolution can alter the structure of predator-prey communities. J. Evol. Biol. 27, 374–380.

Fryxell, J.M., Smith, I.M., Lynn, D.H., 2005. Evaluation of alternate harvesting strategies using experimental microcosms. Oikos 111, 143–149.

Fussmann, G.F., Ellner, S.P., Hairston, N.G., 2003. Evolution as a critical component of plankton dynamics. Proc. R. Soc. B Biol. Sci. 270, 1015–1022.

Gil, D., Graves, J., Hazon, N., Wells, A., 1999. Male attractiveness and differential testosterone investment in zebra finch eggs. Science 286, 126–128.

Ginzburg, L.R., 1998. Assuming reproduction to be a function of consumption raises doubts about some popular predator-prey models. J. Anim. Ecol. 67, 325–327.

Ginzburg, L.R., Taneyhill, D.E., 1994. Population-cycles of forest lepidoptera—a maternal effect hypothesis. J. Anim. Ecol. 63, 79–92.

Hairston, N.G., Lampert, W., Caceres, C.E., Holtmeier, C.L., Weider, L.J., Gaedke, U., Fischer, J.M., Fox, J.A., Post, D.M., 1999. Lake ecosystems—rapid evolution revealed by dormant eggs. Nature 401, 446.

Hairston, N.G., Holtmeier, C.L., Lampert, W., Weider, L.J., Post, D.M., Fischer, J.M., Caceres, C.E., Fox, J.A., Gaedke, U., 2001. Natural selection for grazer resistance to toxic cyanobacteria: evolution of phenotypic plasticity? Evolution 55, 2203–2214.

Hairston, N.G., Ellner, S.P., Geber, M.A., Yoshida, T., Fox, J.A., 2005. Rapid evolution and the convergence of ecological and evolutionary time. Ecol. Lett. 8, 1114–1127.

Hamilton, S.L., Caselle, J.E., Standish, J.D., Schroeder, D.M., Love, M.S., Rosales-Casian, J.A., Sosa-Nishizaki, O., 2007. Size-selective harvesting alters life histories of a temperate sex-changing fish. Ecol. Appl. 17, 2268–2280.

Haridas, C.V., Tuljapurkar, S., 2005. Elasticities in variable environments: properties and implications. Am. Nat. 166, 481–495.

Hasselquist, D., Nilsson, J.-A., 2009. Maternal transfer of antibodies in vertebrates: transgenerational effects on offspring immunity. Philos. Trans. R. Soc. B Biol. Sci. 364, 51–60.

Houck, M.A., Oconnor, B.M., 1991. Ecological and evolutionary significance of phoresy in the astigmata. Annu. Rev. Entomol. 36, 611–636.

Inchausti, P., Ginzburg, L.R., 1998. Small mammals cycles in northern europe: patterns and evidence for a maternal effect hypothesis. J. Anim. Ecol. 67, 180–194.

Isaksson, C., Uller, T., Andersson, S., 2006. Parental effects on carotenoid-based plumage coloration in nestling great tits, parus major. Behav. Ecol. Sociobiol. 60, 556–562.

Jacquin, L., Blottiere, L., Haussy, C., Perret, S., Gasparini, J., 2012. Prenatal and postnatal parental effects on immunity and growth in 'lactating' pigeons. Funct. Ecol. 26, 866–875.

Jorgensen, C., Enberg, K., Dunlop, E.S., Arlinghaus, R., Boukal, D.S., Brander, K., Ernande, B., Gaerdmark, A., Johnston, F., Matsumura, S., Pardoe, H., Raab, K., Silva, A., Vainikka, A., Dieckmann, U., Heino, M., Rijnsdorp, A.D., 2007. Ecology—managing evolving fish stocks. Science 318, 1247–1248.

Kinnison, M.T., Hairston, N.G., 2007. Eco-evolutionary conservation biology: contemporary evolution and the dynamics of persistence. Funct. Ecol. 21, 444–454.

Kinnison, M.T., Palkovacs, E.P., Darimont, C.T., Carlson, S.M., Paquet, P.C., Wilmers, C.C., 2009. Some cautionary notes on fisheries evolutionary impact assessments. Proc. Natl. Acad. Sci. U.S.A. 106, E115.

Kuparinen, A., Merila, J., 2007. Detecting and managing fisheries-induced evolution. Trends Ecol. Evol. 22, 652–659.

Kuparinen, A., Merila, J., 2008. The role of fisheries-induced evolution. Science 320, 47–48.

Lande, R., Saether, B.E., Engen, S., 1997. Threshold harvesting for sustainability of fluctuating resources. Ecology 78, 1341–1350.

Law, R., 1979. Optimal life histories under age-specific predation. Am. Nat. 114, 399–417.

Law, R., 2007. Fisheries-induced evolution: present status and future directions. Mar. Ecol. Prog. Ser. 335, 271–277.

Lenski, R.E., 1984. Coevolution of bacteria and phage—are there endless cycles of bacterial defenses and phage counterdefenses. J. Theor. Biol. 108, 319–325.

Marshall, C.T., Browman, H.I., 2007. Disentangling the causes of maturation trends in exploited fish populations. Mar. Ecol. Prog. Ser. 335, 249–251.

Metcalf, C.J.E., Pavard, S., 2007. Why evolutionary biologists should be demographers. Trends Ecol. Evol. 22, 205–212.

Meylan, S., Miles, D.B., Clobert, J., 2012. Hormonally mediated maternal effects, individual strategy and global change. Philos. Trans. R. Soc. B Biol. Sci. 367, 1647–1664.

Michod, R.E., 1979. Evolution of life histories in response to age-specific mortality factors. Am. Nat. 113, 531–550.

Milner, J.M., Nilsen, E.B., Andreassen, H.P., 2007. Demographic side effects of selective hunting in ungulates and carnivores. Conserv. Biol. 21, 36–47.

Morris, W.F., Pfister, C.A., Tuljapurkar, S., Haridas, C.V., Boggs, C.L., Boyce, M.S., Bruna, E.M., Church, D.R., Coulson, T., Doak, D.F., Forsyth, S., Gaillard, J.-M., Horvitz, C.C., Kalisz, S., Kendall, B.E., Knight, T.M., Lee, C.T., Menges, E.S., 2008. Longevity can buffer plant and animal populations against changing climatic variability. Ecology 89, 19–25.

Morrissey, M.B., Walling, C.A., Wilson, A.J., Pemberton, J.M., Clutton-Brock, T.H., Kruuk, L.E.B., 2012. Genetic analysis of life-history constraint and evolution in a wild ungulate population. Am. Nat. 179, E97–E114.

Nussey, D.H., Postma, E., Gienapp, P., Visser, M.E., 2005. Selection on heritable phenotypic plasticity in a wild bird population. Science 310, 304–306.

Olsen, E.M., Heino, M., Lilly, G.R., Morgan, M.J., Brattey, J., Ernande, B., Dieckmann, U., 2004. Maturation trends indicative of rapid evolution preceded the collapse of northern cod. Nature 428, 932–935.

Olsen, E.M., Carlson, S.M., Gjosaeter, J., Stenseth, N.C., 2009. Nine decades of decreasing phenotypic variability in atlantic cod. Ecol. Lett. 12, 622–631.

Olsen, A.S., Sarras Jr., M.P., Leontovich, A., Intine, R.V., 2012. Heritable transmission of diabetic metabolic memory in zebrafish correlates with DNA hypomethylation and aberrant gene expression. Diabetes 61, 485–491.

Ozgul, A., Tuljapurkar, S., Benton, T.G., Pemberton, J.M., Clutton-Brock, T.H., Coulson, T., 2009. The dynamics of phenotypic change and the shrinking sheep of St. Kilda. Science 325, 464–467.

Ozgul, A., Childs, D.Z., Oli, M.K., Armitage, K.B., Blumstein, D.T., Olson, L.E., Tuljapurkar, S., Coulson, T., 2010. Coupled dynamics of body mass and population growth in response to environmental change. Nature 466, 482–485.

Ozgul, A., Coulson, T., Reynolds, A., Cameron, T.C., Benton, T.G., 2012. Population responses to perturbations: the importance of trait-based analysis illustrated through a microcosm experiment. Am. Nat. 179, 582–594.

Pelletier, F., Clutton-Brock, T., Pemberton, J., Tuljapurkar, S., Coulson, T., 2007. The evolutionary demography of ecological change: linking trait variation and population growth. Science 315, 1571–1574.

Pelletier, F., Garant, D., Hendry, A.P., 2009. Eco-evolutionary dynamics. Philos. Trans. R. Soc. B Biol. Sci. 364, 1483–1489.

Perrin, M.C., Brown, A.S., Malaspina, D., 2007. Aberrant epigenetic regulation could explain the relationship of paternal age to schizophrenia. Schizophr. Bull. 33, 1270–1273.

Persson, L., Amundsen, P.A., De Roos, A.M., Klemetsen, A., Knudsen, R., Primicerio, R., 2007. Culling prey promotes predator recovery—alternative states in a whole-lake experiment. Science 316, 1743–1746.

Pimentel, D., 1961. Animal population regulation by genetic feedback mechanism. Am. Nat. 95, 65–79.

Pimentel, D., Stone, F.A., 1968. Evolution and population ecology of parasite-host systems. Can. Entomol. 100, 655–662.

Pimentel, D., Levin, S.A., Olson, D., 1978. Coevolution and stability of exploiter-victim systems. Am. Nat. 112, 119–125.

Pinder, M., 2009. Interactive Effects of Maternal and Paternal Environments on Maternal Investment in Offspring (UGrad. thesis). University of Leeds, Leeds, UK.

Plaistow, S.J., Benton, T.G., 2009. The influence of context-dependent maternal effects on population dynamics: an experimental test. Philos. Trans. R. Soc. B Biol. Sci. 364, 1049–1058.

Plaistow, S.J., Lapsley, C.T., Beckerman, A.P., Benton, T.G., 2004. Age and size at maturity: sex, environmental variability and developmental thresholds. Proc. R. Soc. Lond. B Biol. Sci. 271, 919–924.

Plaistow, S.J., Lapsley, C.T., Benton, T.G., 2006. Context-dependent intergenerational effects: the interaction between past and present environments and its effect on population dynamics. Am. Nat. 167, 206–215.

Plaistow, S.J., St Clair, J.J.H., Grant, J., Benton, T.G., 2007. How to put all your eggs in one basket: empirical patterns of offspring provisioning throughout a mother's lifetime. Am. Nat. 170, 520–529.

Post, D.M., Palkovacs, E.P., 2009. Eco-evolutionary feedbacks in community and ecosystem ecology: interactions between the ecological theatre and the evolutionary play. Philos. Trans. R. Soc. B Biol. Sci. 364, 1629–1640.

Qvarnstrom, A., Price, T.D., 2001. Maternal effects, paternal effects and sexual selection. Trends Ecol. Evol. 16, 95–100.

Rando, O.J., 2012. Daddy issues: paternal effects on phenotype. Cell 151, 702–708.

Rasanen, K., Kruuk, L.E.B., 2007. Maternal effects and evolution at ecological time-scales. Funct. Ecol. 21, 408–421.

Reznick, D., 1982. The impact of predation on life-history evolution in trinidadian guppies—genetic-basis of observed life-history patterns. Evolution 36, 1236–1250.

Reznick, D.N., Butler, M.J., Rodd, F.H., Ross, P., 1996. Life-history evolution in guppies (poecilia reticulata).6. Differential mortality as a mechanism for natural selection. Evolution 50, 1651–1660.

Roff, D.A., 2002. Life History Evolution. Sinauer, Sunderland, MA.

Rossiter, M., 1994. Maternal effects hypothesis of herbivore outbreak. Bioscience 44, 752–763.

Roth, O., Klein, V., Beemelmanns, A., Scharsack, J.P., Reusch, T.B.H., 2012. Male pregnancy and biparental immune priming. Am. Nat. 180, 802–814.

Savory, F.R., Benton, T.G., Varun, V., Hope, I.A., Sait, S.M., 2014. Stressful environments can indirectly select for increased longevity. Ecol. Evol. 4, 1176–1185.

Schoener, T.W., 2011. The newest synthesis: understanding the interplay of evolutionary and ecological dynamics. Science 331, 426–429.

Shertzer, K.W., Ellner, S.P., 2002. State-dependent energy allocation in variable environments: life history evolution of a rotifer. Ecology 83, 2181.

Stearns, S.C., 1992. The Evolution of Life Histories. Oxford, New York.

Stenson, J.A.E., 1981. The role of predation in the evolution of morphology, behavior and life-history of 2 species of chaoborus. Oikos 37, 323–327.

Stockwell, C.A., Hendry, A.P., Kinnison, M.T., 2003. Contemporary evolution meets conservation biology. Trends Ecol. Evol. 18, 94–101.

Strauss, S.Y., Lau, J.A., Schoener, T.W., Tiffin, P., 2008. Evolution in ecological field experiments: implications for effect size. Ecol. Lett. 11, 199–207.

Tenhumberg, B., Tyre, A.J., Roitberg, B., 2000. Stochastic variation in food availability influences weight and age at maturity. J. Theor. Biol. 202, 257–272.

Trotter, M.V., Krishna-Kumar, S., Tuljapurkar, S., 2013. Beyond the mean: sensitivities of the variance of population growth. Methods Ecol. Evol. 4, 290–298.

Tuljapurkar, S., Horvitz, C.C., Pascarella, J.B., 2003. The many growth rates and elasticities of populations in random environments. Am. Nat. 162, 489–502.

Tuljapurkar, S., Gaillard, J.-M., Coulson, T., 2009. From stochastic environments to life histories and back. Philos. Trans. R. Soc. B Biol. Sci. 364, 1499–1509.

Uller, T., 2008. Developmental plasticity and the evolution of parental effects. Trends Ecol. Evol. 23, 432–438.

Van Leeuwen, A., De Roos, A.M., Persson, L., 2008. How cod shapes its world. J. Sea Res. 60, 89–104.

Van Wijk, S.J., Taylor, M.I., Creer, S., Dreyer, C., Rodrigues, F.M., Ramnarine, I.W., Van Oosterhout, C., Carvalho, G.R., 2013. Experimental harvesting of fish populations drives genetically based shifts in body size and maturation. Front. Ecol. Environ. 11, 181–187.

Via, S., Gomulkiewicz, R., Dejong, G., Scheiner, S.M., Schlichting, C.D., Vantienderen, P.H., 1995. Adaptive phenotypic plasticity—consensus and controversy. Trends Ecol. Evol. 10, 212–217.

Walsh, M.R., Delong, J.P., Hanley, T.C., Post, D.M., 2012. A cascade of evolutionary change alters consumer-resource dynamics and ecosystem function. Proc. R. Soc. B Biol. Sci. 279, 3184–3192.

Wilcox, D.L., Maccluer, J.W., 1979. Coevolution in predator-prey systems—saturation kinetic-model. Am. Nat. 113, 163–183.

Yoshida, T., Jones, L.E., Ellner, S.P., Fussmann, G.F., Hairston, N.G., 2003. Rapid evolution drives ecological dynamics in a predator-prey system. Nature 424, 303–306.

Yoshida, T., Ellner, S.P., Jones, L.E., Bohannan, B.J.M., Lenski, R.E., Hairston, N.G., 2007. Cryptic population dynamics: rapid evolution masks trophic interactions. PLoS Biol. 5, 1868–1879.

> CHAPTER SIX

Individual Trait Variation and Diversity in Food Webs

Carlos J. Melián*,†,1,2, Francisco Baldó‡,§,1, Blake Matthews¶,
César Vilas‖,#, Enrique González-Ortegón§, Pilar Drake§,
Richard J. Williams**

*Department of Fish Ecology and Evolution, Center for Ecology, Evolution and Biogeochemistry, Swiss
Federal Institute of Aquatic Science and Technology, Switzerland
†National Center for Ecological Analysis and Synthesis, University of California, Santa Barbara, California,
USA
‡Instituto Español de Oceanografía, Centro Oceanográfico de Cádiz, Cádiz, Spain
§Instituto de Ciencias Marinas de Andalucía (CSIR), Cádiz, Spain
¶Department of Aquatic Ecology, Swiss Federal Institute of Aquatic Science and Technology, Switzerland
‖IFAPA, Centro El Toruño, El Puerto de Santa María, Cádiz, Spain
#Marine Science Institute, University of California, Santa Barbara, California, USA
**Microsoft Research Ltd., Cambridge, United Kingdom
1Shared first authors.
2Corresponding author: e-mail address: carlos.melian@eawag.ch

Contents

Advances in Ecological Research, Volume 50
ISSN 0065-2504
http://dx.doi.org/10.1016/B978-0-12-801374-8.00006-2

Abstract

In recent years, there has been a renewed interest in the ecological consequences of individual trait variation within populations. Given that individual variability arises from evolutionary dynamics, to fully understand eco-evolutionary feedback loops, we need to pay special attention to how standing trait variability affects ecological dynamics. There is mounting empirical evidence that intra-specific phenotypic variation can exceed species-level means, but theoretical models of multi-trophic species coexistence typically neglect individual-level trait variability. What is needed are multispecies datasets that are resolved at the individual level that can be used to discriminate among alternative models of resource selection and species coexistence in food webs. Here, using one the largest individual-based datasets of a food web compiled to date, along with an individual trait-based stochastic model that incorporates Approximate Bayesian computation methods, we document intra-population variation in the strength of prey selection by different classes or predator phenotypes which could potentially alter the diversity and coexistence patterns of food webs. In particular, we found that strongly connected individual predators preferentially consumed common prey, whereas weakly connected predators preferentially selected rare prey. Such patterns suggest that food web diversity may be governed by the distribution of predator connectivity and individual trait variation in prey selection. We discuss the consequences of intra-specific variation in prey selection to assess fitness differences among predator classes (or phenotypes) and track longer term food web patterns of coexistence accounting for several phenotypes within each prey and predator species.

1. INTRODUCTION

Over the past century, ecologists have proposed numerous theories to explain why natural communities have more species than predicted by theoretical models (Cohen et al., 1990a; Hutchinson, 1959; MacArthur, 1955; May, 1973). In recent years, there has been significant progress towards understanding the structure, dynamics, and coexistence of ecological networks (Brose et al., 2006; Gross et al., 2009; Rooney et al., 2006), yet very little is known about the connection between the drivers of intra–specific variation in resource use observed in the empirical data (Bolnick et al., 2002; Cianciaruso et al., 2009; Post et al., 2008) with the observed patterns of diversity in multi–trophic ecosystems (Abrams, 2010; Bolnick et al., 2011; Chesson, 1978; Ings et al., 2009; May, 2006; Murdoch, 1969; Pachepsky et al., 2007; Stouffer, 2010; Svanbäck et al., 2008; Violle et al., 2012). Competition theory predicts that stable coexistence occurs because of a balance between differences in species niches and competitive abilities (Chesson, 1984). However, intra–specific variation in resource use can exceed the differences in species-level averages within ecological communities, suggesting that individual-level variation might strongly affect community dynamics (Bolnick et al., 2011) and the likelihood of species coexistence (Bolnick et al., 2002; Clark, 2010; Cohen et al., 2005; Lloyd-Smith, 2005; Roughgarden, 1972). Variation in prey selection by predators within natural populations may either promote or limit the diversity of food webs, depending on the number and nature of interactions between individual predators and the abundance of prey populations. Thus, the number of prey an individual predator consumes can be used as a proxy to quantify its effect on prey abundance. On the one hand, where individuals with high feeding rates or fast prey finders (i.e. strongly connected predators) are common and preferentially target rare prey, then we may expect predators to cause declines in prey population size and eventually multiple extinctions in the prey community, increasing the probability of reduction in food web stability (Chesson, 1984; Holt, 1977; Murdoch, 1969). On the other hand, if most individual predators within a population select preferentially common prey, the rare prey species advantage analogous to a Janzen-Connell effect (Janzen, 1970), then most prey populations may coexist in the food web. Thus, the interaction between predator connectivity and prey selection at the individual level can have implications for understanding coexistence of prey and predator species in food webs. There is evidence from both theoretical and empirical work that intra-population variation in prey selection by predators can involve interactions between ecological and evolutionary dynamics. For example, optimal foraging theory with adaptive foragers have shown switching from rare to common prey as prey abundance increase (Beckerman et al., 2010; Kondoh, 2003; Stephens and Krebs, 1986). However, there can be substantial individual variation among predators in the propensity to switch among prey resources (Bolnick et al., 2011) with some individuals switching quickly and others delaying the switch depending on the ecological context (Carnicer et al., 2009). These changes provide intraspecific variation that might include behavioural syndromes associated with individual personalities and may show consistent foraging or mating strategies regardless of the abundance of resources in the environment (Wolf and Weissing, 2012). From the empirical side, individual predator connectivity (i.e. number of interactions) can be estimated from analyzing gut contents, where the number of prey items can give some indicator of an individual's rate of consumption (e.g., fast vs. slow prey finders) and hence degree of connectivity (e.g., strongly vs. weakly connected) to the prey community. Individual connectivity thus could therefore be a signature of different feeding rates driven by intra-population trait variation, plasticity, or even behavioural syndromes (Biro and Stamps, 2010). For a given level of connectivity, the individual prey selection (i.e., the nature of interactions) can be estimated from the difference between a predator's gut contents and independent measures of the empirical abundance distribution of prey in the environment. Previous work has documented extensive individual specialization within predator populations, but the consequences for food web diversity are still unclear (Bolnick et al., 2007; Polis, 1991; Polis and Winemiller, 1996; Tinker et al., 2012; Winemiller, 1990; Winemiller and Layman, 2005). This is partly because obtaining empirical estimations of predator's gut contents and independent measures of empirical abundance of prey species is difficult in the field and so little is known about intra-population variation in the strength of prey selection. Until now, the lack of high-resolution individual-based data with thousands of individual diets and independent measures of resource abundance, has made it difficult to test the role of intra-specific variation in the distribution of predator connectivity and resource selection on species coexistence in large food webs (Dunne, 2006; Polis, 1991). Here, we propose a method to test whether individual predators with different abundance of prey in their gut differ in their propensity to select rare versus common prey in a species rich food web. We use individual trait-based models and propose predator learning behaviour as a trait (Giraldeau and Caraco, 2000; West-Eberhard, 2003) that may explain the connection between individual rate of consumption (e.g., fast vs. slow prey finders), the degree of connectivity (e.g., strongly vs. weakly connected), and prey selectivity. We take a two-step modelling approach to test for intra-population and sampling variation in the strength of prey selection (Fig. 6.1). First, we compare whether a model with or without predator learning better explains the observed distribution of predator connectivity (Fig. 6.1A). We use learning as our trait that serves as a proxy to quantify the strength of past successful feeding encounters in determining individual diets. We then estimate the speed of learning (α) for those occasions where the learning model is the most likely to explain the observed variance in predator connectivity at the individual level (when $\alpha > 0$). Second, to investigate the mechanisms underlying individual variation in the strength of prey selection we used the predicted speed of learning and fit a model of prey selection to test how variation in the number of prey items per individual predator was related to an independent measurement of the observed abundance distribution of resources in the environment (Fig. 6.1B). Overall, the modelling approach uses two-independent sources

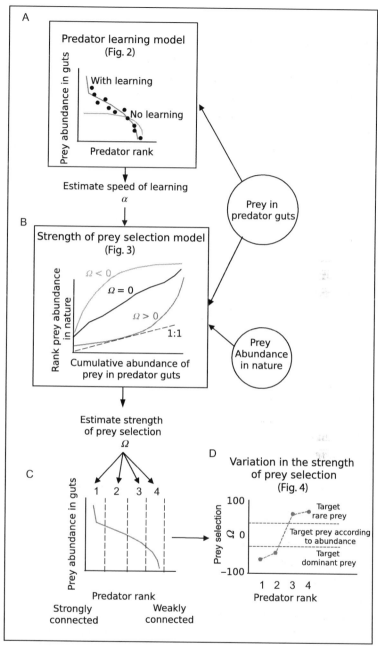

Figure 6.1 *Schematic representation of the approach.* (A) Empirical (black dots) and predicted distribution of prey abundance in predator guts (i.e. predator connectivity) with (red (dark grey in the print version)) or without (orange (light grey in the print version)) predator learning. *x*-Axis and *y*-axis represent the predator rank from strongly to weakly connected and prey abundance in guts, respectively. (B) Rank abundance in nature of prey species *i* (*y*-axis) and predicted cumulative abundance of prey *i* in predator guts

(Continued)

of data, namely the gut contents of predators and abundance of prey in the environment, and estimates two parameters of ecological relevance, namely the rate of predator learning (α) and the strength of prey selection (Ω) (Fig. 6.1).

We tested our method using a community of fish predators from the Guadalquivir estuary community in southwest Spain. We analyzed the stomach contents of 5725 individual fish predators (10 predator species) and identified 88101 prey items (43 taxa). The fish were collected from four daily net hauls at two sites (either oligohaline or polyhaline) once a month for a year ($N=95$ samplings, one missed sampling, Fig. 6.A1). Hence, the data have sufficient intraday, monthly, and seasonal resolution to make inferences at the individual, species, and food web levels simultaneously (Fig. 6.A2).

This combination of temporal and individual resolution for a larger number of species is not available from previous data sets, which are either focused on a few thousands individuals in a few species or on a few individuals per species in food webs with a large number of species (Dunne, 2006; Polis, 1991; Woodward and Warren, 2007). To simplify the fitting of the prey selection model to the high-resolution food web dataset, we divided predator individuals into four connectivity classes based on the number of prey items per individual predator, ranging from just a few prey items (slow prey finders) to a large number of prey items per predator (fast prey finders, Fig. 6.1C). This model simplification allowed us to test, for example, whether individuals within different classes of predator connectivity preferred either rare or dominant prey items (Fig. 6.1D). We followed this procedure for three different levels of aggregation, (i) for each sampling date, (ii) for each predator species pooled across multiple sampling dates with similar environments (Appendix), and (iii) for all the data pooled together. Overall, this approach allows us to combine high-resolution individual trait-level interaction data with resource abundance data in order to reveal the most likely mechanisms underlying predator preference to select the

Figure 6.1—Cont'd (x-axis) in the three scenarios explored: (1) Predator selecting preferentially rare prey (positive density-dependent prey selection, red (dark grey in the print version) with $\Omega > 0$); (2) no preference (neutral density-dependent prey selection, black with $\Omega = 0$), and (3) selecting preferentially common prey (negative density-dependent prey selection, blue (light grey in the print version) with $\Omega < 0$). Dotted black line represents the identity line to show the different trend between the positive and negative density-dependent prey selection scenarios. (C) Four classes of individual predators ranging from strongly connected (1) or fast prey finders to weakly connected (4) or slow prey finders. (D) Strongly to weakly connected predators targeting dominant ($\Omega < 0$) and rare ($\Omega > 0$) prey, respectively.

common or rare prey that are essential for the maintenance of biodiversity in multi-trophic ecosystems. Overall, we find strong evidence for intra-population variation in the strength of prey selection that could influence food web diversity. In particular, our results suggest that the weakly connected individual predators that favour rare prey have the most pronounced negative effects on food web diversity.

2. MATERIAL AND METHODS

2.1. Stochastic individual trait-based predator–prey model

Following the schematic in Fig. 6.1, we present an approach that models individual-level variation in the predator connectivity (i.e. the number of prey items) and the strength in prey selection. Our model is a stochastic, predator–prey model with $N_1, N_2, \ldots, N_S$ the prey population abundance of species-n, and $w_1, w_2, w_3, \ldots, w_T$ the number of prey items in the gut of individual predator-j (Table 6.1). Prey population abundance is sorted from the most common, N_1, to the rarest species, N_S, with S the observed prey species number and T the number of individual predators sampled. In the next section, we expand the model by considering learning as a trait to test whether the model with learning behaviour is a plausible mechanism to connect the degree of connectivity (e.g., strongly connected and fast prey finders vs. weakly connected and slow prey finders) with prey selectivity.

2.1.1 Speed of learning and the distribution of predator connectivity

We model variation in predator connectivity assuming identical learning abilities among individual predators. We model this learning ability using a single parameter, α, which we refer to as the learning rate. To investigate how learning rate affects the distribution of predator connectivity, we consider that the probability to find a prey item is driven by an individual prior experience in catching a prey. The probability to find a prey item is then given by the number of items in the stomach content of each individual predator j, w_j, and the speed of learning, α. For each time step, an individual predator j at time t has probability of catching a prey given by

$$\mathcal{P}_{jt} = \frac{\dfrac{\mathcal{K}}{1 + \mathcal{K}e^{-\alpha w_{j_{t-1}}}}}{\sum_{i=1}^{T} \dfrac{\mathcal{K}}{1 + \mathcal{K}e^{-\alpha w_{i_{t-1}}}}} \tag{6.1}$$

Table 6.1 Glossary of mathematical notation, dimensions, and values used

Variable or parameter [dimension]	Definition	Value
N_n [inds]	Abundance of prey species n	State variable
N [inds]	Abundance all prey species	Empirical value
S [species]	Number prey species	Empirical value
T [inds]	Number of individual predators sampled	Empirical value
ω_{j_t} [prey items]	Number of prey items in stomach content of predator j at time t	State variable
α [dimensionless]	Speed of learning	$[10^{-5}, 10^{-1}]$
$\mathcal{K}$ [dimensionless]	Learning capacity	$[10^2, 10^3]$
τ_c [inds]	Number of individuals in class c obtained by simulations	$[0, T]$
$\mathcal{P}_{ct}$ [dimensionless]	Probability class c catch a prey item at time t	$[0, 1]$
p_{cn} [net energy (time)$^{-1}$]	Profitability of prey species n for individuals in class c	$U\,[1, 101]$
$\mathcal{P}_{cn}$ [dimensionless]	Probability class c eat prey species n with profitability p_{cn}	$[0, 1]$
Ω_c [net energy (time)$^{-1}$]	Prey selection strength individuals in class c	$U\,[-10^2, 10^2]$
Ω_{c-} [net energy (time)$^{-1}$]	Negative prey selection strength individuals in class c	$U\,[-10^2, <-30]$
Ω_{co} [net energy (time)$^{-1}$]	Neutral prey selection strength individuals in class c	$U\,[-30, 30]$
Ω_{c+} [net energy (time)$^{-1}$]	Positive prey selection strength individuals in class c	$U\,[>30, 100]$
μ [ind (time)$^{-1}$]	Prey mortality rate	1
λ [ind (time)$^{-1}$]	Prey birth rate	1

where $\mathcal{K}$ defines the learning capacity for all individual predators, α is the speed of learning equal for all individual predators regardless their number of items, w_{jt-1} is the number of prey items in the stomach content of predator j at time $t-1$, and T is the number of individual predators. A larger value of a implies a higher finding rate of prey (Giraldeau and Caraco,

2000). Note the probability to find a prey item is a function of predator density and what other predators ate in the previous time step. Thus, we assume that with more number of individual predators in a given environment or volume the lower the probability to catch a prey for each of these predators. If there are huge numbers, density-dependent probability can collapse for each individual predator and some individuals predators may starve.

In summary, the number of prey items in the stomach content of each individual j, w_j, and the speed of learning, α, capture the probability to find a prey. We remark that when $\alpha = 0$, that is, when learning is not present, there is no effect of previous successful encounters to find new prey items. We remark that the model without the learning trait may represent a prey population dynamics with predators that are represented as individuals instead of species (Cohen et al., 1990b; Wilson et al., 2003). In the scenario without learning, it is expected that the variance in the number of prey items among individuals is lower than in the model with learning, leading to a homogeneous number of prey items per individual predator. The model with and without learning can be used to compare the expected distribution of predator connectivity with the empirical observations (Fig. 6.1A).

2.1.2 Profitability and the strength of prey selection for individual predators

In order to predict whether strongly (i.e. fast prey finders) or weakly connected predators (i.e. slow prey finders) tend to favour rare versus dominant prey items, respectively, we further expand the learning model to individual predators by considering that the profitability of a given prey varies among individual predators. We model this function as an individual-level trait, and use it to estimate variation in prey selection for a given speed of learning (as estimated above) and an observed prey abundance distribution. By explicitly including individual-level variation in predator profitability functions along with the distribution of prey abundance, we can estimate the strength of the individual predator preference for rare or common prey.

We modelled the profitability function based on the idea that the ratio of energy gained for a given handling time is an essential variable to distinguish random encounters from active searching of common or rare prey by each individual predator. Short handling times have a marked effect on profitability of prey (Catania and Remple, 2005), and variation in each individual predator in handling time (capture, killing, eating, or digesting) can have large consequences for profitability of foraging on a particular prey. We take into account variation in handling time in each individual predator by defining the profitability of prey species n for individual predator j as a function of the ratio of energy gained to handling time, $p_{jn} = (E/h)_{jn}$. This profitability for individual j can be defined as a probability as (Stephens and Krebs, 1986)

$$\mathcal{P}_{jn} = \frac{p_{jn}}{\sum_{i=1}^{S} p_{jn}}, \qquad (6.2)$$

where S is the number of prey species. Because we do not have empirically derived estimates of energy gained and handling time for the sampled prey species, in the simulations we provide each predator with a vector of profitabilities that includes all prey species and this vector of profitabilities is decoupled from abundance. In other words, an individual predator may have a low number of prey items (i.e., slow finder) but this individual may have a set of traits that favour high profitability values for rare prey. The same situation may occur for individual predators with a large number of items (i.e., fast prey finder), who may spend a long time capturing or killing a rare prey species n, while an individual with a low number of items can find the same rare prey highly profitable. In summary, we relax the profitability ranking and decouple it from abundance of prey to allow for variation in handling versus searching time or when digestive abilities vary greatly among individual predators.

We define the strength in prey selection for predator j, by Ω_j, as the difference in profitability for foraging on the rarest, N_S, and the most common prey, N_1. The following three scenarios are possible for each individual predator j

$$\Omega_j \begin{cases} \Omega_{j-}(<0) = p_{jS} - p_{j1} \\ \Omega_{jo}(=0) = p_{jS} - p_{j1} \\ \Omega_{j+}(>0) = p_{jS} - p_{j1}, \end{cases} \qquad (6.3)$$

where Ω_{j-}, Ω_{jo}, and Ω_{j+} represent the negative, neutral, and positive density-dependent prey selection by individual predator j, respectively. The strength of prey selection will be a negative number when predator j choose preferentially common prey (i.e., negative density-dependent prey selection or the fitness of prey population increases as it becomes less common, thus $p_{ji} > p_{jS}$). In the case of no preference, profitability is equal across all the prey species and the individual predator chooses prey according to their abundance. In this scenario, the value will be zero or close to zero (i.e., neutral or nearly-neutral prey selection, respectively, with $p_{j1} \approx p_{jS}$). The strength of prey selection will be a positive number when predator j

preferentially selects rare prey (i.e., the advantage of the common prey satisfying $p_{j1} < p_{jS}$). In this scenario, the cumulative abundance of rare prey in predator guts will grow faster than in the neutral or the negative density-dependent prey selection scenario (Fig. 6.1B). The vector of profitabilities is included in the simulations as a monotonic function with decaying (negative density-dependent prey selection), equal (neutral), or increasing (positive) values from the most common to the rarest species.

2.1.3 Profitability and the strength of prey selection for connectivity classes

At this point, the model is still too complex to recover predator estimates of the strength of prey selection given the food web and abundance data. In order to compare food web and species abundance data with model predictions, we have to divide the predator population into four classes (c) of varying connectivity (Fig. 6.1C), which are (1) strongly connected predators (or fast prey finders); (2) medium-strongly connected predators; (3) medium connected predators, and (4) weakly connected individuals (or slow prey finders).

Equation (6.1) for each individual predator j can be now written for each class, such that, an individual in class c at time t has probability of catching a prey given by

$$\mathcal{P}_{ct} = \frac{\sum_{i=1}^{T_c} \dfrac{\mathcal{K}}{1 + \mathcal{K}e^{-aw_{t-1}}}}{\sum_{i=1}^{T} \dfrac{\mathcal{K}}{1 + \mathcal{K}e^{-aw_{t-1}}}}, \tag{6.4}$$

where T_c is the number of individuals in class c at time t. The profitability of prey species n for an individual predator in class c is now given by $p_{cn} = (E/h)_{cn}$, and the profitability for class c can be written in probabilistic terms as

$$\mathcal{P}_{cn} = \frac{p_{cn}}{\sum_{i=1}^{S} p_{cn}}. \tag{6.5}$$

The strength of prey selection for each class, Ω_c (i.e., Strongly (Ω_1), medium-strongly (Ω_2), medium (Ω_3), and weakly (Ω_4) connected individuals) now follows these scenarios

$$\Omega_c \begin{cases} \Omega_{c-}(<0)=p_{cS}-p_{c1} \\ \Omega_{co}(=0)=p_{cS}-p_{c1} \\ \Omega_{c+}(>0)=p_{cS}-p_{c1}, \end{cases} \tag{6.6}$$

where Ω_{c-}, Ω_{co}, and Ω_{c+} represent the negative, neutral, and positive density-dependent prey selection values for each class c, respectively. We can now study the strength of prey selection in four classes of predators by using the speed of learning, α, and the strength of prey selection for each class, Ω_c.

2.1.4 Dynamics of prey populations driven by predators

Prey dynamics were modelled as a function of predator events. Thus for each predation event a prey population decreases. Suppose the population size of prey species n at some time t is N_n. What will it be after a small increment in δt in time? Call births and deaths "events". Then the population size at the later time $t+\delta t$ depends on which events occur during the small interval δt (Yodzis, 1989). For δt sufficiently small, an individual of prey species n at time t dies by an individual predator in connectivity class c with probability

$$D_{N_n}(t+\delta t)=\mu \mathcal{P}_{ct}\mathcal{P}_{cn}\left(\frac{N_n}{N}\right)\delta t, \tag{6.7}$$

where $\mathcal{P}_{ct}$ is the probability of catching a prey by an individual in class c at time t (Eq. 6.4), $\mathcal{P}_{cn}$ is the profitability of species n for connectivity class c (Eq. 6.5), and the last term is the frequency of prey species n with N the total number of individuals. We scale time to prey dynamics, so mortality rate of prey species, $\mu=1$, and the attack rate of individual predators in class c on prey species n is given by $\mathcal{P}_{ct}\mathcal{P}_{cn}$. At the same time, each birth event produces an offspring in the prey species n. This probability is described as

$$B_{N_n}(t+\delta t)=\lambda\left(\frac{N_n}{N-1}\right)\delta t, \tag{6.8}$$

and we scale time to prey dynamics, so the birth rate of prey species $\lambda=1$. The changes in prey abundance for the individual–learning based model with prey selection are described as

$$\mathcal{P}[N_n-1|N_n]=D_{N_n}B_{N_{\neq n}} \tag{6.9}$$

$$\mathcal{P}[N_n+1|N_n]=D_{N_{\neq n}}B_{N_n}, \tag{6.10}$$

where $B_{N_{\neq n}}$ and $D_{N_{\neq n}}$ are the probabilities that a birth or a death occur in a species other than species n, respectively.

2.2. Guadalquivir estuary food web as a case study: Sampling of gut contents and resource distribution

Gut contents—We analyzed the stomach contents of 5725 fish individuals (total length < 140 mm), including mainly postlarvae and juveniles but also adults of small resident species of the 10 most abundant fish species (*Engraulis encrasicolus, Sardina pilchardus, Pomatoschistus* spp., *Dicentrarchus punctatus, Cyprinus carpio, Pomadasys incisus, Argyrosomus regius, Liza saliens, Liza ramada,* and *Aphia minuta*). We analyzed 95 distinct sample collections (12 months × 2 stations × 4 samplings, minus 1 missing sampling, Fig. 6.A1) and identified a total of 88,101 prey items spanning 43 taxa, mostly zooplankton—including copepods, cladocerans, ostracods, and cirripids as the dominant groups—and hyperbenthic species (e.g. mysids, gammarid amphipods and larval decapods and fishes), but also endobenthes (e.g. polychaetes and gastropods) (Baldó and Drake, 2002). Four connectivity classes were extracted from the empirical variance in the number of prey items per individual predator (Fig. 6.A1): (1) strongly connected predators (or fast prey finders) contain equal or more than 200 prey items; (2) medium–strongly connected predators contain equal or more than 100 and less than 200; (3) medium connected predators contain equal or more than 10 and less than 100; and (4) weakly connected individuals (or slow prey finders) are those individuals with equal or more than 1 and less than 10 prey items in the stomach.

Distribution of prey resources—The species abundances were estimated from samples collected using nets of two different sizes (1 mm, and 250 µm), where the abundance of the different species were given per 10^5 m^3 of filtered water, as estimated either from the current speed measurements taken made a digital flow meter for each net. To estimate the abundance of edobenthic preys, five replicate samples of the macrobenthic community were monthly taken randomly at each sampling site, with an Ekman-Birge grab (15 × 15 cm^2).

Aggregation—Many food web studies aggregate species across samplings and environmental gradients (Dunne, 2006). In order to test for the robustness of the strength of prey selection pattern across environmental gradients, we aggregated each predator species across sampling dates that have similar environmental conditions and analyzed groupings with more than 2000 individual prey and predators sampled (Figs. 6.A1A, 6.A2A, and 6.A3). Similarly, we analyzed each sampling date pooled across all predator species (Figs. 6.A1B, 6.A2B, and 6.A4), and for the entire dataset. Additional sampling details are described in the supplement.

2.3. Fitting the model to the data: Speed of learning (α) and the distribution of predator connectivity

To compare the empirical observations with model predictions of the distribution of predator connectivity across samplings and populations, we used previous experience of the predator given by the number of prey items in the stomach content of each individual predator j, the speed of learning, α, and the learning capacity, $\mathcal{K}$, as the drivers of variability in predator connectivity. The number of prey species and individual predators in the model are the same than in the empirical observations for those observations with more than 2000 individual prey and predators (Figs. 6.2, 6.A3, and 6.A4). Results for Figs. 6.2, 6.A3, and 6.A4 were obtained after running 10^5 replicates. Each replicate was run for the observed number of prey items sampled within the same predator population and temperature–salinity range

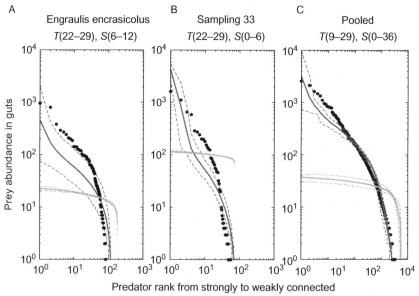

Figure 6.2 *Distribution of predator connectivity. x*-Axis and *y*-axis represent the predator rank from strongly to weakly connected and prey abundance in guts, respectively. (A) For the species *Engraulis encrasicolus* in the temperature and salinity range 22–29 °C (summer) and 6–12 (medium salinity), respectively. (B) For all the predators in the sampling number 33 in the temperature and salinity range 22–29 °C (summer) and 0–6 (low salinity), respectively. (C) For the data pooled in the temperature and salinity range 9–29 °C and 0–36, respectively. Black circles represent the empirical data, solid and dotted red (dark in print version) and orange (light grey in the print version) lines are the mean and confidence interval (CI) from the model with ($\alpha > 0$) and without ($\alpha = 0$) learning, respectively.

(Fig. 6.2A and 6.A3), the observed number of prey items within the same sampling using all the predators sampled (Fig. 6.2B and 6.A4), and for all the data pooled (Fig. 6.2C).

In Figs. 6.2, 6.A3, and 6.A4, we simulated the distribution of the number of items of per individual predator j denoted as $\mathbb{W} = [w_1, w_2, \ldots, w_T]$, with T the number of predators sampled, and compare it with the empirical data, $\mathbb{D} = [d_1, d_2, \ldots, d_T]$. All individual predators have equal w_j at the outset with w_j values larger than 0 for all individual predators. We calculated the distance between model predictions and the empirical data as the sum of the absolute values of the misfits as follows (Tarantola, 2006)

$$\rho(d_1, d_2, \ldots, d_T | \alpha, \mathcal{K}) = \sum_{j=1}^{T} \log\left(P\left(d_j | w_j\right)\right), \tag{6.11}$$

where $\log(P(d_j|w_j))$ captures the difference between the observed (d_j) and simulated (w_j) number of prey items of individual predator j and used the standard rejection-ABC algorithm (Beaumont, 2010; Sunnåker et al., 2013; Toni and Stumpf, 2010) that works as follows

1. Draw θ from the prior where $\theta = [\alpha, \mathcal{K}]$. In absence of field estimations of the speed of learning, α, and the learning capacity, $\mathcal{K}$, the values were randomly chosen from a uniform distribution with range $U[10^{-5}, 10^{-1}]$ and $U[10^2, 10^3]$, respectively.
2. Simulate $\mathbb{W}$ from each run with parameter θ.
3. Calculate distance for $\rho(\mathbb{W}, \mathbb{D})$ between simulated ($\mathbb{W}$) and empirical data ($\mathbb{D}$).
4. Reject θ if $\rho(\mathbb{W}, \mathbb{D}) > \epsilon$.

The tolerance $\epsilon \geq 0$ is the desired level of agreement between the simulations ($\mathbb{W}$) and the empirical data ($\mathbb{D}$) (Toni and Stumpf, 2010). Accepted model replicates to calculate the confidence interval (CI) taking the percentiles 0.05 and 0.95 were generated using a family of $\theta = [\alpha, \mathcal{K}]$ parameter values with tolerance $\epsilon = 2 \times$ minimum log-distance where the minimum log-distance refers to the replicate that minimizes the absolute difference between the empirical and simulated data.

2.4. Choosing among alternative models: Strength of prey selection (Ω)

2.4.1 Negative, neutral, and positive density-dependent prey selection

Results for Figs. 6.3 and 6.4 were obtained by using the information gained from the simulations with the speed of learning. We used the CI range of

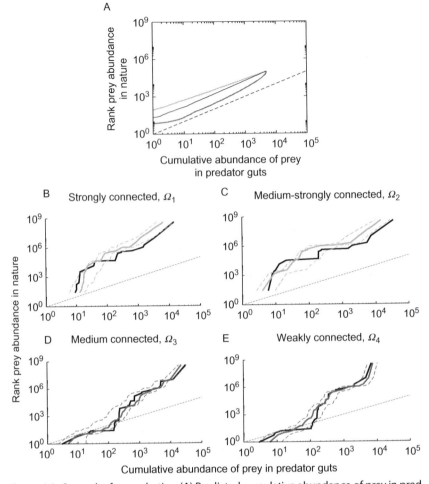

Figure 6.3 *Strength of prey selection.* (A) Predicted cumulative abundance of prey in predator guts (x-axis) with predator selecting preferentially rare prey (red (dark grey in the print version), $\Omega > 0$), no preference (black, $\Omega = 0$), and common prey (orange (light grey in the print version), $\Omega < 0$). (B and C) Predicted (blue (light grey in the print version)) and empirical (black) cumulative abundance of prey in predator guts (x-axis) for the strongly (B) and medium-strongly (C) connected individual predators for the data poled. Strongly (B) and medium-strongly (C) connected predators select preferentially common prey $\left(K_{\mathcal{H}_1, \mathcal{H}_{1_0}} = 14, \Omega_{1-} = -61, \alpha = 0.009, \text{ and } K_{\mathcal{H}_2, \mathcal{H}_{2_0}} = 36, \Omega_{2-} = -41, \alpha = 0.0072, \right.$ respectively). (D and E) Predicted (red (dark grey in the print version)) and empirical (black) cumulative abundance of prey in predator guts (x-axis) for the medium (D) and weakly (E) connected individual predators for the data pooled. Weakly (B) and medium (C) connected individuals select preferentially rare prey $\left(K_{\mathcal{H}_3 + \mathcal{H}_{3_0}} = 82, \Omega_{3+} = 91, \alpha = 0.001, \text{ and } K_{\mathcal{H}_4 + \mathcal{H}_{4_0}} = 107, \Omega_{4+} = 73, \alpha = 0.0062, \text{ respectively} \right)$. Solid and dotted red (dark grey lines in the print version) and orange (light grey lines in the print version) lines are the mean and CI from the best-fit values given by the model with learning. Dotted black lines represent the identity line to show the different trend between the strongly and medium-strongly connected individuals (B and C) and the medium and weakly connected individuals (D and E).

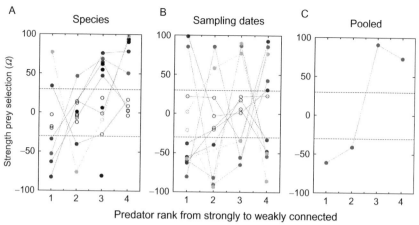

Figure 6.4 *Strength of prey selection (species, sampling dates, and data pooled).* Strength of prey selection, Ω (*y*-axis), as a function of strongly (1) to weakly (4) connected predators (*x*-axis). (A) Strongly to weakly connected predators target preferentially dominant ($\Omega < 0$) and rare ($\Omega > 0$) prey, respectively. (B) Strongly and weakly connected predators target both dominant and rare prey. (C) Strongly to weakly connected predators target preferentially dominant and rare prey, respectively. Red, orange, blue and black (Filled dots in the print version) dots represent the classes for the species, sampling dates and the pooled data outside the range predicted by the neutral density-dependent prey selection model, respectively. Strength of prey selection values plotted in these colours has Bayes factor values >3.

speed of learning values that best fitted the data (see for example the distance profile in sampling 27 in Fig. 6.A4). As in the previous section, results were obtained after running 10^5 replicates for each population and sampling with more than 2000 individual prey and predators. We simulated the diet of each individual predator *j*. The diet is given by the number of items of each prey species *n* in the stomach content of individual predator *j*. We also simulated the number of individuals within each connectivity class *c*. We denote each simulated individual predator diet in class *c* for all prey species $\mathcal{S}$, as $\mathbb{S}_{jc}^n = \left[s_{jc}^1, s_{jc}^2, \ldots, s_{jc}^s\right]$. We compare the simulated diets with the empirical ones, $\mathbb{D}_{jc}^n = \left[d_{jc}^1, d_{jc}^2, \ldots, d_{jc}^s\right]$, across all the empirically observed and simulated predators (i.e., $\mathbb{D}_c = \left[\mathbb{D}_{1c}^n, \mathbb{D}_{2c}^n, \ldots, \mathbb{D}_{T_c}^n\right]$ and $\mathbb{S}_c = \left[\mathbb{S}_{1c}^n, \mathbb{S}_{2c}^n, \ldots, \mathbb{S}_{\tau_c}^n\right]$, respectively). $\mathcal{S}$, $\mathcal{T}_c$, and τ_c are the total number of prey species, the number of individual predators empirically sampled in class *c*, and the number of individual predators in class *c* obtained by simulation, respectively. All individual predators *j* have equal values at the outset with at least one of the $s_{jc}^{\mathcal{S}}$ values larger than 0 for all *j*.

As in the fitting for the speed of learning, we calculated the distance between model predictions and the empirical data as the sum of the absolute values of the misfits for each connectivity class c as follows (Tarantola, 2006)

$$\rho\left(d_{1_c}^1, d_{1_c}^2, \ldots, d_{\mathcal{T}_c}^S | \alpha, \Omega_c\right) = \frac{\left[\sum_{j=1}^{\upsilon}\sum_{n=1}^{S} \log\left(P\left(d_{jc}^n | s_{jc}^n\right)\right)\right] \times |\mathcal{T}_c - \tau_c + 1|}{\xi},$$

(6.12)

where d_{jc}^n and s_{jc}^n are the empirical data and the model prediction of the number of prey items per individual predator j for prey species n for each of the four connectivity classes, respectively. υ is the smallest number after comparing the empirical with the simulated individual predators in class c and ξ represents the number of comparisons when $d_{jc}^n > 0$ and $s_{jc}^n > 0$. The best fit is given by the parameter combination that: (1) maximizes the similarity between the empirical ($\mathcal{T}_c$) and simulated (τ_c) number of individuals in class c, (2) maximizes the number of comparisons between the empirical and simulated values, ξ, and (3) minimizes the absolute differences in the number of prey items per prey species across all prey species and across all the individual predators in class c. We used the standard rejection-ABC algorithm (Beaumont, 2010; Sunnåker et al., 2013; Toni and Stumpf, 2010) that works as follows

1. Generate θ from the prior where $\theta = [\alpha, \Omega_c]$. α values were randomly chosen from the CI range that best fitted the data for each population and sampling. The strength of prey selection values for each class, Ω_c, were chosen from a uniform distribution with ranges $U[-30,30]$, $U[-100,<-30]$, and $U[>30,100]$ for the neutral, ($\mathcal{H}_{co}$, negative, ($\mathcal{H}_{c-}$, and positive, ($\mathcal{H}_{c+}$, density-dependent prey selection models, respectively. For the negative and positive density-dependent prey selection scenarios, we ranked the randomly obtained Ω_c values for each class with the abundance values.

    We remark an strict neutral model would imply to sample values close to 0. We relax the neutral model to allow for a more conservative nearly neutral model and use strength of prey selection values in the range $U[-30,30]$.

2. Simulate the diets, $\mathbb{S}_c$, from each run with parameter θ.

3. Calculate distance for $\rho(\mathbb{S}_c, \mathbb{D}_c)$ between simulated ($\mathbb{S}_c$) and empirical data ($\mathbb{D}_c$).

4. Reject θ if $\rho(\mathbb{S}_c, \mathbb{D}_c) > \epsilon$.

2.4.2 Occam factor

The best-fit distance value gives us a first support for each model. In order to find the evidence supporting each model, we also need to quantify the penalization for each model given the data. That is to say, how much of the prior variance, the one given by the initial uniform distribution, is lost after taking the data into account, the variance in the posterior distribution. One way to find evidence for each model is by taking the best-fit distance that the model can achieve and multiplying it by an "Occam factor" (MacKay, 2003). The "Occam factor" is a term whose value is less than one that penalizes each model for having the strength of selection parameter, Ω_c.

The evidence for each model, $(\mathcal{H}_{ci} = [(\mathcal{H}_{co}, (\mathcal{H}_{c-}, \text{and } (\mathcal{H}_{c+}]$, or the probability to reproduce the empirical data for each connectivity class c, $\mathbb{D}_c$, given each model, $\mathcal{H}_{ci}$, $P(\mathbb{D}_c|\mathcal{H}_{ci})$, reads as follows

$$P(\mathbb{D}_c|\mathcal{H}_{ci}) \simeq P\left(\mathbb{D}_c|\mathcal{L}'\left(\hat{a}, \hat{\Omega}_{c_i}\right), \mathcal{H}_{c_i}\right) \times \frac{\sigma_{\mathcal{H}_{ci}|\mathbb{D}_c}}{\sigma_{\mathcal{H}_{ci}}} \tag{6.13}$$

where $P\left(\mathbb{D}_c|\mathcal{L}'\left(\hat{a}, \hat{\Omega}_{ci}\right), \mathcal{H}_{c_i}\right)$ is the minimum distance from model $\mathcal{H}_{c_i}, \sigma_{\mathcal{H}_{ci}|\mathbb{D}}$ is the posterior variance of the strength of selection parameter, Ω_{ci}, given the data, and $\sigma_{\mathcal{H}_{ci}}$ is the prior variance of the strength of selection parameter, Ω_{ci}. As we have assumed that the prior is uniform on a fixed interval, representing the range of values of the strength of prey selection that were possible a prior, then the Occam factor is

$$\text{Occam factor} = \frac{\sigma_{\mathcal{H}_{ci}|\mathbb{D}_c}}{\sigma_{\mathcal{H}_{ci}}}, \tag{6.14}$$

or the factor by which each model space collapses when the data arrive (Gregory, 2005; MacKay, 2003).

2.4.3 Bayes factor

Once we have the metrics to quantify the best-fit distance values given by each model and the penalization for each model given the data, we can now compare each pair of models to quantify the model that best explain the data. We will use the Bayes factor, $\mathcal{K}_{\mathcal{H}_{ci}}, (\mathcal{H}_{cj}$, to do so and it is given by

$$\mathcal{K}_{\mathcal{H}_{ci}, \mathcal{H}_{cj}} = \frac{P(\mathbb{D}_c|\mathcal{H}_{ci})}{P(\mathbb{D}_c|\mathcal{H}_{cj})}, \tag{6.15}$$

where $\mathcal{K}_{\mathcal{H}_a \mathcal{H}_{cj}} < 1:1$, $\mathcal{K}_{\mathcal{H}_a \mathcal{H}_{cj}} > 3:1$, $\mathcal{K}_{\mathcal{H}_a \mathcal{H}_{cj}} > 10:1$, and $\mathcal{K}_{\mathcal{H}_a \mathcal{H}_{cj}} < 100:1$, supports model ($\mathcal{H}_{cj}$, supports model ($\mathcal{H}_{ci}$, substantial support for model ($\mathcal{H}_{ci}$, and strongly supports model ($\mathcal{H}_{ci}$, respectively (Jeffreys, 1961). Bayes factor values reported were obtained by comparing the positive (or negative) density-dependent prey selection model with the neutral density-dependent prey selection model with the condition $\mathcal{K}_{\mathcal{H}_{c+} \mathcal{H}_{co}}$ (or $\mathcal{K}_{\mathcal{H}_{c-} \mathcal{H}_{co}}) > 3:1$ provided $\mathcal{K}_{\mathcal{H}_{c+} \mathcal{H}_{c-}}$ (or $\mathcal{K}_{\mathcal{H}_{c-} \mathcal{H}_{c+}}) > 3:1$. Otherwise we consider there is no support for the positive or negative density-dependent prey selection model.

3. RESULTS

3.1. Rate of learning and the distribution of predator connectivity

In order to estimate the rate of learning, we used groups with greater than 2000 predator and prey individuals sampled. This sampling size was sufficient to stabilize the variance in connectivity (Fig. 6.A2). A total of 13 of the 62 predator populations (Fig. 6.A2A) and 11 of the 95 samplings (Fig. 6.A2B) have more than 2000 prey and predator individuals sampled and were used to estimate the speed of learning, followed by an analysis to detect individual variation in prey selection. In all of these cases, including for the pooled analysis of all the data, the distribution of predator connectivity obtained from the model without learning (i.e. $\alpha = 0$ in Eq. 6.1) strongly deviates from the empirical data (Fig. 6.2). By contrast, the learning model ($\alpha > 0$ in Eq. 6.1, and red lines in Fig. 6.2) predicts the variance of the distribution of predator connectivity at population level (Figs. 6.2A and 6.A3), across multiple samplings (Figs. 6.2B and 6.A4), and for all the data pooled (Fig. 6.2C). Overall, the distribution of the speed of learning values, α, was always in the range [0.01–0.5]. The learning model consistently captured the strongly and weakly connected individual predators, but it often failed to capture the distribution of the medium and medium-strongly connected predators (Appendix Figs. 6.A3 and 6.A4). Nevertheless the dramatic improvement of the model with versus without learning suggests that this learning trait is an important parameter to model and to carry forward for the analysis of individual variation in prey selection (from Fig. 6.1A to B).

The length of individual predators at intra-specific level explains both the mean length of prey and the total number of prey in 6 of the 13 environmental conditions with more than 2×10^3 prey and predator sampled ($R^2 = [0.22, 0.46]$, and $R^2 = [0.21, 0.61]$ and all $p < 0.01$, respectively) (Appendix Fig. 6.A5). At the sampling level, the length of individual predators explains the mean length of prey and the total number of prey in 3 and 1 samplings of the 11 samplings with more than 2000 prey and predator sampled, respectively. These patterns suggest there is a trend at intra-specific level with individuals with a few or several prey of small and large size, respectively, but there is also substantial variation across individuals with few or several prey of similar length.

3.2. Variation in the strength of prey selection among individuals

Our observation of non-zero rates of learning estimated at the population level, suggests that successful prey captures leading to an increasing probability of future predation success (i.e., learning) is an important mechanism driving predator connectivity. This is true, even in absence of individual-level variation in the speed of learning. However, to discriminate active prey selection from a random encounter model we need to take into account the empirical observations of resource abundance and the profitability function for each predator individual. If prey occurs in the observed gut of individual predators more or less frequently than prey abundance in the environment, then these deviations suggest density-dependent resource selection by individual predators. In such a scenario, diversity in food webs will depend on the preference of predators for rare versus common prey (Fig. 6.1B). To detect such scenarios, we tested for prey selection for each sampling date (pooling across species), for each predator species (pooling across sampling dates in similar environmental conditions), and for the entire dataset.

Analysis of pooled dataset—We found that predators with many prey items in their gut preferentially consume common resources (i.e., negative density-dependent prey selection, best fit and CI in orange lines (light grey lines in the print version), Fig. 6.3B and C), while predators with few prey items preferentially select rare prey (i.e., positive density-dependent prey selection, best fit, and CI in red lines (dark grey lines in the print version), Fig. 6.3D and E). The best fits for the speed of learning, α, did not differ significantly across connectivity classes (Kolmorgorov–Smirnov test for all pairwise comparisons, $p > 0.1$), but the sign and the strength of prey selection, Ω_c, differed significantly between the weakly (and medium) and strongly (and medium-strongly) connectivity classes (Kolmorgorov–Smirnov test for all pairwise comparisons between predator connectivity classes, $p < 0.01$, Figs. 6.3 and 6.4C).

Analysis by predator species—Strongly and weakly connected individual predators selecting preferentially common (i.e., positive prey selection,

$\Omega_1 > 0$) and rare (i.e., negative prey selection, $\Omega_4 < 0$) resources were observed in 22 of the 48 groupings of predator species pooled across samplings. Weakly and medium connected individuals select preferentially rare prey in 12 combinations (positive density-dependent prey selection, Fig. 6.4A) whereas medium-strongly and strongly connected individuals select preferentially common prey in six combinations (negative density-dependent prey selection, Fig. 6.4A).

Analysis by sampling date—Strongly and weakly connected individual predators selecting preferentially common (i.e., positive prey selection) and rare (i.e., negative prey selection) resources were observed in 31 of the 43 sampling dates pooled across species. Weakly and medium connected individuals select preferentially rare resources in 8 samplings (positive density-dependent prey selection, Fig. 6.4B), while medium-strongly and strongly connected individuals select preferentially common resources in 10 samplings (negative density-dependent prey selection). Strongly and weakly connected individuals select preferentially rare and common resources in 4 and 9 combinations, respectively (Fig. 6.4B). Similar to our analysis by predator species, our analysis of each sampling date revealed that predators in different connectivity classes shared both evidence for positive and negative density-dependent prey selection. In contrast, however, the shift from negative to positive prey selection with decreasing predator connectivity was less evident (Fig. 6.4A vs. B).

Summary—Overall, these results suggest that there is ample evidence for intra-population variation in the strength of prey selection by predators. It is equally common to observe such variation within species as across species within a given sampling. Our results also suggest neutral density-dependent prey selection with rare and common resources occurring in the observed gut of the weakly and strongly connected individuals, respectively.

4. DISCUSSION

The patterns of predator connectivity and prey selection presented in this study may be a consequence of strongly and weakly connected predators actively foraging in a system in which prey are distributed in a spatially heterogeneous and patchy manner. Strongly connected predators (or fast prey finders) could result from individuals preferentially searching the most abundant prey species in high-density but species poor patches. A strongly connected predator that finds this kind of patch will then be highly connected to just one or a few common prey species that are highly profitable for this individual predator. Conversely, weakly connected predators could result from individuals preferentially searching low-density patches with a large number of rare prey species. A weakly connected predator will then be weakly connected to more than one rare prey species that are highly profitable for these weakly connected predators. These mechanisms are taken into account in our modelling framework and suggest that individual-level traits in prey selection of spatially heterogeneously distributed resources can drive patterns of species diversity and coexistence in multi-trophic ecosystems. Patterns of coexistence will ultimately depend on the frequency of individual predators that promote (i.e., strength of prey selection $\Omega < 0$) or inhibit diversity (i.e., strength prey selection $\Omega > 0$) (Fig. 6.1).

4.1. Intra-specific variation in prey selection

The patterns of intra-specific variability reported in this study may help to reconcile the mechanisms promoting variation within natural populations with the patterns that promote or inhibit diversity in ecological networks. It has been shown that non-random interactions among species can increase diversity (Bastolla et al., 2009), but non-random interactions among species can also decrease diversity if, as shown here, most weakly connected individuals across predator populations are preferentially selecting rare prey. Nestedness and compartments may increase coexistence and diversity in food webs (Bastolla et al., 2009), but the connection between the empirical evidence of nested diets within populations (Araújo et al., 2010; Bolnick et al., 2007; Cantor et al., 2012; Pires et al., 2011) and food web diversity is still at an incipient stage. Nestedness and compartments at species level can be obtained from at least two scenarios of intra-specific variability with opposite consequences for species diversity: either strongly and weakly connected individuals selecting common and rare prey, respectively, or alternatively strongly and weakly connected individuals selecting rare and common prey, respectively. Because the number of strongly connected individuals in large networks may be orders of magnitude lower than the number of weakly connected individuals, the species-level effects on diversity and stability of these two opposite intra-specific level patterns may differ significantly. We found that highly connected individual predators are preferentially selecting common prey, so they promote rather than limit prey diversity because their overall effect on extinction probability of resource species is lower than we would expect based on their number of interactions. The opposite is true for weakly connected individuals. They preferentially select rare resources and thus their effect on extinction probability of resource species is larger than we expect based on their low number of interactions. Hence, the greater the proportion of weakly connected individuals consuming rare resources, the more pronounced their negative effect on diversity in the network will be.

4.2. Connecting the strength of prey selection and diversity in food webs

In the context of biodiversity theories, current models can explain species abundance patterns for many groups (Allen and Savage, 2007; Rosindell et al., 2011), but most of these approaches do not explain patterns of intra-specific variation in ecological interactions nor their implications for diversity patterns in ecological networks and multi-trophic ecosystems (Bolnick et al., 2011; Violle et al., 2012; Volkov et al., 2009). The models we have developed in this study explore a plausible set of mechanisms that may be useful as a benchmark to predict the empirical patterns of predator connectivity within samplings and populations. In cases where we do not identify significant prey selection, this could occur due to the lack of power or because of stochastic variation in prey selection driven by random encounter processes (Fig. 6.4, open circles). Further work is needed to investigate the temporal consistency of individual diet variation (Bolnick et al., 2003; Sih et al., 2012), for example, by exploring patterns between individual diet variation and morphology (Matthews et al., 2010).

In general, the fit of our models might be improved by considering additional traits. We have studied only learning as a trait involved in prey selection but individuals may differ in diet and prey selectivity for a variety of additional reasons including genetic components, size, sex, morphology, metabolic rates, or physiology (Bolnick et al., 2003). Traits involved in these characteristics of individuals may combine in unexpected ways to improve predictions in resource selection in heterogeneously distributed resources. Analysis combining some of these components like individual size and morphology (Beckerman et al., 2006; Ingram et al., 2011; Woodward et al., 2010), learning in foraging groups (Giraldeau and Caraco, 2000), and genetic and phenotypic characteristics are likely required to improve the predictions we present in this study so as to further disentangle the effect of multiple traits in resource selection at individual level and their effect on diversity in food webs (Bolnick et al., 2011). From our results, it is not entirely straightforward to infer the consequences for food web stability and dynamics. Although the majority of individuals exhibit destabilizing preference for rare prey ($\approx 95\%$ of individuals in our food web are weakly connected), these same individuals consume comparatively few prey per capita, which may mitigate their destabilizing effect. Conversely, highly connected individuals tend to prefer common prey with stabilizing effects on prey populations and although these strongly connected individuals are rare in the samplings they also consume large number of prey. Previous theoretical studies have shown the allocation of species diversity to slow energy channels within food webs and how these slow energy channels result in the skewed distribution of interaction strengths that has been shown to confer stability to food webs (Rooney and McCann, 2012). Further theoretical and empirical research are required to explore the connection between slow and fast prey finders (i.e., weakly and strongly connected individuals, respectively) to slow or fast energy channels and the net effect of these opposing trends within species on interaction strength and food web dynamics and stability. For example, under which conditions does interaction strength driven by intra-specific variation in prey selection embedded in food web dynamics yield different outcomes than a species-mean based approach?

4.3. Connecting trait variation with eco-evolutionary food web dynamics

In the present study, we develop individual trait-based models that contain two basic parameters, namely, the speed of learning and the strength of prey selection. These two parameters drive individual choices of resource selection with fast prey finders preferentially targeting common prey and slow prey finders preferentially targeting rare prey. The model predicts that given a non-zero and equal rate of learning for all the individuals, variation in prey selection among individuals is driven by the different rates in successful experience attacking common and rare prey. This result suggests that accumulation of experience from successful prey captures lead to an increasing probability of predation of similar prey types. Some individuals become highly specialized fast prey finders in a type of prey that is, as observed in the empirical data (Figs. 6.3 and 6.4), a common resource. On the other hand, there are individuals that become highly specialized slow prey finders in a type of prey that is rare in the environment, even if they have the same speed of learning as the fast prey finders (Figs. 6.3 and 6.4).

The observed individual variation in prey selection within and among predator populations is a first step towards understanding the potential for eco-evolutionary dynamics in natural populations. It has been shown that feedbacks between the ecological and evolutionary time scales may be common in prey–predator interactions (Yoshida et al., 2003), but it is an ongoing challenge to obtain datasets to infer both the fitness variation among individual predators so as to estimate subsequent evolutionary effects (Fig. 6.5). Consider a scenario where a population of predators consists of both strongly (Fig. 6.5A, red bars (light grey bars in the print version)) and weakly connected predators (Fig. 6.5A, black bars) selecting preferentially common and rare prey, respectively. In order to test for eco-evolutionary dynamics in such a scenario, we would want to assess fitness differences among predator classes (or phenotypes) and track longer term patterns in class (phenotype) frequency through time (Fig. 6.5C). With the present data and modelling we were able to infer different types of density-dependent prey selection from individual predators for a given ecological time scale, but we were unable to estimate how such variation in prey

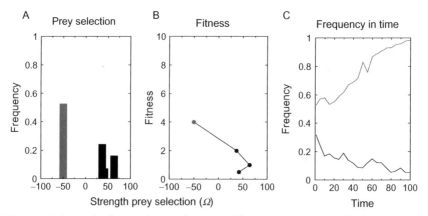

Figure 6.5 *Strength of prey selection, fitness, and frequency in time.* (A) The frequency of strongly (red (light grey in the print version)), medium, and weakly (black) connected individuals as a function of the strength of prey selection, Ω (x-axis), for the species *Engraulis encrasicolus* in the temperature and salinity range 22–29 °C (summer) and 6–12 (medium salinity). (B) A hypothetical scenario where the fitness of each class varies as a function of the strength of prey selection, Ω (x-axis). In this scenario, strongly connected individuals (red dot (light grey dot in the print version)) selecting preferentially the most common prey have higher fitness than weakly connected individuals (black dots). As a consequence, the frequency of the strongly and weakly connected individuals increases and decreases in this population with time, respectively (C).

selectivity might drive frequency-dependent selection pressures experienced by these interacting populations.

4.4. Conclusion

Understanding the mechanisms driving predator connectivity, and how such variation determines the strength of prey selection in high-resolution food web data present several open challenges. In the present study, we have developed individual-based models (Figs. 6.1 and 6.2, orange lines (light grey lines in the print version)) to quantify the speed of learning and the strength in resource selection across predators with different connectivity. By examining these models under different parameter combinations and confronting them with high-resolution individual-level data, we can ascertain the factors driving resource selection in weakly and strongly connected individual predators. The more standard approach of using population level to develop food web and ecological network theory, may fail to anticipate the mechanisms driving species extinction, coexistence, and diversity in multi-trophic ecosystems.

ACKNOWLEDGEMENTS

We thank Daniel Stouffer, Gary Mittelbach, and Martine Maan for useful comments on the manuscript. This study was supported by the Swiss National Science Foundation project 31003A-144162 (to C. M.). C. M. dedicates this chapter to the memory of his mother (1940–2014).

APPENDIX. SAMPLING METHODS

Four daily samples were collected monthly for 1 year period (February 1998 to January 1999) at two sites (oligohaline, o, and polyhaline, p, stations) in the Guadalquivir estuary, southern Spain. The Guadalquivir estuary is a well-mixed temperate estuary with a gradual horizontal change in salinity and a clear seasonal temperature cycle. We selected polyhaline (20 ± 10.1) and oligohaline (3.7 ± 2) sampling stations, situated at 8 and 32 km, respectively, from the river mouth. The samples were collected at the new moon at each sampling site from a traditional fish boat anchored on the left river side. Samples consisted of passive hauls, lasting 2 h, made during the first 2 h of each diurnal and nocturnal flood and ebb tide. Sampled were taken with three nets working in parallel. Nets were made with polyamide gauze of 1 mm and an opening of 2.5 m (width) $\times$ 3 m (depth). At the start of each sampling, water temperature and salinity were measured.

SAMPLINGS AND POPULATIONS ACROSS ENVIRONMENTAL GRADIENTS

To test whether patterns of individual-level variation are robust across environmental gradients, we have categorized the samplings in 9 conditions of salinity and temperature: [0–6 (low salinity), 9–16 °C (winter), green (medium grey in the print version)], [12–36 (high salinity), 9–16 °C (winter), light grey], [0–6 (low salinity), 16–22 °C (spring-fall), blue (dark grey in the print version)], [6–12 (medium salinity), 16–22 °C (spring-fall), cyan (medium grey in the print version)], [>12–36 (high salinity), 16–22 °C (spring-fall), turquoise (medium grey in the print version)], [0–6 (low salinity), 22–29 °C (summer), orange (medium grey in the print version)], [6–12 (medium salinity), 22–29 °C (summer), brown (medium grey in the print version)], [12–36 (high salinity), 22–29 °C (summer), red (dark grey in the print version)], and [6–12 (medium salinity), 9–16 °C (winter), violet (dark grey in the print version)]. At population level, all individuals of the same species were pooled in each of the nine conditions even if individuals

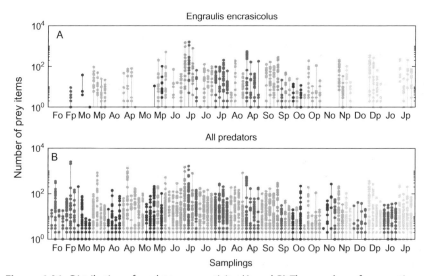

Figure 6.A1 *Distribution of predator connectivity.* (A and B) The number of connections observed in each individual predator (number of prey items, y-axis) in each of the 95 oligohaline, (o), or polyhaline, (p), samplings since February 1998, F_o and F_p, to January 1999, J_o and J_p (x-axis) for the fish species *Engraulis encrasicolus*, a, and for all the fish predators, b. Colours tones (Grey tones in the print version) indicate samplings in each salinity and temperature combination.

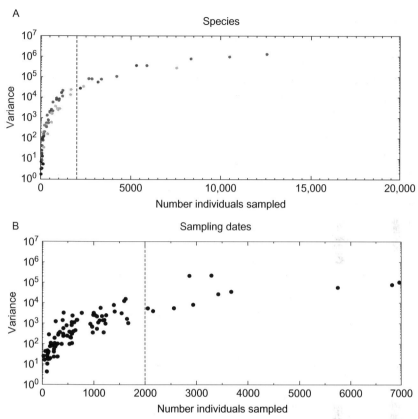

Figure 6.A2 *Sampling effort.* (A) Represents the variance in the number of prey items (y-axis) as a function of the number of prey and predator individuals sampled for all the species (x-axis). Predator species are: *Aphia minuta* (black), *Argyrosomus regius* (violet; dark grey in the print version), *Cyprinus carpio* (blue; dark grey in the print version), *Dicentrarchus punctatus* (orange; grey in the print version), *Engraulis encrasicolus* (red; dark grey in the print version), *Liza ramada* (green; dark grey in the print version), *Liza saliens* (maroon; black in the print version), *Pomatoschistus* spp. (cyan; grey in the print version), *Pomadasys incisus* (magenta; grey in the print version), and *Sardina pilchardus* (indigo; dark grey in the print version). (B) Represents the variance in the number of prey items (y-axis) as a function of the number of prey and predator individuals sampled (x-axis) for all the samples. To detect shifts in the relationship between the number of individuals sampled and the variance in the number of prey items, we used split-line regression. Provided that a shift was detected in the slope of the regression, a threshold of approximately 2000 individuals was detected for both the species and the sampling data. We analyzed the data points that were higher than 2000 individuals (dotted line).

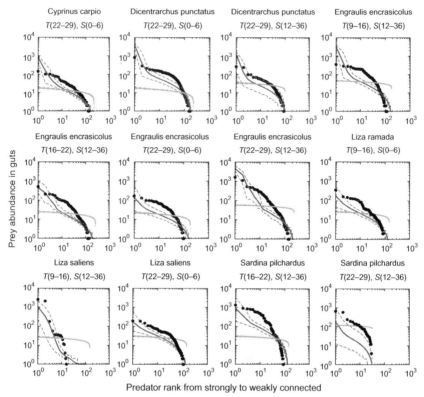

Figure 6.A3 *Distribution of predator connectivity in populations.* x-Axis and y-axis represent the predator rank from strongly to weakly connected and prey abundance in guts, respectively. Black dots represent the empirical distribution for each temperature–salinity combination with more than 2000 individual prey and predators sampled. Solid red (dark grey lines in the print version) and orange lines (light grey lines in the print version) represent the mean and CI from the model with and without learning after 10^5 replicates, respectively.

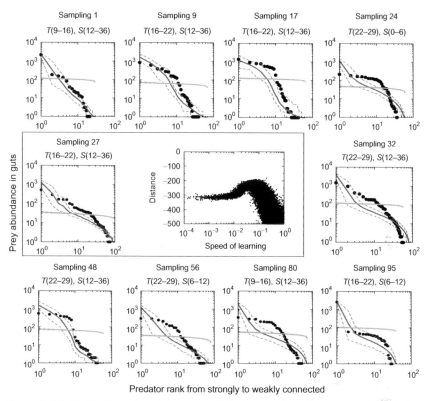

Figure 6.A4 *Distribution of predator connectivity in samplings.* Represents the distribution in individual connectivity for each sampling. *x*-Axis and *y*-axis represent the predator rank from strongly to weakly connected and prey abundance in guts, respectively. Black dots represent the observed species distribution for samplings with a specific temperature–salinity combination and with more than 2000 individual prey and predators sampled. For the Sampling 27, we show the distance profile with the *x*-axis and *y*-axis representing the speed of learning, α, and the distance values (Eq. 11 in the main text), respectively. Solid red (dark grey lines in the print version) and orange lines (light grey lines in the print version) represent the mean and CI from the model with and without learning after 10^5 replicates, respectively.

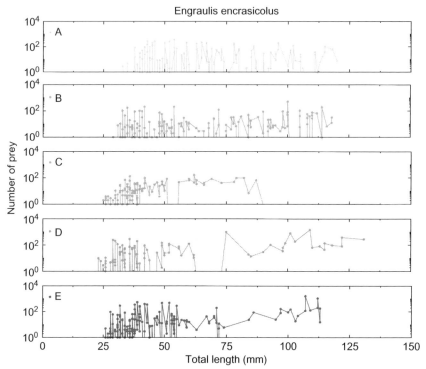

Figure 6.A5 From top to bottom, we represent the number of prey per individual predator for the species *Engraulis encrasicolus* as a function of total length (mm). (A, light grey) Winter, 9–16 °C, high salinity, >12–36, $R^2 = 0.05$, $p > 0.1$. (B, turquoise (medium grey in the print version)) Spring, >16–22 °C, high salinity, >12–36, $R^2 = 0.28$, $p > 0.1$. (C, orange (medium grey in the print version)) Summer, >22–29 °C, low salinity, 0–6, $R^2 = 0.60$, $p < 0.01$. (D, brown (medium grey in the print version)) Summer, >22–29 °C, medium salinity, >6–12, $R^2 = 0.42$, $p < 0.01$, and (E, red (dark grey in the print version)) Summer, >22–29 °C, high salinity, >12–36, $R^2 = 0.33$, $p < 0.01$.

belonging to the same species were collected in different samplings (Figs. 6.2A, 6.4A, and 6.A3). At sampling level, all individuals from the different predator species were pooled (Figs. 6.2B, 6.4B, and 6.A4). Finally we pooled all samplings and all predators (Figs. 6.2C, 6.3, and 6.4C).

REFERENCES

Abrams, P.A., 2010. Implications of flexible foraging for interspecific interactions: lessons from simple models. Funct. Ecol. 24, 7–17.

Allen, A., Savage, V.M., 2007. Setting the absolute tempo of biodiversity dynamics. Ecol. Lett. 10, 637–646.

Araújo, M.S., Martins, M.G., Cruz, L.D., Fernandes, F.R., Linhares, A.X., Dos Reis, S.F., Guimarães, P.R., 2010. Nested diets: a novel pattern of individual level resource use. Oikos 119, 81–88.

Baldó, F., Drake, P., 2002. A multivariate approach to the feeding habits of small fishes in the Guadalquivir estuary. J. Fish Biol. 61, 21–32.

Bastolla, U., Fortuna, M.A., Pascual-García, A., Ferrera, A., Luque, B., Bascompte, J., 2009. The architecture of mutualistic networks minimizes competition and increases biodiversity. Nature 458, 1018–1020.

Beaumont, M.A., 2010. Approximate Bayesian computation in evolution and ecology. Annu. Rev. Ecol. Evol. Syst. 41, 379–406.

Beckerman, A.P., Petchey, O.L., Warren, P.H., 2006. Foraging biology predicts food web complexity. Proc. Natl. Acad. Sci. U. S. A. 103, 13745–13749.

Beckerman, A.P., Petchey, O.L., Morin, P.J., 2010. Adaptive foragers and community ecology: linking individuals to communities and ecosystems. Funct. Ecol. 24, 1–6.

Biro, P.A., Stamps, J.A., 2010. Do consistent individual differences in metabolic rate promote consistent individual differences in behavior? Trends Ecol. Evol. 25, 653–659.

Bolnick, D.I., Yang, L.H., Fordyce, J.A., Davis, J.M., Svanback, R., 2002. Measuring individual-level resource specialization. Ecology 83, 2936–2941.

Bolnick, D.I., Svanbäck, J.A., Fordyce, L.H., Yang, L.H., David, J.M., Hulsey, C.D., Forister, M.L., 2003. The ecology of individuals: incidence and implications of individual specialization. Am. Nat. 161, 1–28.

Bolnick, D.I., Svanbäck, R., Araújo, M.S., Persson, L., 2007. Comparative support for niche variation hypothesis that more generalized populations also are more heterogeneous. Proc. Natl. Acad. Sci. U. S. A. 104, 10075–10079.

Bolnick, D.I., Amarasekare, P., Araujo, M.S., Burger, R., Levine, J.M., Novak, M., Rudolf, V.H.W., Schreiber, S.J., Urban, M.C., Vasseur, D.A., 2011. Why intraspecific trait variation matters in community ecology. Trends Ecol. Evol. 26, 183–192.

Brose, U., Williams, R.J., Martinez, N.D., 2006. Allometric scaling enhances stability in complex food webs. Ecol. Lett. 9, 1228–1236.

Cantor, M., Pires, M.M., Longo, G.O., Guimarães, P.R., Setz, E.Z.F., 2012. Individual variation in resource use by opossums leading to nested fruit consumption. Oikos 122, 1085–1093. http://dx.doi.org/10.1111/j.1600-0706.2012.00070.x.

Carnicer, J., Jordano, P., Melián, C.J., 2009. The temporal dynamics of resource use by frugivorous birds: a network approach. Ecology 90, 1958–1970.

Catania, K.C., Remple, F.E., 2005. Asymptotic prey profitability drives star-nosed moles to the foraging speed limit. Nature 433, 519–522.

Chesson, P.L., 1978. Predator-prey theory and variability. Annu. Rev. Ecol. Evol. Syst. 9, 323–347.

Chesson, P.L., 1984. Variable predators and switching behavior. Theor. Popul. Biol. 26, 1–26.

Cianciaruso, M.V., Batalha, M.A., Gaston, K.J., Petchey, O.L., 2009. Including intraspecific variability in functional diversity. Ecology 990, 81–89.

Clark, J.S., 2010. Individuals and the variation needed for high species diversity in forest trees. Science 327, 1129–1132.

Cohen, J.E., Briand, F., Newman, C.M., 1990a. Community Food Webs: Data and Theory. Springer-Verlag, Berlin Heidelberg.

Cohen, J.E., Luczak, T., Newman, C.M., Zhou, Z.-M., 1990b. Stochastic structure and nonlinear dynamics of food webs: qualitative stability in a lotka-volterra cascade model. Proc. Roy. Soc. Lond. Ser. B 240, 607–627.

Cohen, J.E., Jonsson, T., Müller, C.B., Godfray, H.C.J., Savage, V.M., 2005. Body sizes of hosts and parasitoids in individual feeding relationships. Proc. Natl. Acad. Sci. U. S. A. 102, 684–689.

Dunne, J.A., 2006. The network structure of food webs. In: Pascual, M., Dunne, J.A. (Eds.), Ecological Networks: Linking Structure to Dynamics in Food Webs. Oxford University Press, Oxford, pp. 27–86.

Giraldeau, L.-A., Caraco, T., 2000. Social Foraging Theory. Princeton University Press, Princeton, NJ.

Gregory, P., 2005. Bayesian Logical Data Analysis for the Physical Sciences. Cambridge University Press, Cambridge.

Gross, T., Rudolf, L., Levin, S.A., Dieckmann, U., 2009. Generalized models reveal stabilizing factors in food webs. Science 325, 747–750.

Holt, R.D., 1977. Predation, apparent competition, and the structure of prey communities. Theor. Popul. Biol. 12, 197–229.

Hutchinson, G.E., 1959. Homage to santa rosalia or why are there so many kind of animals? Am. Nat. 93, 145–159.

Ingram, T., Stutz, W.E., Bolnick, D.I., 2011. Does intraspecific size variation in a predator affect its diet diversity and top-down control of prey? PLoS One 6, e20782.

Ings, T.C., Montoya, J.M., Bascompte, J., Blüthgen, N., Brown, L., Dormann, C.F., Edwards, F., Figueroa, D., Jacob, U., Jones, J.I., Lauridsen, R.B., Ledger, M.E., Lewis, H.M., Olesen, J.M., Frank Van Veen, F.J., Warren, P.H., Woodward, G., 2009. Ecological networks - beyond food webs. J. Anim. Ecol. 78, 253–269.

Janzen, D.H., 1970. Herbivores and the number of tree species in tropical forests. Am. Nat. 104, 501–528.

Jeffreys, H., 1961. The Theory of Probability. Oxford University Press, Oxford.

Kondoh, M., 2003. Foraging adaptation and the relationship between food-web complexity and stability. Science 299, 1388–1391.

Lloyd-Smith, J.O., 2005. Superspreading and the impact of individual variation on disease emergence. Nature 438, 293–295.

MacArthur, R., 1955. Fluctuations in animal populations, and a measure of community stability. Ecology 36, 533–536.

MacKay, D., 2003. Information Theory, Inference, and Learning Algorithms. Cambridge University Press, Cambridge.

Matthews, B., Marchino, K.B., Bolnick, D.I., 2010. Specialization of trophic position and habitat use by sticklebacks in an adaptive radiation. Ecology 91, 1025–1034.

May, R.M., 1973. Stability and Complexity in Model Ecosystems. Princeton University Press, Princeton, NJ.

May, R.M., 2006. Network structure and the biology of populations. Trends Ecol. Evol. 21, 394–399.

Murdoch, W.W., 1969. Switching in general predators: experiments on predator specificity and stability of prey populations. Ecol. Monogr. 39, 335–354.

Pachepsky, E., Bown, J.L., Eberst, A., Bausenwein, U., Millard, P., et al., 2007. Consequences of intraspecific variation for the structure and function of ecological communities. Part 2: linking diversity and function. Ecol. Model. 207, 277–285.

Pires, M.M., Guimarães, P.R., Araújo, M.S., Giaretta, A.A., Costa, J.C., dos Reis, S.F., 2011. The nested assembly of individual-resource networks. J. Anim. Ecol. 80, 896–903.

Polis, G.A., 1991. Complex trophic interactions in deserts: an empirical critique of food web theory. Am. Nat. 138, 123–155.

Polis, G.A., Winemiller, K.O., 1996. Food Webs: Integration of Pattern and Dynamics. Chapman and Hall, New York, NY.

Post, D.M., Palkovacs, E., Schielke, E.G., Dodson, S., 2008. Intraspecific phenotypic variation in a predator affects community structure and cascading trophic interactions. Ecology 89, 2019–2032.

Rooney, N., McCann, K.S., 2012. Integrating food web diversity and stability. Trends Ecol. Evol. 27, 40–46.

Rooney, N., McCann, K., Gellner, G., Moore, J.C., 2006. Structural asymmetry and the stability of diverse food webs. Nature 442, 265–269.

Rosindell, J., Hubbell, S.P., Etienne, R.S., 2011. The unified neutral theory of biodiversity and biogeography at age ten. Trends Ecol. Evol. 26, 340–348.

Roughgarden, J., 1972. Evolution of niche width. Am. Nat. 106, 683–718.

Sih, A., Cote, J., Evans, M., Fogarty, S., Pruitt, J., 2012. Ecological implications of behavioural syndromes. Ecol. Lett. 15, 278–289.

Stephens, D.W., Krebs, J.R., 1986. Foraging Theory. Princeton University Press, Princeton, NJ.

Stouffer, D.B., 2010. Scaling from individuals to networks in food webs. Funct. Ecol. 24, 44–51.

Sunnåker, M., Busetto, A.G., Numminen, E., Corander, J., Foll, M., Dessimoz, C., 2013. Approximate Bayesian Computation. PLoS Comput. Biol. 91, e1002803.

Svanbäck, R., Eklöv, P., Fransson, R., Holmgren, K., 2008. Intra-specific competition drives multiple species trophic polymorphism in fish communities. Oikos 117, 114–124.

Tarantola, A., 2006. Popper, Bayes and the inverse problem. Nat. Phys. 2, 492–494.

Tinker, M.T., Guimarães, P.R., Novak, M., Marquitti, F.M.D., Bodkin, J.L., Staedler, M., Bentall, G., Estes, J.A., 2012. Structure and mechanism of diet specialisation: testing models of individual variation in resource use with sea otters. Ecol. Lett. 15, 475–483.

Toni, T., Stumpf, M.P.H., 2010. Simulation-based model selection for dynamical systems in systems and population biology. Bioinformatics 26, 104–110.

Violle, C., Enquist, B.J., McGill, B.J., Jiang, L., Albert, C.H., Hulshof, C., Jung, V., Messier, J., 2012. The return of the variance: intraspecific variability in community ecology. Trends Ecol. Evol. 27, 244–252.

Volkov, I., Banavar, J.R., Hubbell, S.P., Maritan, A., 2009. Inferring species interactions in tropical forest. Proc. Natl. Acad. Sci. U. S. A. 106, 13854–13859.

West-Eberhard, M., 2003. Developmental Plasticity and Evolution. Oxford University Press, Oxford.

Wilson, W.G., Lundberg, P., Vázquez, D.P., Shurin, J.B., Smith, M.D., Langford, W., Gross, K.L., Mittelbach, G.G., 2003. Biodiversity and species interactions: extending lotka-volterra community theory. Ecol. Lett. 6, 944–952.

Winemiller, K.O., 1990. Spatial and temporal variation in tropical fish trophic networks. Ecol. Monogr. 60, 331–367.

Winemiller, K.O., Layman, C.A., 2005. Food web science: moving on the path from abstraction to prediction. In: de Ruiter, P.C., Wolters, V., Moore, J.C. (Eds.), Dynamic Food Webs: Multispecies Assemblages, Ecosystem Development and Environmental Change. Elsevier, Amsterdam, pp. 10–23.

Wolf, M., Weissing, F.J., 2012. Animal personalities: consequences for ecology and evolution. Trends Ecol. Evol. 27, 452–461.

Woodward, G., Warren, P., 2007. Body size and predatory interactions in freshwater: scaling from individuals to communities. In: Hildrew, A.G., Raffaelli, D.G., Edmonds-Brown, R. (Eds.), Body Size: The Structure and Function of Aquatic Ecosystems. Cambridge University Press, Cambridge, pp. 98–117.

Woodward, G., Blanchard, J., Lauridsen, R.B., Edwards, F.K., Jones, J.I., Figueroa, D., Warren, P.H., Petchey, O.L., 2010. Individual-based food webs: species identity, body size and sampling effects. Adv. Ecol. Res. 43, 211–266.

Yodzis, P., 1989. Introduction to Theoretical Ecology. Harper and Row, New York, NY.

Yoshida, T., Jones, L., Ellner, S., Fussmann, G., Hairston Jr., N.G., 2003. Rapid evolution drives ecological dynamics in a predator-prey system. Nature 424, 303–306.

> CHAPTER SEVEN

Community Genetic and Competition Effects in a Model Pea Aphid System

Mouhammad Shadi Khudr*, Tomos Potter*, Jennifer Rowntree*, Richard F. Preziosi*[,1]
*Faculty of Life Sciences, The University of Manchester, Manchester, United Kingdom
[1]Corresponding author: e-mail address: preziosi@manchester.ac.uk

Contents

Advances in Ecological Research, Volume 50
ISSN 0065-2504
http://dx.doi.org/10.1016/B978-0-12-801374-8.00007-4

Abstract

The effect of within-species (intraspecific) genetic variation on associated ecological communities (community genetic effects) has been well documented. However, little is known about the relationship between community genetic effects and other ecological forces. Community genetic effects are likely to be especially important in clonal systems, such as aphids, where individuals experience pressures exerted by both interspecific and intraspecific competitors in the context of variation among host plants. In these systems, the genotypes and genetic variation of both host plants and competitors are likely to be important influences on performance of the other species in the system. Here, we examine the effects of different types of competition and community genetics using a model system of aphids and their faba bean host plants. We observe the effects on a focal aphid experiencing both interspecific and intraspecific competition. We found that among aphids, competition was strongest (i.e. had the greatest negative effect on the fitness of the focal aphid) when competitors were conspecifics and that this relationship persisted in more complex environments. However in all cases, plant genetic environment influenced the outcome of competition, although with seemingly less influence than the aphid competitors themselves. Aphid behaviour (i.e. change in feeding site preference) was strongly influenced by community genetic effects of the host and both interspecific and intraspecific competition, although the patterns of distribution did not seem to follow those of fitness. Our findings highlight the importance of understanding the complex interactions of ecology and evolution in agro-ecosystems.

1. INTRODUCTION

There is a growing recognition that the synthesis of ecology and evolution is essential for understanding how species interact and develop through time and space (Odling-Smee et al., 2003; Pelletier et al., 2009; Schoener, 2011; Thompson, 1994; Whitham et al., 2003, 2006). When ecology and evolution are examined relative to one another, it becomes evident that we need to investigate how multi-level processes influence interacting species within and among communities (Hersch-Green et al., 2011; Hughes et al., 2008; Johnson and Stinchcombe, 2007). The rapidly growing field of community genetics considers that the environment of a species may be significantly defined by the expression of the genomes of other interacting species (community genetic effect) (Agrawal, 2003; Antonovics, 1992; Rowntree et al., 2011b). This perspective presents a clear framework for the integration of ecology, genetics and evolution (Rowntree et al., 2011b). New insights from community genetics research have made clear the need to determine the relative importance of community genetic effects particularly as a force for evolutionary change (Hersch-Green et al., 2011; Rowntree et al., 2011b). As such, the relationship between community genetic effects and other eco-evolutionary processes, such as competition, are of particular interest.

A clear example of community interactions is that of host plants and insect herbivores. Variability in plant hosts, be it architecture, resource quality or resistance to herbivory (Farnsworth et al., 2002; Johnson and Agrawal, 2005; Karban, 1989; Whitham et al., 2003) can have large effects on the fitness of insect herbivores (Haloin and Strauss, 2008; Whitham et al., 2006). In turn, herbivorous insects, through their exploitation, selectivity and manipulation of host resources, undergo a series of co-evolutionary interactions with their hosts, and indeed, with other interacting invertebrates (Dixon, 1977, 1987, 1998; Powell et al., 2006).

1.1. Competition as an eco-evolutionary force

Competition is a significant force that shapes individuals, populations and species (Connell, 1983; Denno et al., 1995; Denno and Kaplan, 2007; Hairston et al., 1960; Hutchinson, 1978; Schoener, 1983, see also Kaplan and Denno, 2007) and, in plant–herbivore communities, can act directly or indirectly via the mediation of the host plant (Denno et al., 1995; Denno and Kaplan, 2007). The magnitude of competition effects, however, may also be influenced by the environment-modifying feedback of the herbivores themselves (Bürger, 2002; Ernebjerg and Kishony, 2011; Fordyce, 2006; Jablonka, 2004; Laland, 2004; Odling-Smee et al., 2003). While many studies have examined effects of plant genetic variation on the form of associated communities, a process that likely influences herbivore competition (see Poelman et al., 2008), only a handful of studies have explicitly examined the influence of plant genetics on insect competition (e.g. Smith et al., 2008). Ideal systems especially useful for the examination of such effects are the low-diversity systems found in agricultural fields. Conventional agricultural systems are usually a mosaic of homogenous genetic patches, that is, monovarietal plots where genetic diversity of the dominant species is highly restricted or entirely absent. In this context, host-plant intraspecific genetic variability will significantly affect the richness, abundance, survivorship and behaviour of associated invertebrate pests (Dungey et al., 2000; Fritz and Price, 1988; Johnson et al., 2006; Utsumi et al., 2011; Whitham et al., 2003; Zytynska and Preziosi, 2011).

1.2. Host-plant effects on aphid fitness and behaviour

For a generalist aphid, decisions about host choice and selection of feeding position on the plant can be influenced by plant chemical cues (Chapman, 2003; Powell et al., 2006; Ranger et al., 2007). The affinity to specific hosts, and to particular parts of the host plant, is based on aphid pre- and post-ingestion probing, and determines herbivore acceptance or discrimination (Powell et al., 2006). In addition, on–plant distribution and among plant distribution of aphids are governed by the availability of suitable locations, varying levels of plant defence, presence of natural enemies and presence and density of competitors (e.g. Ranger et al., 2007; Schuett et al., 2011; Smith et al., 2008; Utsumi et al., 2011).

The literature is rich with research examining the influence of host quality and variation on the performance (reproductive success and survival) of phytophagous insects (e.g. Dungey et al., 2000; Hochwender and Fritz, 2004; Johnson et al., 2006; Ruhnke et al., 2006, 2009; Utsumi et al., 2011; Smith et al., 2008; Whitham et al., 2003; Zytynska and Preziosi, 2011). Moreover, host genotypes are known to vary in their susceptibility and resistance to invertebrate pests, which can result in differential aphid population growth rates among host genotypes (e.g. Hughes et al., 2008; Underwood and Rausher, 2000). However, with regard to bottom–up and cascading effects in multi–trophic systems (Hambäck et al., 2007; Johnson, 2008; Power, 1992), there are few studies exploring the impact of host genotype on the feeding location preference of aphids under intra-specific and interspecific competition (but see Salyk and Sullivan, 1982; Smith et al., 2008; Utsumi et al., 2011; Zytynska and Preziosi, 2011). Understanding the composition of the interactions between plant and aphid genotypes under the conditions of intra- and interspecific competition has relevance to the understanding of pest–crop relationships as well as developing a better comprehension of how ecology affects evolution and vice versa (Johnson and Agrawal, 2007; Johnson and Stinchcombe, 2007; Odling-Smee et al., 2003; Östman, 2011, see also Schoener, 2011; Whitham et al., 2003, 2006; Wimp et al., 2007).

Here, we measure the simultaneous effects of community genetics and competition. We test the effects of interspecific and intraspecific, as well as mixed (inter- plus intraspecific) competition on the fitness (reproductive success) and behaviour (on–plant distribution) of a focal pea aphid genotype. The clonal reproductive nature of aphids allows us to examine intraspecific competition that occurs among specific clones and we use this as our working definition of intraspecific competition, rather than competition within a clone (*sensu* Smith et al., 2008). To our knowledge, this is one of the first studies to examine the influence of the interaction between community genetic (plant genotype) and competition effects at multiple levels on aphids.

2. MATERIALS AND METHODS

2.1. Species and genotype descriptions

We compared community genetic (plant genotype) effects with inter-, intra- and mixed (inter- plus intra-) aphid competition effects on a focal genotype of pea aphid (*Acyrthosiphon pisum* Harris). The focal pea aphid was the pink genotype P127 and was supplied by Imperial College (London). Interspecific competitors were the vetch aphid (*Megoura viciae* Buckton; clonal population established from a single individual collected at the University of Manchester Botanical Grounds, hereafter Meg) and the green peach aphid (*Myzus persicae* Sulzer; clonal population established from a single individual collected near Buxton, UK, hereafter Myz). Intraspecific competitors were green pea aphid genotypes [ORG (=JF01/29), 116, LLO1, BOT]. Genotypes ORG, 116 and LL01 were also supplied by Imperial College (London), while BOT was established from a single aphid progeny of a gravid female collected from a faba bean plant at the University of Manchester Botanical Grounds. We define intraspecific competition as competition among clones of pea aphids. This is distinct from a usual definition of intraspecific competition that would include all competition within a species (both among and within clones), albeit this distinction is probably only relevant for clonally reproducing species. All stocks of aphids were maintained on faba bean plants (*Vicia faba* L. var. major Harz) in growth cabinets at [22 °C and 16:8 (Light:Dark)]. This cultivar (major) was not used in the experimental procedures in order to avoid preconditioning effects.

Three faba bean cultivars were used in the experiments [Masterpiece Green Longpod (LP), Optica (O) and Sutton (Sut)]. All cultivars were procured from commercial seed suppliers. These selectively bred varieties (also referred to as accessions) are genetically distinct synthetics (Duc, 1997; Flores et al., 1998) that do not exist in the wild, but have been maintained for hundreds of generations as inbred lines.

2.2. Experimental conditions

We used three distinct experiments to examine competition effects. Experiment I examined the effects of interspecific competition and we compared the performance and behaviour of P127 in the presence of each of the two interspecific competitors (Myz or Meg) on the three faba bean cultivars. Experiment II examined the effects of intraspecific competition and we compared the performance and behaviour of P127 in the presence of each of the four intraspecific competitors (ORG, 116, BOT, LL01) on the three faba bean cultivars. These two experiments shared a 'competition with self' control treatment where the performance of P127 was assessed without any inter- or intraspecific competitors. However, the 'competition with self' treatment will still include competition among individuals of the same genotype. Experiment III examined the effects of mixed competition and we compared the performance of P127 in the presence or absence of an interspecific competitor (Meg) and each of two intraspecific competitors (LL01, ORG) on the three faba bean cultivars.

For all experiments, Faba beans were planted individually in John Innes no. 3 steam-sterilised compost in 3.5″ plastic pots and covered with 22 ounce plastic cups with two 4 cm × 4 cm fine mesh windows to allow ventilation. Cups were sealed to the pots with PVC tape and plants were randomly placed in a growth cabinet. Plant enclosures were removed from the growth cabinets and watered every 2 days by injecting water directly into the soil through re-sealable pores near the base of the plastic pots. Plants were re-randomised in their placement when returned to the growth cabinets. Plants were standardised in age (1 month) and trimmed to a standardised height (12 cm) at the start of the experiment.

At the start of the experiments, four second-instar aphids were added to each pot. Four P127 individuals were added for the 'competition with self' control treatment, and two P127 individuals with two individuals of each of the competitor aphid lines for the 'inter, intra and mixed competition' treatments. Ten replicates were planted for each treatment group. All P127 aphids on each plant were counted on day 14. Numbers of final aphids for the 'competition with self' treatment in Experiments I and II were divided by two to correct for the doubled number of individuals at the start. Additionally, the positions (whether the aphids were on the plant stems or leaves) of P127 aphids were recorded for Experiments I and II. All experiments were undertaken in September to December 2010.

2.3. Measures of fitness

We used the total number of P127 aphids at day 14 as our measure of aphid fitness. This is equivalent to a measure of population growth rate (e.g. Agrawal et al., 2004) because the starting number of aphids and the number of days was the same for each replicate.

2.4. Aphid behaviour

Position of the aphids on the plants (stems vs. leaves) in Experiments I and II was taken as a measure of aphid behaviour, as it represented a selection of feeding site by the individual aphids.

2.5. Statistical analysis

We analysed the fitness data using generalised linear models with a quasipoisson distribution to account for the expected positively skewed distribution of total aphid numbers. All of these models were run using the glm programme in R version 3.0.2 (R Core Team, 2013). First, using the data from Experiments I and II, we tested the effect of the presence or absence of inter- and intraspecific competition on P127 (total P127 = faba cultivar + competition (inter-/intra-/self) + interaction term). Then, using data from Experiment I, we tested the effect of the different interspecific competitors on P127 (total P127 = faba cultivar + interspecific competitor (Myz/Meg) + interaction term). Next, using data from Experiment II, we tested the effect of different intraspecific competitors on P127 (total P127 = faba cultivar + intraspecific competitor (ORG/116/LL01/BOT) + interaction term). Finally, using data from Experiments II and III, we tested the effect of two different intraspecific competitors with and without a single interspecific competitor (Meg) on P127 (total P127 = faba cultivar + intraspecific competitor (LL01/ORG) + interspecific competition (Y/N) + interaction terms).

We analysed the behavioural (positional) data using generalised linear models with a binary distribution and a logit link function using the glm programme in R version 3.0.2 (R Core Team, 2013). First using data from Experiment I, we tested the effect of the different interspecific competitors on the position of P127 aphids (P127 position = faba cultivar + interspecific competitor (Myz/Meg) + interaction term). Then, using data from Experiment II, we tested the effect of the different intraspecific competitors on the position of P127 (P127 position = faba cultivar + intraspecific competitor (ORG/116/LL01/BOT) + interaction term).

Graphics were produced using the ggplot2 programme (Wickham, 2009) in R version 3.0.2 (R Core Team, 2013).

3. RESULTS

3.1. The effect of competition on P127

There was a significant effect of faba bean cultivar ($F_{2,171} = 3.20$, $p = 0.04$) and a significant effect of competition ($F_{2,171} = 12.44$, $p < 0.0001$) on the number of P127 at day 14, but no significant interaction between host cultivar and competition ($F_{4,171} = 0.975$, $p = 0.42$). Numbers of P127 increased under competition with higher numbers of P127 when the competitor was a different species than when the competitor was a different genotype of the same species. This suggests that the more related to each other the competitors are, the greater the strength of competition among aphids sharing a host plant. It also suggests that the strength of competition among aphids depends on a community genetic effect of the host plant (Fig. 7.1).

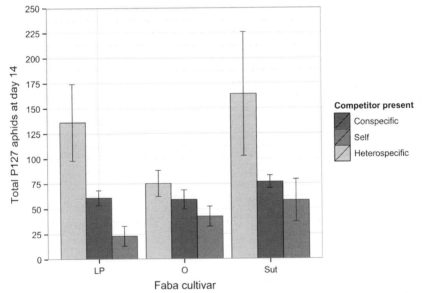

Figure 7.1 Final numbers (mean ± 1 SE) of P127 aphids at day 14 when competing against heterospecifics (red bars; light grey, position 1 in print version), conspecifics (green; dark grey, position 2 in print version) or one's own genotype (blue; mid grey, position 3 in print version) and on three faba bean cultivars.

3.2. Interspecific competition

There was a marginally significant effect of faba bean cultivar ($F_{2,49} = 2.96$, $p = 0.06$), a significant effect of the different interspecific competitors ($F_{1,49} = 12.62$, $p = 0.0009$) and a significant interaction between host cultivar and competitor species ($F_{2,49} = 4.12$, $p = 0.02$) on the number of P127 at day 14. Numbers of P127 were greatest when the host cultivar was Sut and the competitor species Meg, and least when the host cultivar was Sut but the competitor species Myz. This data suggests that the effect of competition among aphid species sharing a host plant depends on a community genetic effect of the host plant (Fig. 7.2).

The position of P127 on the plant depended on the faba bean cultivar ($p < 0.0001$), the competitor species ($p < 0.0001$) and an interaction between the two ($p < 0.0001$). For the cultivars Sut and O, the position of P127 on the plant seemed to be determined by host genotype, rather than competitor species, with P127 on Sut showing a preference for the leaves of the plant and P127 on O showing a preference for the stems. On cultivar LP, the identity of the interspecific competitor was also important, with a greater proportion of P127 found on the stems of the plants when Meg was the

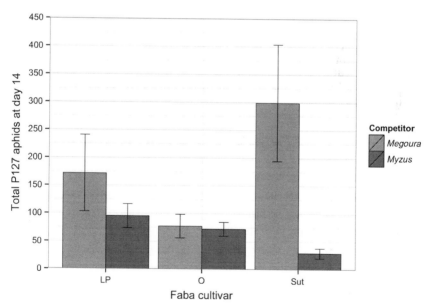

Figure 7.2 Final numbers (mean ± 1 SE) of P127 aphids at day 14 when competing against heterospecifics on three faba bean cultivars. Red (light grey in print version) bars are when competing against Meg and blue (dark grey in print version) bars are when competing against Myz.

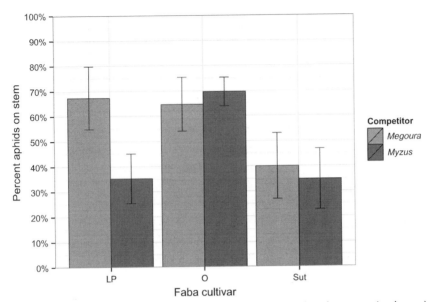

Figure 7.3 Percent (mean ± 1 SE) of P127 aphids found on the plant stem (vs. leaves) when on three different faba cultivars. Red (light grey in print version) bars are when P127 was competing against Meg and blue (dark grey in print version) bars are when P127 was competing against Myz.

competitor species and the opposite when Myz was the competitor species. This shows that there is a community genetic effect of the plant genotype on aphid behaviour that also determines the impact of competition among species of aphids sharing a host plant (Fig. 7.3).

3.3. Intraspecific competition

There was no significant effect of faba bean cultivar ($F_{2,95} = 2.08$, $p = 0.13$), but a significant effect of intraspecific competitor ($F_{3,95} = 5.52$, $p = 0.002$) and a significant interaction between host cultivar and competitor genotype ($F_{6,95} = 2.32$, $p = 0.04$). On cultivar LP, the population size of P127 was greatest when aphid genotype BOT was the competitor, while on cultivar O, the population size of P127 was greatest when aphid genotype 116 was the competitor. Numbers of P127 on cultivar Sut were similar for all genotypes of competitor aphids. Thus, an aphid's ability to compete with different genotypes of the same species also depends on a community genetic effect of the host plant (Fig. 7.4).

The position of P127 on the plant depended on the faba bean cultivar ($p < 0.0001$), the genotype of competitor ($p < 0.0001$) and an interaction between the two ($p < 0.0001$). In the majority of cases, there were a greater

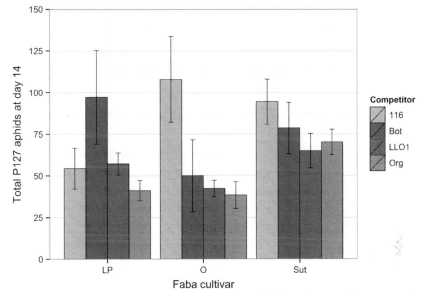

Figure 7.4 Final numbers (mean±1 SE) of P127 aphids at day 14 when competing against conspecifics on three faba bean cultivars. The four conspecific clones were 116 (red bars; light grey, position 1 in print version), Bot (green bars; dark grey, position 2 in print version), LLO1 (blue bars; dark grey, position 3 in print version) and Org (purple bars; mid grey, position 4 in print version).

proportion of P127 on the stems of the plants rather than the leaves. However, this was not always the case and the actual proportion of P127 individuals on stems versus leaves depended on the specific combination of host-plant cultivar and aphid genotype. On host cultivar LP, P127 showed a preference for feeding on the stem if the competitor species was Myz, but a preference for feeding on the leaves if the competitor was Meg. On host cultivar O, P127 showed a slight preference for feeding on the stem with both competitor species. On host cultivar Sut, P127 showed a slight preference for feeding on the leaves with both competitor species. This suggests that the choice of feeding position in the light of intraspecific competition was influenced by a genotype by genotype interaction between the host plant and the aphid herbivores (Fig. 7.5).

3.4. Mixed competition

When both an inter- and an intraspecific competitor were present, the number of P127 on day 14 only depended on a significant interaction between the intraspecific competitor and faba bean cultivar ($F_{2,198} = 5.66$, $p = 0.004$). Numbers of P127 differed across host cultivars and among competitor

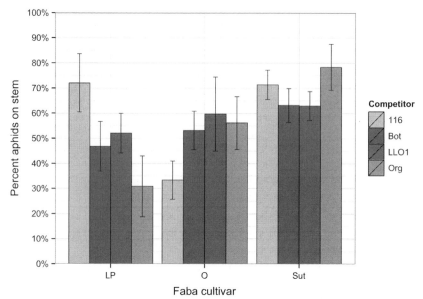

Figure 7.5 Percent (mean ± 1 SE) of P127 aphids found on the plant stem (vs. leaves) when on three different faba cultivars. P127 aphids were in competition with four conspecific clones; 116 (red bars; light grey, position 1 in print version), Bot (green bars; dark grey, position 2 in print version), LLO1 (blue bars; dark grey position 3 in print version) and Org (purple bars; mid grey, position 4 in print version).

genotypes of the same species, but were similar with and without the presence of an interspecific competitor. The exception to this was when the host cultivar was Sut and the intraspecific competitor was LL01. This suggests that, in general, the relationships among genotypes of the same aphid species competing for a shared resource hold in a more complex environment. That is, the strength of competition among genotypes of the same species of aphids is greater than competition among species of aphids sharing a host plant, and that intraspecific competition depends on a community genetic effect of the host (Fig. 7.6; see Table 7.1 for full test details).

4. DISCUSSION

This work is one of the few studies that investigate the effect of intraspecific genetic variation in a host plant on competition amongst con- and heterospecific sap-feeding insects. Our findings also address a current question about the relative importance of community genetic effects compared to other ecological forces.

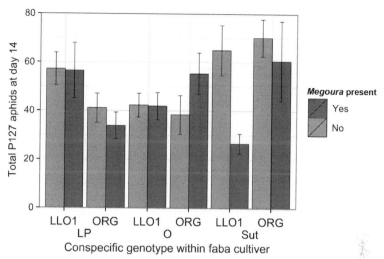

Figure 7.6 Final numbers (mean ± 1 SE) of P127 aphids at day 14 when competing on three faba bean cultivars. P127 aphids were competed against two conspecific clones (LLO1 and Org) and in the absence (red (light grey in print version) bars) or presence (blue (dark grey in print version) bars) of a heterospecific competitor (Meg).

Table 7.1 ANOVA results for 3-way analysis with cultivar, presence or absence of a heterospecific competitor and genotype of the conspecific competitor as factors

Factor	df	F	P
Cultivar	2196	0.0595	0.9423
Heterospecific (Y/N)	1195	0.7266	0.3951
Conspecific (LLO1/Org)	1194	0.4619	0.4976
Cultivar * heterospecific	2192	2.2140	0.1121
Cultivar * conspecific	2190	5.6626	0.0041
Heterospecific * conspecific	1189	1.4391	0.2318
Cultivar * Het * Con	2187	0.9708	0.3807

4.1. Comparative strength of competition

We measured the effect of inter- and intraspecific competition and host-plant genotype on the fitness and behaviour of a focal aphid clone. We show that intraspecific competition has a greater effect than interspecific competition. That is, the population size of our focal aphid (P127) was greater when competing against heterospecifics than when competing against multiple conspecific genotypes, or itself. Theory predicts that for there to be coexistence among competitors sharing a common resource, intraspecific competition (the sum of within and among clonal competition) should be greater than interspecific competition (Chesson, 2000), therefore our data would support the coexistence of the aphid species we tested sharing a faba bean host.

At first glance, these results differ from the work of Smith et al. (2008), who showed that the strength of interspecific competition among aphids colonising milkweed plants (*Asclepias syriaca*) was greater than intraspecific competition across a variety of complex environments. However, in their 'plant genotype by competition' experiment (one of three experiments in that study) the overall response to competition varied among the three aphid species examined and the response within each species varied widely based on host–plant genotype. We note that Smith et al. (2008) used a single genotype of each species, thus, their intraspecific competition treatment would be equivalent to our 'competition with self' treatment. Overall, our results do agree with Smith et al. (2008), in that plant genotype influences competitive interactions.

4.2. Relatedness and competition

It is perhaps counterintuitive that our focal aphid clone had the lowest reproductive rate when in competition with itself, a greater reproductive rate when competing against conspecific clones, and its highest reproductive rate when competing against heterospecifics. One possible explanation for this is that the specific combination of clones utilised reduced plant defences and thus increased the fitness of the focal clone (Dicke and Hilker, 2003; Hays et al., 1999; Sharon Zytynska et al., unpublished data). An alternative explanation is Darwin's hypothesis that more closely related individuals will be in greater competition and competitive success will be greater when the competitor is more distantly related (Darwin, 1859). This 'competition–relatedness hypothesis' (*sensu* Cahill et al., 2008) would suggest, as we find, that our focal clone would have its lowest success when all individuals are of the same clone, higher success when some of the local population is of other clones of the same species and the greatest fitness when some of the population is a different species.

4.3. Interspecific competition

The outcome of interspecific competition among aphids depended upon the environment provided by the host plant, in combination with the identity of the heterospecific competitors. From our data, it is clear that plant genotype can strongly influence the environment provided by the host. This reflects the outcome of numerous previous studies (e.g. Fritz and Price, 1988; Hughes et al., 2008; Johnson, 2008; Rowntree et al., 2011a; Underwood, 2009; Whitham et al., 2006; Zytynska et al., 2011), and, in particular, concurs with the findings of Smith et al (2008), who found that host-plant genotype-mediated competition among aphid species on individual milkweed plants. We can conclude that the abundance and distribution of the host genotypes, which is influenced by natural selection on the host species, influences, in turn, the ecological interactions amongst aphids.

We used different cultivars of a horticultural crop (faba bean) as our host plants. Although these do represent genetically distinct lines of faba beans (Duc, 1997), they are not found in natural or semi-natural environments. However, as much of available land is used for the cultivation of agricultural crops (The World Bank, 2013), our experimental system is representative of much of the environment. In agricultural settings, where the distribution and abundance of host genotypes is fixed, the outcomes of competition among aphids and other herbivores, as well as the potential extent of damage these herbivores inflict on the host plant, is largely determined by management practise, in particular, the use or otherwise of intercropping techniques (Zhu et al., 2000).

There was a strong effect of all factors tested on the behaviour of P127 in the presence of heterospecific competitors. Behaviour (selection of stem or leaf as a feeding site) was strongly influenced by the specific combination of host genetic environment and competing species. The choice of leaves or stems as a feeding site for aphids on faba beans has been linked to host resistance or susceptibility to aphid infestation (Bond and Lowe, 1975), with more resistant plants having a greater proportion of aphids feeding on the stems and more susceptible plants a greater proportion of aphids feeding on the leaves. Additional work, however, has also shown that competition amongst two heterospecific aphids (*Aphis fabae* and *A. pisum*) feeding on a single cultivar of faba bean (Windsor) can change their preference for stem or leaves as a feeding site (Salyk and Sullivan, 1982). More recent studies have demonstrated that previous exposure to aphid herbivory can change the attractiveness of the host plant to subsequent herbivores (Brunissen et al., 2009) and similar effects may have occurred in our system. We suggest that the specific combination of host-plant genotype and the identity of the competitor species may be influencing the susceptibility of the host to P127, which in turn modifies aphid behaviour and choice of feeding site. Although we demonstrate that there is plasticity in feeding site preference, there was no obvious association between preference and fitness of P127 on the plant cultivars used. This leads us to believe that while the impact on aphid fitness of interspecific competition may be the result of exploitation competition, the influence on behaviour is likely to be the result of interference competition.

4.4. Intraspecific competition

Similar to the results for interspecific competition, the fitness outcome of intraspecific competition was influenced by the interaction of host genotype and competitor, and by the genetic identity of the competitor. There was no main effect of host genotype on P127 fitness, in contrast, the focal aphid was influenced by intraspecific competition, a process that results in evolutionary change, combined with the environment it finds itself in, in this case defined by the host-plant genotype. In this experiment, we have broken down the overall host-plant effect into genotype-specific effects, revealing the potential for selection and evolutionary change in host plants to alter the environment and ecology of the competing aphids. Quite literally, the abundance and distribution (aka ecology) of the focal aphid is influenced by genetically based variation that is subject to natural selection. This pattern is consistent with other community genetic studies where the genotype of the dominant species influences both the presence and abundance of species in the plant-associated community (e.g. Johnson and Agrawal, 2005; Johnson et al., 2006; Whitham et al., 2003, 2006, see also Barbour et al., 2009).

The intraspecific competitor in this study represents a portion of the focal aphid's environment that is defined by a single competing genotype of the same species. Thus, changes in the relative frequencies of the aphid genotypes (clones) over generations are, by definition, evolution. In studies competing aphid clones (*M. persicae*) across a range of initial aphid densities, Turcotte et al. (2013) showed a clear effect of starting density on evolutionary trajectory (i.e. change in the relative density of clones) over a very short period of time. Here, in our study, the effect host genotype further modifies the evolutionary trajectory of the focal aphid and, if there were movement of aphid clones among host plants of differing genotypes, it would likely to result in a complex feedback between the ecology and evolution of the aphids and plants. While evolution in a commercial crop is unlikely, the planting patterns of plant cultivars are often subject to yearly change. Such changes in the numbers and distributions of plant cultivars year to year will drastically alter the interactions between plant hosts and their herbivores.

We note here that evolution in the host plant by natural selection or artificial selection is entirely possible (indeed, has been used to shape the phenotypes being planted) but is unlikely to occur in agricultural fields for two reasons: (1) host plants are often planted as genetic monocultures, thus, the lack of genetic variation means there can be no evolutionary response to phenotypic selection; (2) most agricultural systems harvest plants within one generation and replant with a predetermined genotype(s), therefore, there will be no offspring generation of the individuals that have undergone phenotypic selection (instead, we usually consume the propagules or use them as feed for livestock).

Across all treatment groups, P127 were found more often on the stems of the host plants compared to the leaves, but behaviour of P127 was also influenced by the specific interaction of host genotype and conspecific aphid genotype. The pattern of feeding site preference is a mix of both subtle changes in the strength of preference and more drastic switches amongst feeding sites. Interestingly, while the overall strength of competition is greater when the competitor is heterospecific, the pattern of behavioural outcomes remains strongly context dependent on the specific combination of competitor and host genotype. As with the interspecific competitors, there was no strong association between fitness and behaviour despite the fact that both were being influenced by the interaction between the host and competitor genotypes.

4.5. Mixed competition

The response of P127 to mixed competition with both an intraspecific and an interspecific aphid competitor mirrored the patterns observed for intraspecific competition alone. That is, the number of P127 depended on the specific combination of host-plant genotype and the genotype of the intraspecific competitor, and not the presence of an interspecific competitor. This lends weight to our previous supposition that, in this system, the effects of intraspecific competition on a focal aphid are greater than those of interspecific competition. It also supports the theory of coexistence of competitors (Chesson, 2000) and provides a mechanism pertinent to the empirical observation that the different species or clones of aphids can coexist on faba beans. In an experiment testing the effect of genetic diversity within the aphid (*A. pisum*) on interspecific competition (with *M. viciae*), Hazell et al. (2006) showed that aphid genotype could change the outcome of interspecific competition. This highlights further the importance that within-species genetic variation can have on the outcome of species interactions (Mooney, 2011). Had we incorporated genetic variation of the interspecific competitors in our study (we only used a single genotype of the interspecific competitors), we may well have revealed additional complexity as a result of interactions among genotypes of all three species.

4.6. Consequences for agriculture

Ideas for the improvement of agricultural crops have recently re-focused on the incorporation of natural selection, eco-evolutionary dynamics and community interactions (Denison et al., 2003; Loeuille et al., 2013; Thrall et al., 2011). We suggest that these can be taken a step further by including the genetic variation in herbivores. By and large, aphid population growth in agricultural studies is considered a function of plant-defence resistance. Our results indicate that understanding the outcome of plant–insect interactions requires an appreciation of the genetics of both species, that is, the 'resistance' of a plant (and the subsequent consequences for productivity) is only meaningful in the context of the insect species and genotype(s) for which resistance has been observed.

4.7. Relative importance of effects

Recent work in the field of community genetics has emphasised the importance of determining not only the presence of genotypic effects in multiple systems but also their relative importance in the face of other environmental variables (Hersch-Green et al., 2011). In an experiment testing the relative effect of plant genotype and soil fertilisation on competition among plant species, Johnson et al. (2008) found the effect of soil fertilisation regime to be much more important than plant genotype, and that a plant genotype effect could only be seen within fertilisation treatments. Similarly, Tack et al. (2010) found spatial location to have a more profound effect on arthropod community assemblage of oak trees (*Quercus robur*) than plant genotypic effects. However in a meta-analysis of multiple experiments, Bailey et al. (2009) concluded that although, in general, the importance of genetic effects declined with increasing levels of organisation, this was not always the case. In this study, we found the effects of both inter- and intraspecific competition among aphids to be mediated by the host-plant genotype. Therefore, while competition among aphids appears to have a larger effect on aphid fitness than plant genotype *per se*, the consistency of significant interactions with genetic variation in the plants shows that, in fact, the outcome of competitive interactions among aphids depends intimately on community genetic effects of the plant.

4.8. Conclusion

Using a community genetics approach we found that both interspecific and intraspecific competition were context dependent on the genetic variation in host plant. Thus, the ecological dynamics in this system were dependent on heritable traits in two of the species involved. For pea aphids, we directly assessed fitness, and the interplay of ecology and evolution is clear. For faba beans we did not assess fitness but it is a reasonable assumption that plant fitness will be affected by herbivore pressure. As our system is based on an agricultural host plant, the plants are not likely to evolve. However, similar effects will be present in natural systems, where aphid herbivores would act as a selective force on host plants. Studies that clearly demonstrate all components of the interplay between ecology and evolution in multi-species systems are rare, but further studies on the effects of genetic variation, and thus evolution, on ecological dynamics will be central to understanding eco-evolutionary dynamics.

ACKNOWLEDGEMENTS

We thank Mrs. Lisa Andrejczak for her kind assistance in system maintenance. Our sincere thanks go to Dr. Glen Powell (Imperial College, London) for supplying pea aphid genotypes. We also thank Dr. Yvonne Golding for providing green peach aphids. We are grateful to Dr. Daniel Rozen for his constructive criticism and supportive insights. Our hearty thanks also go to Dr. David Penney for his invaluable remarks and proofreading the chapter. This work has been funded via the HECBPS scheme in cooperation between Damascus University and The British Council.

REFERENCES

Agrawal, A.A., 2003. Community genetics: new insights into community ecology by integrating population genetics. Ecology 84, 543–544.

Agrawal, A.A., Underwood, N., Stinchcombe, J.R., 2004. Intraspecific variation in the strength of density dependence in aphid populations. Ecol. Entomol. 29, 521–526.

Antonovics, J., 1992. Toward community genetics. In: Fritz, R.S., Simms, E.L. (Eds.), Ecology and Evolution of Plant Resistance to Herbivores and Pathogens: Ecology, Evolution, and Genetics. University Of Chicago Press, Chicago, USA, pp. 426–449.

Bailey, J.K., Schweitzer, J.A., Úbeda, F., Koricheva, J., LeRoy, C.L., Madritch, M.D., Rehill, B.J., Bangert, R.K., Fischer, D.G., Allan, G.J., Whitham, T.G., 2009. From genes to ecosystems: a synthesis of the effects of plant genetic factors across levels of organisation. Philos. Trans. R. Soc. B 364, 1607–1616.

Barbour, R.C., Forster, L.G., Baker, S.C., Steane, D.A., Potts, B.M., 2009. Biodiversity consequences of genetic variation in bark characteristics within a foundation tree species. Conserv. Biol. 23 (5), 1146–1155.

Bond, D.A., Lowe, H.J.B., 1975. Tests for resistance to *Aphis fabae* (Hom. Aphididae) in field beans (*Vicia faba*) (Leguminosae). Ann. Appl. Biol. 81, 21–32.

Brunissen, L., Cherqui, A., Pelletier, Y., Vincent, C., Giordanengo, P., 2009. Host-plant mediated interactions between two aphid species. Entomol. Exp. Appl. 132, 30–38.

Bürger, R., 2002. Additive genetic variation under intraspecific competition and stabilizing selection: a two-locus study. Theor. Popul. Biol. 61, 197–213.

Cahill, J.F., Kembel, S.W., Lamb, E.G., Keddy, P.A., 2008. Does phylogenetic relatedness influence the strength of competition among vascular plants? Perspect. Plant Ecol. Evol. Syst. 10 (1), 41–50.

Chapman, R.F., 2003. Contact chemoreception in feeding by phytophagous insects. Annu. Rev. Entomol. 48, 455–484.

Chesson, P., 2000. Mechanisms and maintenance of species diversity. Annu. Rev. Ecol. Syst. 31, 343–366.

Connell, J.H., 1983. On the prevalence and relative importance of interspecific competition: evidence from field experiments. Am. Nat. 122 (5), 41–50.

Core Team, R., 2013. R: A Language and Environment for Statistical Computing. R Foundation for Statistical Computing, Vienna, Austria.http://www.R-project.org/.

Darwin, C.R., 1859. The Origin of Species. John Murray, London.

Denison, R.F., Kiers, E.T., West, S.A., 2003. Darwinian agriculture: when can humans find solutions beyond the reach of natural selection? Q. Rev. Biol. 78 (2), 145–168.

Denno, R.F., Kaplan, I., 2007. Plant-mediated interactions in herbivorous insects: mechanisms, symmetry, and challenging the paradigms of competition past. In: Ohgushi, T., Craig, T.P., Price, P.W. (Eds.), Ecological Communities: Plant Mediation in Indirect Interaction webs. Cambridge University Press, Cambridge, UK, pp. 19–50.

Denno, R.F., McClure, M.S., Ott, J.R., 1995. Interspecific interactions in phytophagous insects: competition re-examined and resurrected. Annu. Rev. Entomol. 40, 297–331.

Dicke, M., Hilker, M., 2003. Induced plant defences: from molecular biology to evolutionary ecology. Basic Appl. Ecol. 4 (1), 3–14.

Dixon, A.F.G., 1977. Aphid ecology: life cycles, polymorphism, and population regulation. Annu. Rev. Ecol. Evol. Syst. 8, 329–353.

Dixon, A.F.G., 1987. Parthenogenetic reproduction and the rate of increase in aphids. In: Minks, A.K., Harrewijn, P. (Eds.), Aphids, Their Biology, Natural Enemies and Control. Elsevier, Amsterdam, pp. 269–287.

Dixon, A.F.G., 1998. Aphid Ecology. Chapman and Hall, London.

Duc, G., 1997. Faba bean (*Vicia faba* L.). Field Crop Res 53, 99–109.

Dungey, M., Potts, B.M., Whitham, T.G., Li, H., 2000. Plant genetics affects arthropod community richness and composition: evidence from a synthetic eucalypt hybrid population. Evolution 54 (6), 1938–1946.

Ernebjerg, M., Kishony, R., 2011. Dynamic phenotypic clustering in noisy ecosystems. PLoS Comput. Biol. 7 (3), e1002017.

Farnsworth, K.D., Focardi, S., Beecham, J.A., 2002. Grassland-herbivore interactions: how do grazers coexist? Am. Nat. 159, 24–39.

Flores, F., Moreno, M.T., Cubero, J.J., 1998. A comparison of univariate and multivariate methods to analyze G × E interaction. Field Crop Res 56, 271–286.

Fordyce, J.A., 2006. The evolutionary consequences of ecological interactions mediated through phenotypic plasticity. J. Exp. Biol. 209, 2377–2383.

Fritz, R.S., Price, P.W., 1988. Genetic variation among plants and insect community structure: willows and sawflies. Ecology 69, 845–856.

Hairston, N., Smith, F., Slobodkin, L., 1960. Community structure, population control and competition. Am. Nat. 94, 421–425.

Haloin, J.R., Strauss, S.Y., 2008. Interplay between ecological communities and evolution: a review of feedbacks from microevolutionary to macroevolutionary scales the year in evolutionary biology. Ann. N. Y. Acad. Sci. 1133, 87–125.

Hambäck, P.A., Vogt, M., Tscharntke, T., Thies, C., Englund, G., 2007. Top-down and bottom-up effects on the on the spatiotemporal dynamics of cereal aphids: testing scaling theory for local density. Oikos 116, 1995–2006.

Hays, D.B., Porter, D.R., Webster, J.A., Carver, B.F., 1999. Feeding behavior of biotypes E and H greenbug (Homoptera: Aphididae) on previously infested near-isolines of barley. J. Econ. Entomol. 92 (5), 1223–1229.

Hazell, S.P., McClintock, I.A.D., Fellowes, M.D.E., 2006. Intraspecific heritable variation in life-history traits can alter the outcome of interspecific competition among insect herbivores. Basic Appl. Ecol. 7, 215–223.

Hersch-Green, E.I., Turley, N.E., Johnson, M.T.J., 2011. Community genetics: what have we accomplished and where should we be going? Philos. Trans. R. Soc. B 366 (1569), 1453–1460.

Hochwender, C.G., Fritz, R.S., 2004. Plant genetic differences influence herbivore community structure: evidence from a hybrid willow system. Oecologia 138, 547–557.

Hughes, A.R., Inouye, B.D., Johnson, M.T.J., Underwood, N., Vellend, M., 2008. Ecological consequences of genetic diversity. Ecol. Lett. 11, 609–623.

Hutchinson, G.E., 1978. An Introduction to Population Ecology. Yale University Press, New Haven.

Jablonka, E., 2004. From replicators to heritably varying phenotypic traits. The extended phenotype revisited. Biol. Philos. 19, 353–375.

Johnson, M.T.J., 2008. Bottom-up effects of plant genotype on aphids, ants and predators. Ecology 89, 145–154.

Johnson, M.T.J., Agrawal, A.A., 2005. Plant genotype and environment interact to shape a diverse arthropod community on evening primrose (Oenothera biennis). Ecology 86, 874–885.

Johnson, M.T.J., Agrawal, A.A., 2007. Covariation and composition of arthropod species across plant genotypes of evening primrose (Oenothera biennis). Oikos 116, 941–956.

Johnson, M.T.J., Stinchcombe, J.R., 2007. An emerging synthesis between community ecology and evolutionary biology. Trends Ecol. Evol. 22, 250–257.

Johnson, M.T.J., Lajeunesse, A.A., Agrawal, A.A., 2006. Additive and interactive effects of plant genotypic diversity on arthropod communities and plant fitness. Ecol. Lett. 9, 24–34.

Johnson, M.T.J., Dinnage, R., Zhou, A.Y., Hunter, M.D., 2008. Environmental variation has stronger effects than plant genotype on competition among plant species. J. Ecol. 96, 947–955.

Kaplan, I., Denno, R.F., 2007. Interspecific interactions in phytophagous insects revisited: a quantitative assessment of competition theory. Ecol. Lett. 10 (1), 977–994.

Karban, A., 1989. Fine-scale adaptation of herbivorous thrips to individual host plants. Nature 340, 60–61.

Laland, K.N., 2004. Extending the extended phenotype. Biol. Philos. 19, 313–325.

Loeuille, N., Barot, S., Georgelin, E., Kylafis, G., Lavigne, C., 2013. Eco-evolutionary dynamics of agricultural networks: implications for sustainable management. Adv. Ecol. Res. 49, 339–435.

Mooney, K.A., 2011. Genetically based population variation in aphid association with ants and predators. Arthropod Plant Interact. 5 (1), 1–7.

Odling-Smee, F.J., Laland, K.N., Feldman, M.W., 2003. Niche Construction: The Neglected Process in Evolution. Princeton University Press, Princeton, NJ.

Östman, Ö., 2011. Interspecific competition affects genetic structure but not genetic diversity of daphnia magna. Ecosphere 2 (3), Article 34, 1–9.

Pelletier, F., Garant, D., Hendry, A.P., 2009. Eco-evolutionary dynamics. Philos. Trans. R. Soc. B 364 (1523), 1483–1489.

Poelman, E.H., van Loon, J.J.A., Dicke, M., 2008. Consequences of variation in plant defense for biodiversity at higher trophic levels. Trends Plant Sci. 13 (10), 534–541.

Powell, G., Tosh, C.R., Hardie, J., 2006. Host plant selection by aphids: behavioral, evolutionary, and applied perspectives. Annu. Rev. Entomol. 51, 309–330.

Power, M.E., 1992. Top-down and bottom-up forces in food webs: do plants have primacy? Ecology 73, 733–746.

Ranger, C.M., Singh, A.P., Johnson-Cicalese, J., Polavarapu, S., Vorsa, N., 2007. Intraspecific variation in aphid resistance and constitutive phenolics exhibited by the wild blueberry *Vaccinium darrowi*. J. Chem. Ecol. 33, 711–729.

Rowntree, J.K., Cameron, D.D., Preziosi, R.F., 2011a. Genetic variation changes the interactions between the parasitic plant-ecosystem engineer *Rhinanthus* and its hosts. Philos. Trans. R. Soc. B 366 (1569), 1380–1388.

Rowntree, J.K., Shuker, D.M., Preziosi, R.F., 2011b. Forward from the crossroads of ecology and evolution. Philos. Trans. R. Soc. B 366 (1569), 1322–1328.

Ruhnke, H., Schädler, M., Matthies, D., Klotz, S., Brandl, R., 2006. Are sawflies adapted to individual host trees? A test of the adaptive deme formation hypothesis. Evol. Ecol. Res. 8, 1039–1048.

Ruhnke, H., Schädler, M., Klotz, S., Matthies, D., Brandl, R., 2009. Variability in leaf traits, insect herbivory and herbivore performance within and among individuals of four broadleaved tree species. Basic Appl. Ecol. 10, 726–736.

Salyk, R., Sullivan, D.J., 1982. Comparative feeding behavior of two aphid species: Bean aphid (*Aphis fabae* Scopoli) and pea aphid (*Acyrthosiphon pisum* Harris) (Homoptera: Aphididae). J. N. Y. Entomol. Soc. 90 (2), 87–93.

Schoener, T.W., 1983. Field experiments on interspecific competition. Am. Nat. 122, 240–285.

Schoener, T.W., 2011. The newest synthesis: understanding the interplay of evolutionary and ecological dynamics. Science 331, 426–429.

Schuett, W., Dall, S.R.X., Baeumer, J., Kloesener, M.H., Nakagawa, S., Beinlich, F., Eggers, T., 2011. 'Personality' variation in a clonal insect: the pea aphid, *Acyrthosiphon pisum*. Dev. Psychobiol. 53, 631–640.

Smith, R.A., Mooney, K.A., Agrawal, A.A., 2008. Coexistence of three specialist aphids on common milkweed, *Asclepias syriaca*. Ecology 89 (8), 2187–2196.

Tack, A.J.M., Ovaskainen, O., Pulkkinen, P., Roslin, T., 2010. Spatial location dominates of host plant genotype in structuring a herbivore community. Ecology 91 (9), 2660–2672.

The World Bank (2013) World Development Indicators: Agricultural Inputs Table 3.2 online resource. http://wdi.worldbank.org/table/3.2 (accessed February 2014).

Thompson, J.N., 1994. The Coevolutionary Process. University of Chicago Press, Chicago.

Thrall, P.H., Oakeshott, J.G., Fitt, G., Southerton, S., Burdon, J.J., Sheppard, A., Russell, R.J., Zalucki, M., Heino, M., Denison, R.F., 2011. Evolution in agriculture: the application of evolutionary approaches to the management of biotic interactions in agro-ecosystems. Evol. Appl. 4 (2), 200–215.

Turcotte, M.M., Reznick, D.N., Hare, J.D., 2013. Experimental test of an eco-evolutionary dynamic feedback loop between evolution and population density in the green peach aphid. Am. Nat. 181 (S1), S46–S57.

Underwood, N., 2009. Effect of genetic variance in plant quality on the population dynamics of a herbivorous insect. J. Anim. Ecol. 78, 839–847.

Underwood, N., Rausher, M.D., 2000. The effects of host-plant genotype on herbivore population dynamics. Ecology 81, 1565–1576.

Utsumi, S., Ando, L., Craig, T.P., Ohgushi, T., 2011. Plant genotypic diversity increases population size of a herbivorous insect. Proc. R. Soc. B 278 (1721), 3108–3115.

Whitham, T.G., Young, W.P., Martinsen, G.D., Gehring, C.A., Schweitzer, J.A., Shuster, S.M., Wimp, G.M., Fischer, D.G., Bailey, J.K., Lindroth, R.L., Woolbright, S., Kuske, C.R., 2003. Community and ecosystem genetics: a consequence of the extended phenotype. Ecology 84, 559–573.

Whitham, T.G., Bailey, J.K., Schweitzer, J.A., Shuster, S.M., Bangert, R.K., LeRoy, C.J., Lonsdorf, E.V., Allan, G.J., DiFazio, S.P., Potts, B.M., Fischer, D.G., Gehring, C.A., Lindroth, R.L., Marks, J.C., Hart, S.C., Wimp, G.M., Wooley, S.C., 2006. A framework for community and ecosystem genetics: from genes to ecosystems. Nat. Rev. Genet. 7, 510–523.

Wickham, H., 2009. ggplot2: Elegant Graphics for Data Analysis. Springer, New York.

Wimp, G.M., Wooley, S., Bangert, R.K., Young, W.P., Martinsen, G.D., Keim, P., Rehill, B., Lindroth, R.L., Whitham, T.G., 2007. Plant genetics predicts intra-annual variation in phytochemistry and arthropod community structure. Mol. Ecol. 16, 5057–5069.

Zhu, Y., Chen, H., Fan, J., Wang, Y., Li, Y., Chen, J., Fan, J., Yang, S., Hu, L., Leung, H., Mew, T.W., Teng, P.S., Wang, Z., Mundt, C.C., 2000. Genetic diversity and disease control in rice. Nature 406 (6797), 718–722.

Zytynska, S.E., Preziosi, R.F., 2011. Genetic interactions influence host preference and performance in a plant-insect system. Evol. Ecol. 25, 1321–1333.

Zytynska, S.E., Fay, M.F., Penney, D., Preziosi, R.F., 2011. Genetic variation in a tropical tree species influences the associated epiphytic plant and invertebrate communities in a complex forest ecosystem. Philos. Trans. R. Soc. B 366 (1569), 1329–1336.

Genetic Correlations in Multi-Species Plant/Herbivore Interactions at Multiple Genetic Scales: Implications for Eco-Evolutionary Dynamics

Julianne M. O'Reilly-Wapstra*,[†],[1], Matthew Hamilton*,[†],
Benjamin Gosney*,[†], Carmen Whiteley*, Joseph K. Bailey[‡],
Dean Williams[§], Tim Wardlaw[§], René E. Vaillancourt*, Brad M. Potts*,[†]

*School of Biological Sciences, University of Tasmania, Hobart, Tasmania, Australia
[†]National Centre for Future Forest Industries, University of Tasmania, Hobart, Tasmania, Australia
[‡]Department of Ecology and Evolutionary Biology, University of Tennessee, Knoxville, Tennessee, USA
[§]Forestry Tasmania, Hobart, Tasmania, Australia
[1]Corresponding author: e-mail address: joreilly@utas.edu.au

Contents

Abstract

In plant/herbivore systems, elucidating the hierarchical genetic correlations that exist between enemies to a host plant (e.g., in the magnitude of damage) and determining how stable these effects are across environments is crucial for our understanding of potential eco-evolutionary dynamics in these systems. This sort of information would allow us to better know how plant populations have evolutionarily diverged in their phenotypic traits, which organisms are driving the evolutionary change and how rapid evolutionary change in one enemy or plant species can feedback to affect other herbivore and pathogen species showing genetically correlated responses to the host plant. Here, we investigate consistency in patterns of the genetic correlations within and among populations in preferences among multiple enemies for a globally planted species, *Eucalyptus globulus*, and explore how stable these genetically based correlations are across environments. We show plant enemies respond to underlying host genetic variation at two genetic hierarchical scales and that relationships between plant enemies are both independent and correlative. Our finding of a significant positive genetic correlation in damage to the host plant between a sawfly and a fungal leaf pathogen species suggests a potential indirect eco-evolutionary feedback loop mediated by the genetic correlation. We also demonstrate that among population divergence patterns are not constrained by within population correlations, and this decoupling between among and within population patterns suggests that diffuse relationships between two enemies are not constraining the evolutionary diversification in resistance of populations of their host plant. In addition, when genetic effects are present, there is clear stability in the genetic influences of this host tree on its enemies across trials. This indicates that in an eco-evolutionary dynamic setting, the strength and consistency of any selective force would be maintained across environments.

1. INTRODUCTION

A recent surge in evolutionary ecology research has contributed some exciting findings showing that evolutionary processes affect ecological dynamics (e.g. population size, population growth rate and community organisation), and that these processes are occurring on timescales that we can measure (a field of research coined eco-evolutionary dynamics; Ellner, 2013; Hiltunen et al., 2014, chapter 2 of this volume; Post and Palkovacs, 2009; Schoener, 2011; Turcotte et al., 2011; Turcotte et al., 2013). For example, eco-evolutionary feedback loops have been demonstrated in the green peach aphid (*Myzus persicae*) where intraspecific density of aphids alters aphid genotype frequencies through time, which in turn, alters aphid concurrent population growth rate and density (Turcotte et al., 2013). In this system, the effects of evolution on aphid population growth and density could be measured in only three generations (16 days). Eco-evolutionary processes can also affect ecological interactions among species and some of the best systems for highlighting these patterns are found in the field of plant/herbivore interactions. In a willow/herbivore community system, variation in the composition of herbivore communities leads to variation in induced re-growth of willow. This variation in re-growth results in divergent trait evolution of a subsequent herbivore, the leaf beetle *Plagiodera versicolora* and, hence, there is an indirect effect of the original herbivore community on the subsequent colonising leaf beetle (Utsumi et al., 2013). This latter example highlights the fact that plants are attacked by a succession of herbivores and pathogens, often more than one at a time throughout their life cycle (Khudr et al., 2014, chapter 7 of this volume; Strauss and Irwin, 2004). A major challenge in evolutionary ecology is to understand the role that each of these enemies plays, either independently or through interactions with other enemies, in the evolution of the host plant. Within host plant species, there is often clear variation among different genetically and geographically distinct populations in plant resistance to herbivores (Thompson, 2005). Determining the role of organisms in driving population divergence in the host plant through their potential selective impacts has broad implications for how we understand the evolution of plant defensive traits, co-evolution between plants and enemies, multi-trophic interactions and complex community and ecosystem networks.

Placing ecological studies in a genetic framework is necessary to investigate the possible evolutionary ramifications of ecological interactions. To date, it is commonplace for studies in the field of plant/herbivore interactions to focus on only one level of genetic variation within the plant system. For example, informative studies of the role of enemies as selective agents on plant resistance and/or the genetic correlations that may exist in resistance to multiple herbivores have tended to focus on within plant population genetic variation (Agrawal, 2005; Johnson and Agrawal, 2007; Mauricio and Rausher, 1997; Stinchcombe and Rausher, 2001; Wise, 2007), while studies examining the geographic mosaic of evolutionary relationships across the species range have concentrated on among plant population level variation (Benkman, 1999; Gómez et al., 2009; Muola et al., 2010). To better understand how plant populations have diverged (e.g. what are the possible selective factors driving divergence), one needs to consider both levels (Armbruster, 1991). However, robust experimental systems where multiple levels of the genetic hierarchy are available, along with a large collection of genetic material to allow broad inference space to investigate underlying factors influencing plant population divergence (Colautti and Barrett, 2011) in plant/enemy interactions, and are still quite rare (Andrew et al., 2010; Laine, 2004; Leimu and Fischer, 2010).

Exploring the hierarchy (among and within populations) of genetic correlations that may exist in plant resistance to more than one herbivore or pathogen provides a framework to investigate aspects of population divergence and the eco-evolutionary dynamics between them (Armbruster, 1991). For example, genetic correlation analysis (estimating the degree to which two or more traits are affected by the same genes; Conner and

Hartl, 2004; Falconer and Mackay, 1996) offers insight into how enemies may affect the evolution of the plant. Enemies may act independently (a lack of correlation; pairwise selection) and hence create variable selective forces on the plant by potentially selecting on a variety of traits throughout the life of the host plant. Alternatively, the evolutionary effect of an enemy on the plant may be influenced by other organisms and this will be signalled by significant genetic correlations between resistance to multiple enemies (diffuse selection; Rausher, 1996; Stinchcombe and Rausher, 2001). In this case, multiple enemies could be selecting on either the same traits or genetically correlated traits of their host plants (Leimu and Koricheva, 2006). This could result in increased or more consistent directional change in the host plant if the correlation is positive or it could restrict the rate of directional change if the correlation is negative. How these correlations differ among and within populations could provide information on how enemies, as possible selective agents, might be influencing population divergence in their host plant (Armbruster, 1991). If the within and among population level genetic correlations in damage from different enemies are consistent, then population divergence is expected to be constrained by the within population genetic structure (Armbruster, 1991; Pruitt et al., 2010; Schluter, 1996). In this case, population genetic divergence in a host plant may be due to strong diffuse selective impacts of more than one plant enemy (if no other factors are affecting the relationship). However, if the among population correlation pattern is not consistent with a significant within population genetic correlation in damage from different enemies, then this signals that population divergence is not constrained by the within population genetic correlations between multiple enemies (Colautti and Barrett, 2011).

Eucalypts are proving to be an important model species to address fundamental questions relating to plant/enemy interactions (Andrew et al., 2010; Barbour et al., 2009; Miller et al., 2007; Moore and Foley, 2005; Wiggins et al., 2006). In particular, the quantitative genetics and ecology of the widespread *Eucalyptus globulus* have been well studied. Previous research has examined the univariate biotic interactions between *E. globulus* and some of its key enemies finding significant genetically based variation in susceptibility and associated impacts on fitness to mammalian herbivores (O'Reilly-Wapstra et al., 2010, 2012), invertebrates (Jordan et al., 2002; Rapley et al., 2004a) and leaf fungal pathogens (Hamilton et al., 2013; Milgate et al., 2005). Here, we used over 26,000 trees from up to 500 open-pollinated families, sourced from up to 16 different populations, grown across several environments (field trials) to examine the potential role of these multiple enemy interactions in the evolutionary diversification of *E. globulus*. We specifically examined the genetic correlations in damage caused by these enemies at both the within and among plant population level in both the juvenile and adult life-stage of the host plant. We have six main aims: (1) document variation in the response of each plant enemy (herbivore/pathogen) to underlying host plant genetic variation at two genetic scales, (2) test if the response by enemies at two host plant genetic scales is consistent across environments (field trials), (3) test if preferences by organisms are consistent across juvenile and adult foliage, (4) test whether these enemies are acting independently by estimating the within plant population genetic correlation in response of enemies, (5) examine if among plant population divergence is likely constrained by within population genetic correlations in responses of the herbivores/pathogens and (6) examine if correlations in enemy response to genetically based variation in the plant are stable across different environments.

2. METHODS

2.1. Genetic hierarchy of *E. globulus*

E. globulus is a dominant species in south-eastern Australian native forests. It can range in height from a short dwarf coastal tree (2–5 m) to a medium woodland tree (15–20 m) and a tall open forest tree (up to 60–70 m). The native range of this species has been classified into a hierarchy of genetically distinct classes. We define the hierarchy of genetic classes of *E. globulus* as follows: "population" refers to genetically differentiated broad geographical grouping of trees (previously defined as sub-races) (Dutkowski and Potts, 1999); while "family" refers to the progeny derived from the open-pollinated seed collected from a single parent tree within a population. There are 20 populations recognised across the whole geographic range of *E. globulus*, which are defined based on geography and quantitative genetically based differences in 35 morphological traits amongst trees growing in base-population progeny trials derived from families of wild parental trees (Dutkowski and Potts, 1999). The parent trees from which families were grown for the present study where from across the geographic range of each population as indicated in Fig. 8.1.

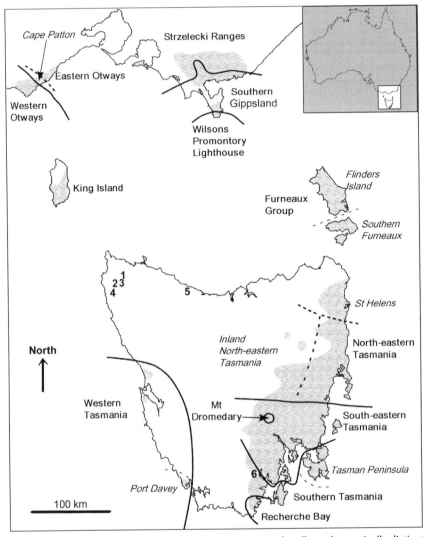

Figure 8.1 Map illustrating the location of the 20 geographically and genetically distinct native populations of *Eucalyptus globulus* across its range in south-eastern Australia. Non-italic font indicates populations that can also be classified as broader genetically distinct groupings of trees known as races (we have not analysed our data as races in this chapter). The Kuark population located near Strzelecki Ranges is not indicated on the map. This is an intergrade population that is part of a continuous gene pool in the Gippsland and Strzelecki areas (Jones et al., 2013). Numbers 1–6 on the map illustrate the location of the six common environment field trials: (1) *Togari*, (2) *Salmon River 1*, (3) *Salmon River 2*, (4) *Temma*, (5) *Massy Greene* and (6) *Oigles Road*. Modified from Dutkowski and Potts (1999).

2.2. Field trials and damage assessments

2.2.1 Summary of field trials and damage assessments

In this chapter, we studied foliar damage symptoms on *E. globulus* trees growing in six different common garden field trials in Tasmania (Togari, Salmon River 1, Salmon River 2, Temma, Massy Greene and Oigles Road; Fig. 8.1). These trials were planted at different times for different experimental purposes between 1989 and 2006. Consequently, the damage assessments on these trials were also performed at different times. The specific details of each trial and the assessments undertaken can be found below. In brief, all trees in each common garden are planted 2.5 m apart in straight rows, which were themselves 3–4 m apart (1000–1667 stems ha^{-1}). Each trial was divided into a number of replicates and randomised incomplete blocks within replicates, with each family (and thus population) represented in each replicate. The number of replicates, incomplete blocks within replicates and the number of trees planted for each population and family in each replicate, differ between trials. These details are found below for each field trial.

    E. globulus is heteroblastic, changing from juvenile to adult foliage at about 3–4 years of age in plantations (Jordan et al., 1999). Damage by organisms was assessed on juvenile foliage in all field trials and on adult foliage in only two field trials (Temma and Salmon River 2), due to time and logistical constraints. Analyses for juvenile and adult foliage assessments were performed separately. We studied the damage on juvenile foliage from four different groups of organisms: mammals, where browse damage by the common brushtail possum, *Trichosurus vulpecula*, and Tasmanian pademelon, *Thylogale billardierii*, was scored as one collective score (O'Reilly-Wapstra et al., 2002); sawfly, *Pergus affinis* ssp. *insularis* (Jordan et al., 2002); eucalypt leaf beetle, *Paropsisterna agricola* (Rapley et al., 2004b; referred to a leaf beetle agricola) and a fungal leaf pathogen complex, *Teratosphaeria* spp. (Hamilton et al., 2013; referred to as fungal leaf pathogen 1). On the adult foliage, we present results for damage symptoms from six organisms: the fungal leaf pathogen complex of the genus *Teratosphaeria* (but most likely different species to that found on the juvenile foliage) referred to as fungal leaf pathogen 2; a gum moth *Plesanemma fucata*; a leaf beetle, *Paropsisterna bimaculata*, referred to as leaf beetle bimaculata; an unidentified leaf fungal pathogen referred to as fungal leaf dots; an unidentified petiole fungal pathogen referred to as petiole fungus and finally an unidentified leaf miner referred to as leaf miner.

    All six common gardens were naturally infested by the enemies (organisms were not artificially introduced to the trials) and, hence, not all organisms were present in all common gardens. In some cases, organisms were present in the same garden but separated temporally. What was common among field trials, however, were the populations and families that were planted and, hence, what the organisms damaged. Because not all populations and families were represented in each field trial, statistical analysis of damage was only done using populations and families of *E. globulus* that were common among all field trials (i.e. the same families were present on multiple trials). Below, we detail the common gardens and damage assessments.

2.2.2 Four far north-western Tasmania field trials (Togari, Salmon River 1, Salmon River 2 and Temma): Mammalian herbivore and Teratosphaeria damage on juvenile foliage and damage by six organisms on adult foliage

Damage by the fungal leaf pathogen complex *Teratosphaeria* spp. (fungal leaf pathogen 1) was assessed on juvenile foliage of four *E. globulus* progeny trials in north-western Tasmania and browsing damage by marsupial herbivores (the common brushtail possum, *Trichosurus vulpecula*, and the Tasmanian pademelon, *Thylogale billardierii*) was assessed on juvenile foliage at one of these trials. As we cannot separate damage by the two marsupial species, we have kept them as a pooled result and called it 'mammal' damage. These four trials are referred to as Togari (where mammal damage was assessed), Salmon River 1, Temma and Salmon River 2. Togari (40°57′S, 144°55′E) and Salmon River 1 (41°02′S, 144°49′E) were planted in 2005, whereas Temma (41°07′S, 144°45′E) and Salmon River 2 (41°01′S, 144°52′E) were planted a year later. The minimum distance between these trials is 4.6 km between the two Salmon River trials and the greatest distance between the trials is 24 km between the Togari and Temma trials. The two Salmon River and Temma sites are on yellow–brown mottled clay on Precambarian mudstone while the Togari site is on a loamy soil over clay on Cambrian mud, silt and sandstone geology.

The trials were established according to incomplete block designs with families allocated into 16–25 replicates, with each replicate comprising 12–15 incomplete blocks and each family represented as a single-tree plot within each replicate. The families in these trials were open–pollinated families from throughout the native range of *E. globulus*, as well as advanced-generation breeding-program germplasm (Potts et al., 2008). Only native-forest families from 13 populations were included in analyses of the juvenile foliage. The presence/absence of mammal browsing on foliage was recorded for each tree in the Togari trial in 2006 ($n = 138$ families and 2153 individual trees used in the analysis). Fungal leaf pathogen 1 infestation was scored by estimating the percentage of necrotic leaf area (Carnegie et al., 1994) in the entire juvenile-crown of each tree after a widespread outbreak in 2007 in all four trials ($n = 139$ families and 2044 individual trees for Togari, 146 families and 2532 individual trees for Salmon River 1, 140 families and 2734 individual trees for Salmon River 2 and 124 families and 2501 individual trees for Temma were used in the analysis) (Hamilton et al., 2013).

Adult foliage assessments of a subset of trees from the Salmon River 2 and Temma field trials were performed in April 2012 (trees were 6 years old). A branch from the mid-canopy on the eastern side of each tree was cut and 20 representative leaves of the branch were picked for organism damage assessments. Twenty leaves constituted approximately a third of the leaves on the branch. Foliage was sampled from ten families per population and three trees per family (these three trees were from the open-pollinated seed collected from the same wild parent tree) in each trial ($n = 30$ per population). We sampled across 5 of 25 replicates at Salmon River 2 and 6 of 25 replicates at Temma. At Salmon River 2, 138 families and 408 individual trees, and at Temma, 128 families and 368 individual trees were used in the analysis. Leaves were returned to the laboratory and dried at 30 °C prior to assessments. Arthropod and fungal damage assessments were scored using a symptom based approach. Distinct symptoms were scored on each of the 20 leaves as presence/absence and quantitative abundance data for each species was based on the proportion of 20 leaves affected (presence/absence) by each species. In total, damage symptoms from 49 organisms were scored on foliage across the two field trials. Most symptoms were rare (<10% of leaves infected) and, here, we present results for six abundant organisms (listed above in the methods; Plate 1).

2.2.3 Massy Greene field trial: Sawfly damage

Sawfly damage was assessed on adult foliage in an *E. globulus* family trial near Burnie, in north-western Tasmania, Australia (41°05′S 145°54′E). The trial site was on basalt-derived krasnozem soil. The trial was one of five base-population trials established in 1989 from progenies grown from open-pollinated seed. The trial contained nearly 600 open-pollinated families collected from 16 populations throughout Tasmania and southern Victoria, Australia. The trial comprised 5 replicates, each of 25 incomplete

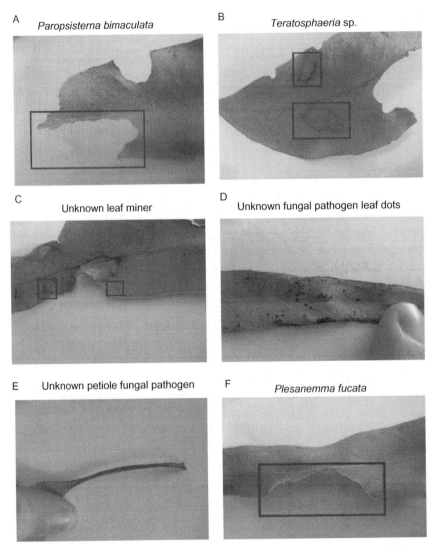

Plate 1 Photos of the damage symptoms by the six organisms identified on adult *Eucalyptus globulus* foliage.

blocks of 24 families. Each family was represented in a replicate as a two–tree plot, giving a total of 10 individuals per family. In brief, the trial was damaged by sawflies during several growing seasons and was scored for this damage in January and November 1995. Damage was scored as presence/absence and was pooled for both time periods (damage between these two time periods was highly correlated). Further details on the trial and sawfly assessments are given in Jordan et al. (2002). In this chapter, data for 16 populations, 573 families and 4841 individual trees was used in the analysis.

2.2.4 Oigles Road field trial: Leaf beetle agricola oviposition and larvae damage

The Oigles Road *E. globulus* family trial was established in 1999 near Geeveston (43°10′S, 146°54′E) in southern Tasmania. This site was on dolerite derived clay loam krasnozem soil. The families in the trial were derived from open-pollinated seed collected from trees in native stands representing nine of the populations from throughout the range of *E. globulus*. There were 10 replicates in the trial, each with 20 incomplete blocks (except in three of the replicates where only 15 or 18 incomplete blocks were planted). There were 10 families in each incomplete block, each represented by a four-tree row plot. A survey of leaf beetle *Paropsisterna agricola* (referred to as leaf beetle agricola in results section) oviposition was conducted in December 2001, coinciding with the peak activity of this species. The eastern side of each tree was assessed for 15 s and the number of egg batches (eggs) and larval clutches (larvae) counted during that time were recorded. Infestation level of each tree was determined by summing the number of egg batches and larval clutches. For full details on this trial and leaf beetle assessments, refer to Rapley et al. (2004b). In this chapter, data for 9 populations, 225 families and 7041 individual trees were used in the analysis.

3. STATISTICAL ANALYSIS
3.1. Juvenile foliage analysis

Univariate mixed models were fitted using ASReml (Gilmour et al., 2009) to examine genetic variation in the prevalence of *E. globulus* enemies within and among plant populations. For the four far north-western Tasmanian trials (Togari 1, Salmon River 1, Salmon River 2 and Temma), the models fitted replicate, assessor (person who scored damage assessments) and population effects as fixed factors and incomplete block within replicate, row within replicate and family within population effects as random factors. Similar models were fitted for the analysis of data from the Massy Greene and Oigles Road trials, details of which are outlined in Jordan et al. (2002) and Rapley et al. (2004b), respectively. Mammalian herbivore damage (Togari), sawfly damage (Massy Greene) and leaf beetle agricola presence (Oigles Road) were treated as binary data and analysed with binomial models using a probit link function (Gilmour et al., 2009) and the significance of family within population variance was estimated using a Z-test. In the case of fungal leaf pathogen 1 damage, a likelihood ratio test was used to test the significance from zero of the family within population variance at each trial. For each damage indicator and trial, the significance of the population effect was gauged with a Wald F-test (Gilmour et al., 2009). The population least squares mean was estimated with standard error based on the family within population variance.

To test within population family correlations (r_f) between organisms on the juvenile foliage, bivariate models that extended the above univariate models were used. Unlike the univariate models, these bivariate models did not fit a link function (i.e. it was assumed that genetic correlations were equivalent on the binomial and underlying scales; Chambers et al., 1996). These models were used to estimate family correlations among mammal browsing, sawfly and leaf beetle agricola (refer to Costa et al., 2006, for details of similar analyses). Separate multivariate models were used to estimate equivalent correlations between fungal leaf pathogen 1 damage and each of the other enemies based on the assumption that inter-trial (among field trials) family correlations for fungal leaf pathogen 1 damage were equal and inter-trait family correlations were equal across the trials assessed for fungal leaf pathogen 1 damage.

To estimate the correlation between fungal leaf pathogen 1 and mammalian herbivore damage, only fungal leaf pathogen 1 data from three of the four far north-western trials that did not experience substantial mammalian browsing prior to the assessment of fungal damage were used (i.e. fungal leaf pathogen 1 damage data from the browsed Togari trial were excluded). This ensured that early age mammal browsing did not interfere with the fungal leaf pathogen 1 and maintained consistency in our data analysis where we analysed data for fungal leaf pathogen 1 and mammal browsing that only occurred on different sites.

Population correlations among mammalian herbivore, sawfly, leaf beetle agricola and fungal leaf pathogen 1 damage/presence were estimated in the same manner as genetic correlations by fitting population as a random, rather than fixed, factor in the same models. Significance tests of population and family correlation differences from zero and one were undertaken using a two- and one-tailed likelihood ratio tests, respectively (Gilmour et al., 2009). Table 8.1 shows the number of common families and number of common populations for each correlation for the juvenile foliage analysis. While families may express different phenotypes temporally and across

Table 8.1 Number of families and populations that were common for each correlation analysis between each organism assessed on the juvenile foliage of *Eucalyptus globulus* across six field trials

Organisms	No. families	No. populations
Sawfly versus leaf beetle agricola	39	8
Sawfly versus mammals	109	13
Sawfly versus fungal leaf pathogen 1	140	13
Leaf beetle agricola versus mammals	26	7
Leaf beetle agicola versus fungal leaf pathogen 1	35	7
Mammal versus fungal leaf pathogen 1	138	13

common gardens, the strength of having common families across multiple gardens is that any correlation observed in damage across trials must be due to a common genetic control as organisms have not interfered with each other's activities.

3.2. Adult foliage analysis

We used a simplified model for adult foliage analysis where incomplete block within replicate was not included in the model (as it was for the juvenile foliage) because we used a subset of trees in each replicate, due to time and logistical constraints. To determine the genetic response of each organism to population and family within population level genetic variation, restricted maximum likelihood mixed model analyses were undertaken fitting population as a fixed factor, and replicate and family within population as random factors. The significance of the population and family within population variances were tested with 'one-tailed' likelihood ratio tests (Gilmour et al., 2009).

For organisms where univariate analyses revealed significant ($P < 0.05$) variation in response to genetically based differences in *E. globulus* at both trials (i.e. for three organisms at the population level and none at the family within population level; Table 8.4), a bivariate model was fitted to (i) test if genetically based variances were equal across trials and (ii) estimate inter-trial correlations. The bivariate model extended the univariate model to include the same explanatory variables, but with population fitted as a random term and with levels of replicate considered independent across trials. 'Two-tailed' likelihood ratio tests were used to test if variances for population level genetic responses were significantly different across trials, by fitting a constrained model in which population variances were forced to be equal across trials (Hamilton et al., 2013). 'Two-tailed' likelihood ratio tests were also conducted to determine if population correlations were significantly different from zero and 'one-tailed' likelihood ratio tests were used to determine if these correlations were significantly different from one.

To determine if inter-organism population correlations were significantly different from each other across Temma and Salmon River 2 trials, a four-variate model (i.e. two organisms by two trials) was fitted and the difference tested with a likelihood ratio test by fitting a constrained model in which inter-trial population correlations were forced to be equal across trials. Standard errors of parameters were estimated from the average information matrix, using a standard truncated Taylor series approximation (Gilmour et al., 2009). Analyses were conducted using ASReml (Gilmour et al., 2009).

4. RESULTS

The first aim of this work was to determine whether there was genetically based variation in damage by each enemy at two hierarchical levels, within and among plant populations. If we look at juvenile foliage first, the mean damage response on juvenile foliage of each organism to each population is illustrated in Table 8.2 and damage for all species on juvenile foliage differed significantly among these *E. globulus* populations and also among families within populations (Table 8.3). Similarly to the juvenile foliage analysis, all six organisms on the adult foliage showed a response to the genetic variation of *E. globulus* at the population level, but this response was not evident in both trials for all six organisms (Table 8.4). Only three organisms (fungal leaf pathogen 2, fungal leaf dots and leaf miner) showed a response in both trials. A formal test of the population variance for each organism (results not shown in table) showed that the population variance for leaf beetle bimaculata was significantly different between the two trials [tested with a likelihood ratio test by constraining population variances to be equal across sites; $X^2(1)$; $P < 0.05$], was on the margins of being significantly different for leaf miner ($P = 0.06$) and was not significantly different for all other organisms. Only two organisms—fungal leaf pathogen 2 and the petiole fungus—exhibited significant variation in their damage to adult foliage at the family within population level (Table 8.4). Again, these responses were not evident on both trials with a formal test of the family

Table 8.2 Population least square mean damage values (±standard error) for all organisms assessed on *Eucalyptus globulus* juvenile foliage

Population	Sawfly (proportion of trees with damage)	Fungal leaf pathogen sp. 1 (average % foliage damaged per tree)	Mammal browsing (proportion of trees with damage)	Leaf beetle agricola (proportion of trees with damage)
Eastern Otways	0.68 (0.04)	4.49 (0.03)	0.27 (0.05)	–
Flinders Island	0.45 (0.04)	5.63 (0.02)	0.13 (0.04)	0.30 (0.04)
Inland North Eastern Tasmania	0.60 (0.06)	–	–	0.11 (0.03)
King Island	0.23 (0.04)	5.64 (0.02)	0.14 (0.03)	–
Kuark	–	–	–	0.44 (0.07)
North Eastern Tasmania	0.48 (0.06)	6.32 (0.02)	0.17 (0.04)	0.27 (0.03)
Port Davey	0.27 (0.11)	–	–	–
Recherche Bay	0.24 (0.11)	7.48 (0.02)	0.21 (0.04)	–
South–eastern Tasmania	0.41 (0.04)	7.05 (0.02)	0.16 (0.04)	0.32 (0.04)
Southern Furneaux	0.55 (0.04)	5.93 (0.02)	0.24 (0.04)	0.33 (0.04)
Southern Gippsland	0.18 (0.05)	4.94 (0.02)	0.32 (0.05)	–
Southern Tasmania	0.13 (0.03)	7.05 (0.02)	0.21 (0.04)	0.21 (0.02)
St Helens	0.39 (0.08)	5.30 (0.02)	0.43 (0.06)	0.24 (0.04)
Strzelecki Ranges	0.39 (0.03)	5.37 (0.02)	0.29 (0.05)	0.48 (0.04)
Tasman Peninsula	0.29 (0.11)	–	–	–
Western Otways	0.49 (0.03)	4.92 (0.02)	0.25 (0.05)	–
Western Tasmania	0.37 (0.05)	5.61 (0.02)	0.18 (0.04)	–

NB: Sawfly, mammal and agricola data were treated as binary (presence/absence) data for analysis.

Table 8.3 Genetic variation in damage to *Eucalyptus globulus* juvenile foliage by the four enemy species at two genetic levels; among populations (fixed effect) and among families within populations (random effect)

Site	Species	Data collected	Among populations		Among families within populations		
					Variance component		
			F_{df}	P	(n)	SE	P
Togari	Mammal browsing	Presence/absence	$3.06_{12,115}$	<0.001	0.053 (138)	0.0269	0.05
Togari	Leaf fungal pathogen 1	% Leaf area affected	$2.57_{12,120}$	<0.01	0.003 (140)	0.0001	<0.0001
Temma	Leaf fungal pathogen 1	% Leaf area affected	$10.43_{12,111}$	<0.001	0.005 (125)	0.0009	<0.0001
Salmon River 1	Leaf fungal pathogen 1	% Leaf area affected	$5.22_{12,124}$	<0.001	0.002 (147)	0.0004	<0.0001
Salmon River 2	Leaf fungal pathogen 1	% leaf area affected	$4.80_{12,116}$	<0.001	0.001 (141)	0.0002	<0.001
Massy Greene	Sawfly larvae	Presence/absence	$10.26_{15,574}$	<0.001	0.209 (176)	0.0286	<0.0001
Oigles Road	Leaf beetle agricola oviposition and larvae	Presence/absence	$10.19_{8,137}$	<0.001	0.065 (47)	0.0135	<0.0001

Table 8.4 Genetic variation in damage to *Eucalyptus globulus* adult foliage by the six enemy species at two genetic levels; among populations (fixed effect) and among families within populations (random effect)

Species	Site	Among populations		Among families within populations	
		F_{df}	P	Z	P
Gum moth	Salmon River 2	$2.76_{13,124}$	<0.001	0.02	0.50
	Temma	$1.80_{13,114}$	0.051	1.03	0.14
Fungal leaf pathogen 2	Salmon River 2	$4.28_{13,124}$	<0.001	2.3	<0.01
	Temma	$4.08_{13,111}$	<0.001	1.01	0.15
Leaf beetle bimaculata	Salmon River 2	$1.23_{13,392}$	0.25	0	–
	Temma	$3.36_{13,112}$	<0.001	1.24	0.10
Fungal leaf dots	Salmon River 2	$2.96_{13,121}$	<0.001	0.03	0.50
	Temma	$2.01_{13,114}$	<0.05	0.51	0.30
Petiole fungus	Salmon River 2	$2.03_{13,123}$	<0.05	2.06	<0.05
	Temma	$1.67_{13,350}$	0.06	0	–
Leaf miner	Salmon River 2	$2.19_{13,393}$	<0.01	0	–
	Temma	$3.09_{13,351}$	<0.001	0	–

Results from Salmon River 2 (138 families) and Temma (128 families) are presented.

within population variances showing a significant difference between trials for the petiole fungus ($P<0.05$; results not shown in table).

We used the adult foliage data to address the second aim of whether highly damaged populations or families on one site were also highly damaged on a different site. We tested this for the three organisms that responded to variation in *E. globulus* populations on both trials (Table 8.4) and these responses were consistent across the populations (that is, population differences were similar) on both trials. This is evidenced by the positive among population correlations (r_s) across sites for all three organisms, but this correction was only significantly different from zero at the 0.05 level for fungal leaf pathogen 2 (fungal leaf pathogen 2, $r_s=0.91$ $P<0.01$; fungal leaf dots, $r_s=0.81$ $P=0.07$; leaf miner, $r_s=0.74$ $P=0.08$). Despite not showing a statistically significant response to the family within population *E. globulus* genetic variation at both trials (Table 8.4), the fungal leaf pathogen 2 response to the family within population host variation was also consistent across trials

(r_f=0.79, P=0.04). Results from a previous study (Hamilton et al., 2013) of fungal leaf pathogen 1 responses to juvenile foliage on the same trials we used in this present study, also shows consistent responses of this organism to plant populations and families within populations across multiple trials.

Teratosphaeria (fungal leaf pathogen 1 and 2) was the only organism(s) that we assessed for damage on both juvenile and adult foliage (aim 3). We found significant responses by Teratosphaeria to E. globulus population level variation were evident in all trials and for both foliage types (Tables 8.3 and 8.4). The response of Teratosphaeria to E. globulus population level variation was consistent for both adult and juvenile foliage, where the population level correlation across foliage types was positive on both trials, but only significantly different from zero at one of the trials (Salmon River 2 r_s=0.52, P=0.001; Temma r_s=0.45, P=0.24). The preference of Teratosphaeria to adult and juvenile foliage did not show a significant correlation at the family within population level for either trial (Salmon River 2 r_f=0.28, P=0.51; Temma r_f=0.31, P=0.20).

To test the fourth aim of whether enemies are acting independently, we examined the family within population genetic correlation of enemies in response to juvenile foliage (Fig. 8.2). While there was significant variation in the response of all enemies to underlying host plant genetic variation (Table 8.3), the relationship among enemies appeared to be generally uncorrelated (Fig. 8.2). This result argues that most organisms are responding independently to the within population genetic variation in this host tree. However, there was a significant positive family–level genetic correlation in damage between the leaf fungal pathogen 1 and sawflies (r_f level in Fig. 8.2). This pattern appeared to be uniform within all E. globulus populations as we could not detect heterogeneity in the family genetic correlations for the different populations (likelihood ratio test, χ^2=10.12, DF=12, P=0.61). When we examined the among population level genetic correlation in response to E. globulus by these two enemies, we detected no significant correlation in damage among populations (r_s; Fig. 8.2) indicating that the patterns of genetic correlation within and among populations are not consistent (aim 5).

The final aim of this chapter was to test if genetically based correlations in damage among organisms were consistent across trials. We tested this aim using data from the adult foliage (as we did not have equivalent data in the juvenile foliage across trials). Only two organisms (of the three organisms that showed a population level genetically based response to the adult foliage on both trials) exhibited a correlation in their preferences among plant

Between families within populations (r_f)

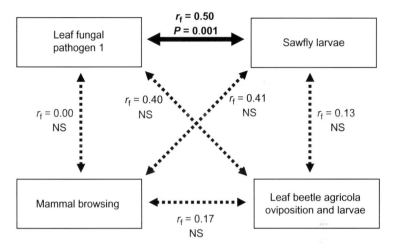

Between populations (r_s)

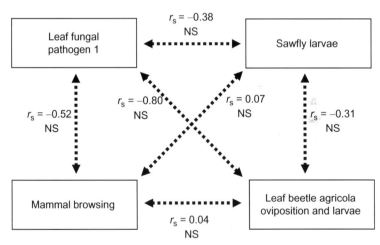

Figure 8.2 Correlations in damage on *Eucalyptus globulus* juvenile foliage between each enemy species among families within populations (an estimate of the additive genetic correlation; r_f) and among populations (r_s). *NS* denotes non-significance at the 0.05 level.

populations in the adult foliage (Table 8.5). There was a significant, highly negative correlation (r_s) between fungal leaf dots and the leaf miner. This negative genetically based relationship is also evident in Fig. 8.3B and C where it is clear that preferences of the leaf miner and fungal leaf dots to

Table 8.5 Correlations among the responses of three organisms to population level genetic variation in *E. globulus* adult foliage and their significances from zero

	Fungal leaf pathogen 2			Fungal leaf dots		
	Population correlation (*r*)	*P* (from zero)	*P* (across sites)	Population correlation (*r*)	*P* (from zero)	*P* (across sites)
Fungal leaf dots	−0.04	1	0.41			
Leaf miner	−0.18	0.58	0.44	−0.92	<0.05	0.78

Correlations were constrained to be equal across sites. The significance of whether the population correlations were significantly different across sites is also shown. Correlations were estimated using data from 14 populations.

population level variation in the host plant are effectively opposite. The other responses exhibited by these three organisms at the plant population level were largely independent of each other (Table 8.5). Importantly, results show that irrespective of whether genetically based responses of organisms on the adult foliage are independent or correlated, both of these patterns are very stable across trials (Table 8.5).

5. DISCUSSION

Overall, there were six main findings to this study. First, across multiple trials, damage by a diverse range of plant enemies (mammalian herbivores, invertebrates and fungal leaf pathogens) was affected by underlying host genetic variation at two distinct genetic scales; family within populations and among populations (aim 1). Responses of organisms to both levels of the genetic hierarchy were particularly pronounced in the juvenile foliage, with less consistency in the responses of organisms to family within population variation in our studies with adult foliage. Second, the genetically based influences of *E. globulus* on its enemies, when they do occur, appear to be relatively consistent across trials (aim 2)—highlighted by the highly significant positive correlations of the fungal leaf pathogens (*Teratosphaeria*) responses to population- and family-level variation in *E. globulus* across the trials. In addition, and thirdly, at the plant population level, there is consistency in *Teratosphaeria* response to juvenile and adult foliage. In some cases, organism damage did not consistently exhibit a detectable response to host genetic variation across trials, as shown for the leaf beetle bimaculata on the adult foliage (Table 8.4). However, in this

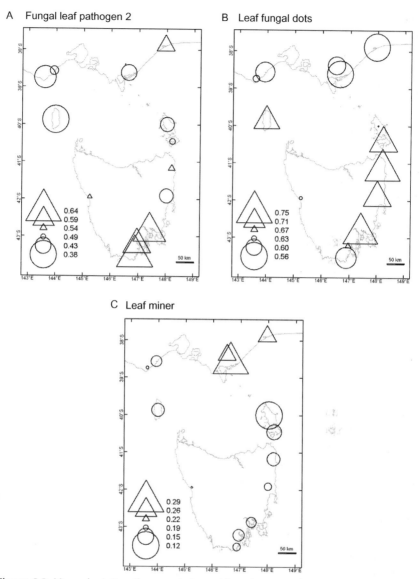

Figure 8.3 Maps depicting the population arithmetic mean damage values on adult *Eucalyptus globulus* foliage by a fungal leaf pathogen 2 (a *Teratosphaeria* species) (A), a second pathogen referred to as 'leaf fungal dots' (B) and a leaf miner (C). Damage scores represent the proportion of 20 leaves affected (presence/absence) by the species. Large triangles represent greater damage while large circles represent less damage.

specific case, the absence of a detectable response was most likely due to over-saturation of beetle infestation at one trial (Salmon River 2; data not shown) and hence, the expression of population divergence in the beetles response was lost, although it may have been evident at an earlier stage in the infestation of the trial. This result highlights the transient nature of detecting plant genetic effects in experimental systems and most likely reflects that the expression of genetic effects in natural plant systems can also fluctuate through temporal scales. For example, in the forest tree *Pinus elliottii*, heritability estimates for fungal rust infection (*Cronartium quercuum*) increased linearly with increasing incidence of rust infection (Dieters et al., 1996) and the detection of correlations in invertebrate responses to genetic variation in the host plant, *Oenothera biennis*, changed through time (Johnson and Agrawal, 2007). The appearance of temporal variation in genetic effects can be due to plasticity in relevant traits influencing the ecological interactions (Fordyce, 2006).

Our fourth major finding was that apart from two genetic correlations between organisms, the majority of the relationships in damage to the genetically based variation in *E. globulus* among enemies were uncorrelated and, hence, organisms primarily exhibit pairwise relationships with the host tree (aim 4). In a recent study, Wise and Rausher (2013) suggested that the common finding of pairwise relationships of herbivores feeding on a host plant (e.g. Simms and Rausher, 1989) may be due to the small number of plant genotypes used in studies. In this study, we tested a relatively large number of families and populations and, consequently, our system may provide some generality in supporting this question of common, independent relationships. Additionally, a lack of correlation between organisms in their response to host genetics could be due to a variable response of the same organism to host family differences (within populations) depending upon, for example, trial or season. However, this would not appear to be the case in the present study as when genetic differences do exist there is consistency in the response of specific organisms to underlying host plant genetics across multiple trials, times and leaf stage (leaf fungal pathogen 2 in this study and leaf fungal pathogen 1 and mammal responses in previous studies). These findings gives us additional confidence in the observed independent response patterns and in our conclusion that many of the organisms studied exhibit pairwise interactions with the host at the genetic level. As many of these organisms have documented fitness consequences in this system (Jordan et al., 2002; Milgate et al., 2005; O'Reilly-Wapstra et al., 2012), evolutionary diversification in *E. globulus* is most likely occurring in the context of variable selective forces with pairwise interactions occurring between *E. globulus* and these enemies. For genetic divergence in resistance to occur in response to selection by enemies, the presence of enemies would not only need to have clear fitness consequences but also be relatively consistent (e.g. temporally consistent). This would need to be particularly true in plant systems with variable selective pressures (e.g. multiple herbivores/pathogens) which could impede or slow down directional selection (Strauss and Irwin, 2004). Alternatively, if the effects of two or more co-occurring species are positively genetically correlated then this could also result in stronger and consistent selection on traits.

Our finding of significant within population genetic correlations (aim 4) between sawfly damage and leaf fungal pathogen 1 (positive relationship; Fig. 8.2) and between fungal leaf dots and a leaf miner (negative relationship) indicates that the host plants relationships with these organisms is operating in a diffuse manner (Stinchcombe and Rausher, 2001; Wise and Rausher, 2013). We will discuss each of these relationships separately. As the sawfly damage and leaf fungal pathogen 1 correlation were detected when organisms were assessed on different trials, at different times and in the face of potential genotype by environment interactions affecting plant phenotype, our results indicate a true genetic correlation in the response of these two organisms. The only links between trees were through the tree pedigree at the family level. Thus, this non-independent response suggests pleiotropy, resulting from either common genetic control of two traits to which the organisms are independently responding or both organisms responding to the same trait. Currently, the mechanisms driving genetic variation in resistance of *E. globulus* to both sawfly and the pathogen are largely unknown. It has been suggested that bark thickness, where the thinner, smoother bark of some trees may hinder sawfly larval movement up the tree (Dutkowski and Potts, 1999), and that leaf width and glaucousness may also affect oviposition (Carne, 1965). Alternatively, constitutive leaf mesophyll density is believed to play a role in the variation in resistance of eucalypts to *Teratosphaeria* attack (Smith et al., 2007). Nevertheless, regardless of the mechanism, the significant family within population correlation between these two organisms (this present study) and their documented negative fitness consequences for *E. globulus* (Jordan et al., 2002; Milgate et al., 2005) suggests that any selection pressure imposed by one will result in a correlated genetic response by the plant population, which will impact the other organism (that is, a potential indirect eco-evolutionary feedback loop mediated by a genetic correlation).

To date, very little is known about the leaf miner and fungal leaf dot symptoms in our *E. globulus* system. As opposed to the sawfly damage and leaf fungal pathogen 1 discussed above, these symptoms were assessed in the same field trials and thus the strong, negative relationship we detected at the population level could be due to ecological interference (Wise, 2009) between the causal organisms where, for example, the miner avoided leaves with fungal lesions. Alternatively, this relationship could be due to the organisms responding in opposite directions to the same genetic variation. We tested these two questions by analysing the correlation in the residuals (accounting for the remaining variance—genetic and environmental—after the host population variance has been accounted for). In our previous analysis for these traits, there was no significant genetic variance at the family within population level, consequently, if the residual correlation follows the population correlation (a negative relationship) then this provides evidence that the correlation between the organisms is due to ecological interference. In contrast, the correlation of residuals on each trial was positive and significant for only one trial ($r=0.14$, $P<0.001$; $r=0.03$, $P>0.05$ for Salmon River 2 and Temma, respectively). This contrasting pattern of a correlations amongst these symptoms at the among and within population levels argues against interference and suggests that the organisms are responding in an opposite manner to population variation in a host trait(s). This may lead to a possible evolutionary trade-off for the host plant, restricting the rate of directional change (Strauss and Irwin, 2004; Wise and Rausher, 2013).

The fifth aim in this study was to determine if among population divergence in trait evolution is constrained by within population genetic correlation in responses of organisms. If genetic correlations represent constraints on population divergence, then the genetic correlation within populations should be a reasonable predictor of trait correlations among populations (Colautti and Barrett, 2011). The significant positive within population genetic correlation for sawflies and the fungal leaf pathogen was not detected among populations (Fig. 8.2) and, consequently, the response of the organisms at the population level is not constrained by the within population patterns. Our fifth conclusion here is that the diffuse relationships between these two organisms are not constraining the evolutionary diversification in resistance of populations of their host plant. A breakdown in genetic correlations can occur through strong correlational selection on favoured traits in only one of the species; or sexes, as elegantly demonstrated in the dioecious plant *Silene latifolia* (Delph et al., 2011). In our study, we suggest the decoupling may occur in several ways. For example, population divergence in a trait to which both organisms are responding could pass a threshold for one organism but not the other. Another possibility is if multiple traits are affecting the response of the organisms to the plant and one or both of the organisms are responding differently to another trait. In other words, a third factor(s) has overridden the correlated response of these two organisms at the population level, which could arise with strong natural selection on population divergence in traits which indirectly affect biotic interactions. For example, photo-damage has been proposed as being an important abiotic factor in influencing the evolution of foliar traits that also influence biotic interactions in eucalypt species (Close and McArthur, 2002). Additionally, drought has proven to be a powerful selective agent in other systems such as pinyon pine (*Pinus edulis*) (Sthultz et al., 2009) and *E. globulus* (Dutkowski and Potts, 2012), where it can affect plant resistance traits such as foliar phenolic levels (McKiernan et al., 2014).

The final finding from this study is that regardless of whether the genetic effects of *E. globulus* are acting independently or in a correlative manner between organisms, when the genetic effect is evident, it is consistent across multiple environments (aim 6). This conclusion is reflected in the strong, stable responses of the *Teratosphaeria* species to host genetic variation on juvenile and adult foliage across trials and the lack of significant difference between trials in both the correlative and independent population level relationships in Table 8.5. Two recent studies (O'Reilly-Wapstra et al., 2013a,b) have also highlighted the stability (consistency) in the *E. globulus* genetic signal across trials in the expression of plant secondary metabolites and in the holistic physicochemical profile of the species. This lack of genetic by environment interaction in these chemical traits was also evident at both among and within populations of the *E. globulus* genetic hierarchy. Additionally, another recent study (McKiernan et al., 2012) examining the effects of climate change factors on the genetically based quantitative expression of plant secondary metabolites in two eucalypt species (*E. globulus* and *E. pauciflora*) also showed stability in the population level ranking across CO_2 environments. Consequently, it appears that the genetic ranking of many quantitative traits we have examined in our plant/herbivore/pathogen investigations in eucalypts are consistent across environments.

In conclusion, we have used this forest tree system to highlight some important points that are relevant to the field of eco-evolutionary dynamics. A challenge in eco-evolutionary dynamics of plant/herbivore systems is to take a multi-species approach (e.g. Hendry, 2013; Moya-Laraño et al., 2014, chapter 3 of this volume; Turcotte et al., 2011) and incorporate knowledge of genetically based correlations between enemy responses to host plants at different levels of the host plants genetic hierarchy. This information would better inform us of how plant populations have evolutionarily diverged in their phenotypic traits, which organisms are driving the evolutionary change and how rapid evolutionary change in one enemy or plant species can feedback to affect other genetically correlated species. This chapter demonstrates that genetically based correlations between natural enemies and how they respond to a host tree may not be consistent across the hierarchical levels of a genetic variation within a tree species. Here, within plant population correlations in enemy response were not present among plant populations, indicating that population divergence in plant resistance to two enemies is not constrained by the within population genetic correlations. Additionally, the consistency we showed across different trials in the correlated enemy response to the genetic variation in host plants (environments) indicates that in an eco-evolutionary dynamic setting, the strength and consistency of any selective force would be maintained across environments (Hallsson and Björklund, 2012).

ACKNOWLEDGEMENTS

We thank Greg Jordan, Erik Wapstra and anonymous reviewers for constructive comments on earlier drafts of this chapter. We thank Hugh Fitzgerald, Paul Tilyard, Justin Bloomfield, Naomi Glancy, Alison Miller, James Marthick and Forestry Tasmania staff for assistance with field work. We thank Lynn Forster for assistance with the symptom identification and Peter Harrison for assistance with processing leaf samples. We thank Forestry Tasmania and Gunns Limited for access to land. Funding was provided by the University of Tasmania, CRC for Forestry, Forestry Tasmania and ARC Discovery (DP0451533 and DP120102889) and Linkage (LP0562415) grants.

REFERENCES

Agrawal, A.A., 2005. Natural selection on common milkweed (*Asclepias syriaca*) by a community of specialized insect herbivores. Evol. Ecol. Res. 7, 651–667.
Andrew, R.L., Wallis, I.R., Harwood, C.E., Foley, W.J., 2010. Genetic and environmental contributions to variation and population divergence in a broad-spectrum foliar defence of *Eucalyptus tricarpa*. Ann. Bot. 105, 707–717.
Armbruster, W.S., 1991. Multilevel analysis of morphometric data from natural plant populations: insights into ontogenetic, genetic, and selective correlations in *Dalechampia scandens*. Evolution 45, 1229–1244.
Barbour, R.C., O'Reilly-Wapstra, J.M., De Little, D.W., Jordan, G.J., Steane, D.A., Humphreys, J.R., Bailey, J.K., Whitham, T.G., Potts, B.M., 2009. A geographic mosaic of genetic variation within a foundation tree species and its community-level consequences. Ecology 90, 1762–1772.
Benkman, C.W., 1999. The selection mosaic and diversifying coevolution between crossbills and Lodgepole pine. Am. Nat. 153, S75–S91.
Carne, P.B., 1965. Distribution of the eucalypt defoliating sawfly *Perga affinis affinis* (Hymenoptera). Aust. J. Zool. 13, 593–612.

Carnegie, A.J., Keane, P.J., Ades, P.K., Smith, I.W., 1994. Variation in susceptibility of *Eucalyptus globulus* provenances to *Mycosphaerella* leaf disease. Can. J. Forest Res. 24, 1751–1757.

Chambers, P.G.S., Borralho, N.M.G., Potts, B.M., 1996. Genetic analysis of survival in *Eucalyptus globulus* ssp. *globulus*. Silvae Genet. 45, 107–112.

Close, D.C., McArthur, C., 2002. Rethinking the role of many plant phenolics-protection from photodamage not herbivores? Oikos 99, 166–172.

Colautti, R.I., Barrett, S.C.H., 2011. Population divergence along lines of genetic variance and covariance in the invasive plant *Lythrum salicaria* in eastern North America. Evolution 65, 2514–2529.

Conner, J.K., Hartl, D.L., 2004. A Primer of Ecological Genetics. Sinauer Associates, Inc, Sunderland.

Costa e Silva, J., Potts, B.M., Dutkowski, G.W., 2006. Genotype by environment interaction for growth of *Eucalyptus globulus* in Australia. Tree Genet. Genomes 2, 61–75.

Delph, L.F., Steven, J.C., Anderson, I.A., Herlihy, C.R., Brodie, E.D., 2011. Elimination of a genetic correlation between the sexes via artificial correlational selection. Evolution 65, 2872–2880.

Dieters, M.J., Hodge, G.R., White, T.L., 1996. Genetic parameter estimates for resistance to rust (*Cronartium quercuum*) infection from full-sib tests of slash pine (*Pinus elliottii*), modelled as functions of rust incidence. Silvae Genet. 45, 235–242.

Dutkowski, G.W., Potts, B.M., 1999. Geographic patterns of genetic variation in *Eucalyptus globulus* ssp. *globulus* and a revised racial classification. Aust. J. Bot. 47, 237–263.

Dutkowski, G.W., Potts, B.M., 2012. Genetic variation in the susceptibility of *Eucalyptus globulus* to drought damage. Tree Genet. Genomes 8, 1–17.

Ellner, S.P., 2013. Rapid evolution: from genes to communities, and back again? Funct. Ecol. 27, 1087–1099.

Falconer, D.S., Mackay, T.F.C., 1996. Introduction to Quantitative Genetics, fourth ed. Longman, Harlow.

Fordyce, J.A., 2006. The evolutionary consequences of ecological interactions mediated through phenotypic plasticity. J. Exp. Biol. 209, 2377–2383.

Gilmour, A.R., Gogel, B.J., Cullis, B.R., Thompson, R., 2009. ASReml User Guide Release 3.0. VSN International Ltd, Hemel Hempstead.

Gómez, J.M., Perfectti, F., Bosch, J., Camacho, J.P.M., 2009. A geographic selection mosaic in a generalized plant-pollinator-herbivore system. Ecol. Monogr. 79, 245–263.

Hallsson, L.R., Björklund, M., 2012. Selection in a fluctuating environment leads to decreased genetic variation and facilitates the evolution of phenotypic plasticity. J. Evol. Biol. 25, 1275–1290.

Hamilton, M.G., Williams, D.R., Tilyard, P.A., Pinkard, E.A., Glen, M., Vaillancourt, R.E., Potts, B.M., 2013. A latitudinal cline in disease resistance of a host tree. Heredity 110, 372–379.

Hendry, A., 2013. Eco-evolutionary dynamics: community consequences of (mal)adaptation. Curr. Biol. 23, R870.

Hiltunen, T., Ellner, S.P., Hooker, G., Jones, L.E., Hairston, N.G., 2014. Eco-evolutionary dynamics in a three-species food web with intraguild predation: intriguingly complex. In: Moya-Laraño, J., Rowntree, J., Woodward, G. (Eds.), Advances in Ecological Research, Vol. 50: Eco-evolutionary Dynamics. Elsevier, Amsterdam, pp. 41–73.

Johnson, M.T.J., Agrawal, A.A., 2007. Covariation and composition of arthropod species across plant genotypes of evening primrose, *Oenothera biennis*. Oikos 116, 941–956.

Jones, R.C., Steane, D., Lavery, M., Vaillancourt, R.E., Potts, B.M., 2013. Multiple evolutionary processes drive patterns of genetic differentiation in a forest tree species complex. Ecol. Evol. 3, 1–17.

Jordan, G., Potts, B.M., Wiltshire, R., 1999. Strong, independent quantitative genetic control of vegetative phase change and first flowering in *Eucalyptus globulus* ssp. *globulus*. Heredity 83, 179–187.

Jordan, G.J., Potts, B.M., Clarke, A.R., 2002. Susceptibility of *Eucalyptus globulus* ssp. *globulus* to sawfly (*Perga affinis* ssp. *insularis*) attack and its potential impact on plantation productivity. For. Ecol. Manage. 160, 189–199.

Khudr, M.S., Potter, T., Rowntree, J., Preziosi, R.F., 2014. Community genetic and competition effects in a model pea aphid system. In: Moya-Laraño, J., Rowntree, J., Woodward, G. (Eds.), Advances in Ecological Research, Vol. 50: Eco-evolutionary Dynamics. Elsevier, Amsterdam, pp. 241–263.

Laine, A.-L., 2004. Resistance variation within and among host populations in a plant pathogen metapopulation: implications for regional pathogen dynamics. J. Ecol. 92, 990–1000.

Leimu, R., Fischer, M., 2010. Between-population outbreeding affects plant defence. PLoS One 5, e12614.

Leimu, R., Koricheva, J., 2006. A meta-analysis of genetic correlations between plant resistances to multiple enemies. Am. Nat. 168, E15–E37.

Mauricio, R., Rausher, M.D., 1997. Experimental manipulation of putative selective agents provides evidence for the role of natural enemies in the evolution of plant defence. Evolution 51, 1435–1444.

McKiernan, A.B., O'Reilly-Wapstra, J.M., Price, C., Davies, N.W., Potts, B.M., Hovenden, M.J., 2012. Stability of plant defensive traits among populations in two *Eucalyptus* species under elevated carbon dioxide. J. Chem. Ecol. 38, 204–212.

McKiernan, A.B., Hovenden, M.J., Brodribb, T.J., Potts, B.M., Davies, N.W., O'Reilly-Wapstra, J.M., 2014. Effect of limited water availability on foliar plant secondary metabolites of two *Eucalyptus* species. Environ. Exp. Bot. 105, 55–64.

Milgate, A.W., Potts, B.M., Joyce, K., Mohammed, C., Vaillancourt, R.E., 2005. Genetic variation in *Eucalyptus globulus* for susceptibility to *Mycosphaerella nubilosa* and its association with tree growth. Australas. Plant Pathol. 34, 11–18.

Miller, A.M., McArthur, C., Smethurst, P.J., 2007. Effects of within-patch characteristics on the vulnerability of a plant to herbivory. Oikos 116, 41–52.

Moore, B.D., Foley, W., 2005. Tree use by koalas in a chemically complex landscape. Nature 435, 488–490.

Moya-Laraño, J., Bilbao-Castro, J.R., Barrionuevo, G., Ruiz-Lupión, D., Casado, L.G., Montserrat, M., Melián, C.J., Magalhães, S., 2014. Eco-evolutionary spatial dynamics: rapid evolution and isolation explain food web persistence. In: Moya-Laraño, J., Rowntree, J., Woodward, G. (Eds.), Advances in Ecological Research, Vol. 50: Eco-evolutionary Dynamics. Elsevier, Amsterdam, pp. 75–143.

Muola, A., Mutikainen, P., Lilley, M., Laukkanen, L., Salminen, J.-P., Leimu, R., 2010. Associations of plant fitness, leaf chemistry, and damage suggest selection mosaic in plant-herbivore interactions. Ecology 91, 2650–2659.

O'Reilly-Wapstra, J.M., McArthur, C., Potts, B.M., 2002. Genetic variation in resistance of *Eucalyptus globulus* to marsupial browsers. Oecologia 130, 289–296.

O'Reilly-Wapstra, J.M., Bailey, J.K., McArthur, C., Potts, B.M., 2010. Genetic- and chemical-based resistance to two mammalian herbivores varies across the geographic range of *Eucalyptus globulus*. Evol. Ecol. Res. 12, 1–16.

O'Reilly-Wapstra, J.M., McArthur, C., Potts, B.M., 2012. Selection for anti-herbivore plant secondary metabolites: a Eucalyptus system. In: Iason, G.I., Hartley, S., Dicke, M. (Eds.), The Integrative Role of Plant Secondary Metabolites in Ecological Systems. Cambridge University Press, London, pp. 10–33.

O'Reilly-Wapstra, J., Freeman, J.S., Barbour, R.C., Vaillancourt, R., Potts, B., 2013a. Genetic analysis of the near infrared spectral phenome of a global *Eucalyptus* species. Tree Genet. Genomes 9, 943–959.

O'Reilly-Wapstra, J.M., Miller, A.M., Hamilton, M., Williams, D., Glancy-Dean, N., Potts, B., 2013b. Chemical variation in a dominant tree species: population divergence, selection and genetic stability across environments. PLoS One 8, e58416.

Post, D.M., Palkovacs, E.P., 2009. Eco-evolutionary feedbacks in community and ecosystem ecology: interactions between the ecological theatre and the evolutionary play. Philos. Trans. Biol. Sci. 364, 1629–1640.

Potts, B.M., McGowen, M.H., Williams, D.R., Suitor, S., Jones, T.H., Gore, P.L., Vaillancourt, R.E., 2008. Advances in reproductive biology and seed production systems of *Eucalyptus*: the case of *Eucalyptus globulus*. South. For. 70, 145–154.

Pruitt, J.N., Riechert, S.E., Iturralde, G., Vega, M., Fitzpatrick, B.M., Aviles, L., 2010. Population differences in behaviour are explained by shared within-population trait correlations. J. Evol. Biol. 23, 748–756.

Rapley, L., Allen, G.R., Potts, B.M., 2004a. Genetic variation of *Eucalyptus globulus* in relation to autumn gum moth *Mnesampela privata* (Lepidoptera: Geometridae) oviposition preference. For. Ecol. Manage. 194, 169–175.

Rapley, L.P., Allen, G.R., Potts, B.M., 2004b. Genetic variation in *Eucalyptus globulus* in relation to susceptibility from attack by the southern eucalypt leaf beetle, *Chrysophtharta agricola*. Aust. J. Bot. 52, 747–756.

Rausher, M.D., 1996. Genetic analysis of coevolution between plants and their natural enemies. Trends Genet. 12, 212–217.

Schluter, D., 1996. Adaptive radiation along genetic lines of least resistance. Evolution 50, 1766–1774.

Schoener, T.W., 2011. The newest synthesis: understanding the interplay of evolutionary and ecological dynamics. Science (New York, N.Y.) 331, 426–429.

Simms, E.L., Rausher, M.D., 1989. The evolution of resistance to herbivory in *Ipomoea purpurea*. II. Natural selection by insects and costs of resistance. Evolution 43, 573–585.

Smith, A.H., Gill, W.M., Pinkard, E.A., Mohammed, C.L., 2007. Anatomical and histochemical defence responses induced in juvenile leaves of *Eucalyptus globulus* and *Eucalyptus nitens* by *Mycosphaerella* infection. For. Pathol. 37, 361–373.

Sthultz, C.M., Gehring, C.A., Whitham, T.G., 2009. Deadly combination of genes and drought: increased mortality of herbivore-resistant trees in a foundation species. Glob. Chang. Biol. 15, 1949–1961.

Stinchcombe, J.R., Rausher, M.D., 2001. Diffuse selection on resistance to deer herbivory in the ivy leaf morning glory, *Ipomoea hederacea*. Am. Nat. 158, 376–388.

Strauss, S.Y., Irwin, R.E., 2004. Ecological and evolutionary consequences of multispecies plant-animal interactions. Annu. Rev. Ecol. Syst. 35, 435–466.

Thompson, J.N., 2005. The Geographic Mosaic of Coevolution, first ed. University of Chicago Press, Chicago, IL.

Turcotte, M.M., Reznick, D.N., Hare, J.D., 2011. The impact of rapid evolution on population dynamics in the wild: experimental test of eco-evolutionary dynamics. Ecol. Lett. 14, 1084–1092.

Turcotte, M.M., Reznick, D.N., Hare, J.D., 2013. Experimental test of an eco-evolutionary dynamic feedback loop between evolution and population density in the green peach aphid. Am. Nat. 181, S46–S57.

Utsumi, S., Ando, Y., Roininen, H., Takahashi, J.-i., Ohgushi, T., 2013. Herbivore community promotes trait evolution in a leaf beetle via induced plant response. Ecol. Lett. 16, 362–370.

Wiggins, N.L., McArthur, C., Davies, N.W., McLean, S., 2006. Spatial scale of the patchiness of plant poisons: a critical influence on foraging efficiency. Ecology 87, 2236–2243.

Wise, M.J., 2007. Evolutionary ecology of resistance to herbivory: an investigation of potential genetic constraints in the multiple-herbivore community of *Solanum carolinense*. New Phytol. 175, 773–784.

Wise, M.J., 2009. Competition among herbivores of *Solanum carolinense* as a constraint on the evolution of host-plant resistance. Evol. Ecol. 23, 347–361.

Wise, M.J., Rausher, M.D., 2013. Evolution of resistance to a multiple-herbivore community: genetic correlations, diffuse coevolution, and constraints on the plants response to selection. Evolution 67, 1767–1779.

CHAPTER NINE

When Ranges Collide: Evolutionary History, Phylogenetic Community Interactions, Global Change Factors, and Range Size Differentially Affect Plant Productivity

**Mark A. Genung*,[1], Jennifer A. Schweitzer*, John K. Senior[†],
Julianne M. O'Reilly-Wapstra[†], Samantha K. Chapman[‡],
J. Adam Langley[‡], Joseph K. Bailey***

*Department of Ecology and Evolutionary Biology, University of Tennessee, Knoxville, Tennessee, USA
[†]School of Biological Sciences, University of Tasmania, Hobart, Tasmania, Australia
[‡]Department of Biology, Villanova University, Villanova, Pennsylvania, USA
[1]Corresponding author: e-mail address: mgenung@utk.edu

Contents

Advances in Ecological Research, Volume 50
ISSN 0065-2504
http://dx.doi.org/10.1016/B978-0-12-801374-8.00009-8

Abstract

Humans are extensively changing the global environment, both by altering abiotic conditions through increases in carbon dioxide (CO_2) and reactive nitrogen (N), and by driving patterns of extinctions and introductions that shift community composition and affect the biotic environment. Evolutionary history may play an important role in determining plant responses to global change, if evolution has selected for certain traits (biomass, allocation strategies, range size, etc.) that determine plant responses to rising CO_2 and N. Additionally, the evolutionary history of interacting plants (i.e. phylogenetic relationships within plant communities) may determine how plants respond to global change, as closely related species might be expected to compete more for limiting resources than distantly related species. Using 26 Australian eucalypt species in two subgenera (*Eucalyptus* and *Symphyomyrtus*) of the genus *Eucalyptus*, we conducted the first experiment, to our knowledge, that simultaneously integrated contemporary range size, phylogenetic identity, phylogenetic similarity, and global change factors (CO_2 and N). We showed that plant biomass responded to two three-way interactions: (1) subgenus identity, N fertilization, and phylogenetic similarity and (2) subgenus identity, CO_2 enrichment, and N fertilization. Our results indicate that eco-evolutionary dynamics are linked in diverse and non-intuitive ways where evolutionary history (i.e. subgenus-level differences) mediates how plant productivity responds to resource manipulation, and that the nature of this response depends on the phylogenetic composition of plant communities. Overall, these findings have significant implications for how we understand the ecosystem-level consequences of climate change.

1. INTRODUCTION

Human alteration of the environment is resulting in extensive biodiversity loss and dramatic changes in atmospheric chemistry and nutrient availability (Galloway et al., 2004; Hooper et al., 2012; IPCC, 2007; Vitousek et al., 1997) and is threatening the sustainability of many global ecosystems. These global patterns are a pressing concern for ecologists, evolutionary biologists, and conservationists as different species contribute uniquely to ecosystem services (i.e. provisioning of food and fuel, soil formation, carbon sequestration, among others; Díaz and Cabido, 2001), and it is unclear to what extent these services will operate in the future. While phylogenetic information is now more commonly incorporated into global change studies (Davis et al., 2010; Edwards et al., 2007; Senior et al., 2013; Willis et al., 2010), most focus on the response of species to abiotic environmental factors. Much less attention has been paid to the role of intrinsic factors such as evolutionary history (in this chapter, we use evolutionary history interchangeably with subgenus identity), or biotic factors like the relatedness of interacting species, which has been shown to be a key element for plant productivity and the provisioning of ecosystem services (Smith et al., 2013).

To develop a better understanding of how evolutionary history, phylogenetic relationships, CO_2 enrichment, and N fertilization interact to affect plant responses to a changing climate, it is important to know the properties of plants that may also be related to evolutionary history. For example, it is well established that more productive species have larger ranges, and that productivity is commonly under genetic control, and yet it is unclear whether contemporary range size may be related to evolutionary history (Bailey et al., 2014; Gaston, 2003; Morueta-Holme et al., 2013; Ohlemüller et al., 2008). Although we have limited knowledge of what determines range size in plants (Brown et al., 1996; Morueta-Holme et al., 2013), it is known that species with larger ranges face less extinction risk (Gaston, 2003), and a recent meta-analysis showed that habitat area and climate stability are co-determinants of range size in New World plants (Morueta-Holme et al., 2013). Beyond the abiotic determinants of range size, it may be important to consider how plant traits vary between species with large and small ranges. For example, we know that species that occupy larger ranges are more productive, but it is less clear if variation in productivity determines how plants will respond to anthropogenic changes to CO_2 and N. Also, if productivity varies predictably between plants with large and small ranges, is it related to evolutionary history, and how do phylogenetically based species interactions alter these patterns of productivity in a global change context?

Determining how phylogenetic similarity (in this chapter, phylogenetic similarity refers to the degree of genetic distance between interacting plants) affects competition and/or facilitation between interacting species may be particularly useful in the context of global change research, as phylogenetic similarity at the species level (McKinney and Lockwood, 1999) and genetic similarity at the population level (Luck et al., 2003) have increased, and are predicted to continue increasing, in response to human activities. This suggests that most species are declining in abundance while a few "winning" species increase (McKinney and Lockwood, 1999), reducing the potential for facilitative interactions between phylogenetically distant plants; these declines may be even more significant when measured at the population instead of species level (Luck et al., 2003). Therefore, it is important to consider that changes in phylogenetic similarity and environmental conditions (driven by global changes including climate change and N eutrophication due to intensive agricultural practices) may interact to affect communities and ecosystems; however, we know of no studies that have explicitly manipulated phylogenetic similarity in the context of global change. While there exists a well-developed set of predictions for how a range of abiotic factors will change over the next 100 years (IPCC, 2007), the extent to which the phylogenetic similarity of communities will change is less well projected and it may be difficult to determine how biodiversity loss (at the species or phylogenetic level) and ecosystem functioning can be measured in meaningful ways. Tilman et al. (2012) approached this issue by comparing 16-species plant mixtures with 4-, 2-, and 1-species mixtures (thought to be similar to human-created grasslands) to determine how the conversion of native grasslands to cropland may affect community-level diversity and consequently ecosystem functioning. The effect size of species richness was then compared with ecologically relevant manipulations of global change factors, and the authors determined that, for plant productivity, an increase in species richness from 4 to 16 species was equivalent to adding 54 kg ha^{-1} of N per year (Tilman et al., 2012). For other study systems, different manipulations of species richness or phylogenetic diversity may be necessary to represent reasonable future conditions based on predicted changes to biodiversity over the next 100 years, but studies such as the one described above indicate that biodiversity loss, and biotic interactions and global change factors can have effects of similar magnitude.

In addition to changes in the phylogenetic structure (i.e. not all phylogenetic groups will respond similarity to global change; Senior et al., 2013) of communities due to biodiversity changes, two of the most important elements of global environmental change over the next 100 years will be increases in atmospheric CO_2 concentrations and N eutrophication. Atmospheric CO_2 concentrations are predicted to double pre-industrial concentrations in the next century (IPCC, 2007) from 360 to approximately 700 ppm (predictions range from 550 to 950 ppm; IPCC, 2007). Rates of N fixation are currently twice those observed before pre-industrial times and are expected to continue to rise due to anthropogenic activities (Galloway et al., 2004; McLauchlan et al., 2013; Vitousek et al., 1997). Changes in CO_2 and N do not always operate in isolation, however, as productivity responses to elevated CO_2 can be limited by N availability (Norby et al., 2010). The response of plants to these abiotic factors (CO_2 and N) may by unpredictable due to plant–plant interactions that vary across different communities (Langley and Megonigal, 2010). Plants may respond less to CO_2 and N in the presence of a heterospecific neighbour compared with a conspecific neighbour, if that heterospecific neighbour is a superior competitor and acquires the bulk of the additional resources.

Another approach to predicting how plant–plant interactions between similarly and distantly related plants will change under elevated CO_2 and N involves considering whether phylogenetic niches are conserved or not. Phylogenetic niche conservatism occurs when similar niches are occupied by phylogenetically related individuals (e.g. Cooper et al., 2011; Webb et al., 2002; Wiens and Graham, 2005). However, if phylogenetic niches are not conserved, then distantly related individuals may have similar ecological functions due to convergent evolution giving rise to similar traits in different evolutionary lineages. It is possible that across larger or smaller sections (i.e. clades or series vs. subgenera) of the phylogeny, we could see different effects of niche conservatism and convergent evolution; this would be indicated by an unpredictable relationship between phylogenetic similarity and plant productivity. Also, closely related individuals may have different ecological functions, due to mechanisms such as the competitive exclusion principle (Gause, 1934; Hardin, 1960) that drive divergence when two similar species occupy the same habitat. In the context of altered nutrient availability under climate change, phylogenetic niche conservatism could mean that distantly related focal and neighbour plants may be able to partition nutrient acquisition, resulting in a beneficial interaction for both partners, or a lack of phylogenetic niche conservatism could make it difficult to predict how phylogenetic similarity will mediate the responses of plant communities to climate change.

Here we integrate the evolutionary basis (i.e. subgenus-level differences) to contemporary range size, the phylogenetic similarity of interacting species, CO_2, and N to examine how phylogenetic similarity and global change drivers interact to influence the performance of plants in genus *Eucalyptus*. We hypothesized that (1) contemporary range size has an evolutionary basis; (2) plant biomass would negatively respond to increasing phylogenetic similarity and positively respond to increasing concentrations of CO_2 and N; and that (3) the effects of phylogenetic similarity, CO_2, and N on the productivity of plants would be contingent on evolutionary history (subgenus *Eucalyptus* and subgenus *Symphyomyrtus*) due to the different traits of these groups (e.g. faster growth rates and less herbivore resistance traits in *Symphyomyrtus* relative to *Eucalyptus*; Senior et al., 2013; Stone et al., 1998; Wallis et al., 2010) and to the unpredictable way plant–plant interactions may change under increased CO_2 and N. We also compared the relative effect sizes of subgenus identity, phylogenetic similarity, CO_2 enrichment, and N fertilization to determine whether evolutionary factors (subgenus identity, phylogenetic similarity) or abiotic factors (CO_2, N) have larger effects on plant productivity. We found that evolutionary history impacted plant responses to climate change in multiple ways, including subgenus variation and phylogenetically based interactions within plant communities, and evolutionary factors had a larger impact on plant productivity than did abiotic factors, suggesting that more climate change studies should incorporate an evolutionary perspective.

2. METHODS

2.1. Experimental methods and study species

2.1.1 Design, study species, and global change manipulations

To determine how evolutionary history and phylogenetic similarity interact with abiotic factors (such as atmospheric CO_2 and soil N) to affect plant productivity and biomass allocation, 26 of the 29 species of Tasmanian *Eucalyptus* species were utilized; 16 are in subgenus *Symphyomyrtus* and 10 are in subgenus *Eucalyptus*. Within each subgenus, species are further placed within a series, based on genetic relatedness; there are three series within *Symphyomyrtus* and two within *Eucalyptus* (see Senior et al., 2013 for phylogeny). These phylogenetic classifications are based on a framework created by Brooker (2000) with molecular data (Diversity Arrays Technology; Jaccoud et al., 2001) supporting the most recent subgenus- and series-level classifications (McKinnon et al., 2008; Senior et al., 2013; Steane et al., 2011). The life histories of species with *Eucalyptus* and *Symphyomyrtus* differ in several ways. For example, *Symphyomyrtus* species are generally found at higher elevations than *Eucalyptus* species (average elevation of ~500 m vs. ~325 m) but both subgenera occur over large ranges of elevation and range maps suggest that almost all species pairs are capable of co-occurring in the wild (Williams and Potts, 1996). Continuous and categorical classifications of range size based on geographic range maps and presence/absence counts for all species of eucalypts on the island of Tasmania (Williams and Potts, 1996) were used for analyses that required the ability to nest other factors within range size. We used the presence/absence counts as a continuous variable when creating some figures.

2.1.2 Plant propagation and greenhouse methods

To enhance germination, seeds were folded in a paper towel and soaked overnight in a solution of water and dishwashing detergent and then refrigerated for 30 days at 4 °C. After refrigeration, seeds were sprinkled over 28 separate $30 \times 35 \times 3$ cm trays filled with commercial potting mix, which consisted of eight parts composted fine pine bark and three parts coarse river sand with added macro- and micro-nutrients from Nutricote Grey (Langley Australia Pty Ltd., Welshpool, WA). The added nutrients included nitrogen (N), phosphorus, and potassium in the weight ratio of 19:2.6:10, at an approximate concentration of 3 kg m^{-3}. The surface of the potting mix was covered with vermiculite for water retention. Germinants were grown in a climate-controlled greenhouse for 3 weeks until the majority of seedlings of each species had developed the first pair of leaves. While in the greenhouse, the plants experienced normal day length for Tasmanian winters (9–10 hour days). For each species, 48 seedlings of uniform size were selected for use in the global change/phylogenetic similarity manipulations. After seedlings were selected, they were grown in standard forestry tubes (50^2 cm surface area $\times$ 125 cm deep) for the remainder of the experiment. Biomass was destructively sampled before plants became root-bound.

Global change treatments were created by factorially manipulating CO_2 and N concentrations to yield four different treatments: control, $+CO_2$, $+N$, and $+CO_2+N$ (details on global change treatments can be found in the following paragraph). Within each of the global change treatments, four phylogenetic similarity treatments were established that manipulated the degree of relatedness between neighbouring plants by planting different species together in the same pot according to the following configuration: (1) species monocultures, in which both individuals belonged to the same species; (2) within-series mixtures, in which both individuals belonged to the same series; (3) among-series mixtures; and (4) between-subgenus mixtures that contained species from both the subgenus *Eucalyptus* and subgenus *Symphyomyrtus*. Treatments were established by randomly sampling species to create the appropriate level of phylogenetic similarity and the pool from which neighbour species were randomly drawn was restricted based on the desired phylogenetic similarity for a given pot (for a list of species combinations present in each phylogenetic similarity treatment, see table in Appendix A). The four global change treatments and four phylogenetic similarity treatments (described above) were factorially combined to yield 16 different "global change by phylogenetic similarity" treatments (e.g. species monoculture in $+CO_2$, or within-series mixture in $+CO_2+N$). As mentioned above, 48 individuals of each species were selected; this was done to allow for 3 replicates of each species in each of the 16 "global change by phylogenetic similarity" treatments. The range of phylogenetic similarity captured by these treatments is consistent with what Tasmanian eucalypts experience; in other words, we did not create any phylogenetic similarity treatments (monocultures, between-subgenus mixtures, etc.) that could not theoretically occur in natural systems (Williams and Potts, 1996). In total, 2496 plants were planted (26 species and 48 individuals per species for focal plants, doubled to account for neighbour plants). Because two plants were present in each pot, this yields a total of 1248 pots, with 78 pots in each of the 16 "global change by phylogenetic similarity" treatments. Of the original plants, 1793 (882 focal and 911 neighbour) survived to the end of the experiment. Of the original pots, 688 had both plants (focal and neighbour) survive to the end of the experiment, and these were the only pots used in our analyses (to prevent bias due to plant density per pot). We examined how many of those 688 pots belonged to each of the 16 "global change by phylogenetic similarity" treatments; the treatment with the least pots had 31 (62 plants) and the treatment with the most pots had 55 (110 plants).

For the elevated CO_2 treatments, carbon dioxide was elevated to 720 ppm using a CO_2 control unit (Thermoline Scientific Equipment, Smithfield, Australia) and compressed CO_2. The low CO_2 greenhouse chamber was maintained at ambient CO_2 and frequently monitored for leakage of CO_2 from the neighbouring high CO_2 cell. To avoid greenhouse chamber effects, both the CO_2 treatments and their respective seedlings were exchanged between two greenhouse chambers each week by physically moving the CO_2 tank regulator and all plants as well as monitoring concentrations with an infra-red gas analyzing device (LiCor 6200, LiCor Inc., Lincoln, NE). Pots were randomly repositioned during these periods to avoid positional effects in the greenhouse. The N fertilization treatment was initiated at the same time as the CO_2 manipulation and consisted of the monthly addition of pellets of urea at an approximate concentration of 30 kg N ha^{-1}; this replicated the approximate N addition of forestry practices and a 10-fold increase in N availability compared to the low N treatments. Seedlings in each treatment were uniformly watered daily until the 8th week of the study at which point water was applied as needed. Following 6 months of growth, all plants were harvested and separated into aboveground and belowground components. The aboveground biomass samples were oven-dried for 48 h at 60 °C before weighing. The belowground biomass of each seedling was assessed after it was carefully rinsed separately, over 2 and 0.5 mm sieves, to remove soil while retaining as much of the fine root biomass as possible. The rinsed belowground biomass samples were then dried for 48 h at 60 °C and weighed.

3. STATISTICAL ANALYSES

3.1. Evolutionary basis to range size

To address the hypothesis that contemporary range size has an evolutionary basis, we used a general linear mixed model using restricted estimated maximum likelihood (REML) to estimate variance components. The output of these models gives the degrees of freedom that provide the closest match between the F distribution and the distribution of the test statistic (Kenward and Roger, 1997). All statistical models were carried out in JMP 9.0.2. Two nearly identical models were compared that either included or ignored subgenus-level variation. Both models included the same response variable (total plant biomass) and the same fixed effects: range size (categorical), CO_2, N, and phylogenetic similarity in a fully factorial design. One model included pot identity (to control for potential non-independence within pots), and species identity nested within range size as random effects. The second model included pot identity, species identity nested within subgenus identity, and subgenus identity nested within range size as random effects. We then examined and compared the fixed effects that were significant for the two models. We ask whether the relationship between range size and productivity disappears when subgenus is added as a random effect. If the relationship does break down, this suggests that the relationship between range size and productivity has a phylogenetic basis. A categorical measurement of range size (taken from Williams and Potts, 1996) was used so that variation explained at the level of species and subgenus could be nested within range size. The categorical measurement includes three categories: widespread (12 species), regional (9 species), and localized (5 species) (Williams and Potts, 1996).

3.2. Interactive effects of evolutionary history, phylogenetic similarity, CO_2, and N

To address the hypothesis that plant biomass would positively respond to decreasing phylogenetic similarity, as well as increasing concentrations of CO_2 and N, we used general linear mixed models using REML to estimate variance components. In these models, the factors were subgenus identity, phylogenetic similarity, CO_2, and N in a fully factorial design. Statistical models were carried out in JMP 9.0.2. To account for variation due to neighbour effects and species identity (which were not part of our hypotheses), we included species identity and neighbour species identity as random effects. To account for variation between pots (potentially causing neighbouring plants within a pot to be non-independent), we also included pot number as a random effect in our models. We used total biomass, above-ground biomass, belowground biomass, and the ratio of aboveground to belowground biomass (hereafter A:B) as response variables to examine changes in productivity and biomass allocation in response to manipulations of phylogenetic similarity, CO_2, and N. We analyzed aboveground, below-ground, and total biomass and A:B at the individual level (i.e. not summed, mixture-level responses). Because phylogenetic similarity and $CO_2 * N$ had multiple levels, we conducted post hoc contrasts to determine which levels of phylogenetic similarity differed. These contrasts were conducted using the LSMeans option in JMP 9.0.2 after running the linear mixed models. Instead of using very conservative Bonferroni corrections, we controlled the false discovery rate (FDR) at 0.05, an approach considered appropriate for ecological research when many simultaneous tests are employed (Pike, 2010; Verhoeven et al., 2005). We controlled FDR separately for each interaction term within each response variable.

3.3. Relative effect sizes of phylogenetic similarity, CO_2 manipulation, and N fertilization

Additionally, using our raw data, we examined the relative effect size (Cohen's d; $d = M_1 - M_2/s_{pooled}$, where M_x and s_x indicate the mean value and standard deviation, respectively, for a given group, and $s_{pooled} = ((s_1^2 + s_2^2)/2)^{1/2}$) of subgenus identity, CO_2, N, and phylogenetic similarity on overall plant productivity. We compared the effect sizes of these four main effects. For phylogenetic similarity, which had more than two categories, we calculated all possible pairwise effect sizes and took the average of these values. We also calculated effect sizes for the interactive effects: subgenus identity * phylogenetic similarity (evolutionary history effects) and $CO_2 * N$ (abiotic effects); these effect sizes were calculated as the average pairwise effect size for all possible combinations within the interaction (e.g. ambient CO_2/ambient N, elevated CO_2/ambient N, ambient CO_2/elevated N, elevated CO_2/elevated N).

4. RESULTS

4.1. Evolutionary basis to range size

There is an evolutionary basis to contemporary range size in the genus *Euca-lyptus*. After controlling for variation in productivity that was related to species identity, CO_2, N, and phylogenetic similarity, species with larger ranges were nearly 50% more productive than those with small ranges (Fig. 9.1A and B; Table 9.1A). Furthermore, species with "widespread" ranges experienced the

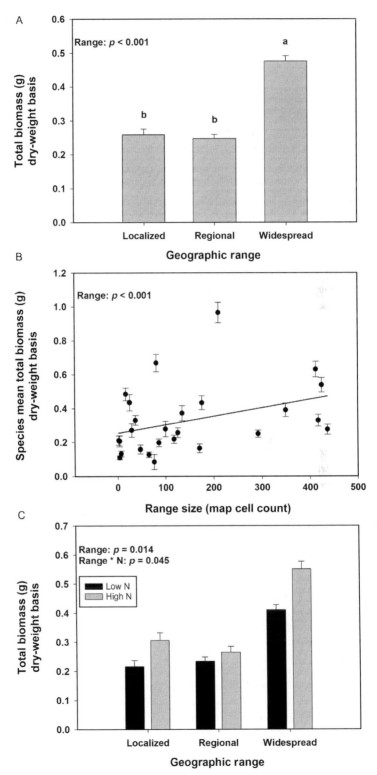

Figure 9.1 See legend on next page.

largest increases in productivity following N fertilization (Fig. 9.1C), suggesting that range size is a trait that influences how species will respond to abiotic environmental changes. When the variation in productivity that was related to species identity was nested within subgenera and subgenera were nested within range size, range size was not a significant predictor of total plant biomass (Table 9.1B), indicating that the variation in productivity associated with range size is due to subgenus-level phylogenetic divergence. These results are important for two reasons: they clearly indicate that evolutionary history is an important determinant of contemporary range size and also a driver of how plants are likely to respond to future changes in global change factors.

4.2. Interactive effects of evolutionary history, phylogenetic similarity, CO_2, and N

In addition to the effects of evolutionary history on plant productivity and contemporary range size, there were significant effects on productivity that were related to the phylogenetic similarity between the interacting species (Fig. 9.2). Total aboveground biomass and total plant biomass (marginally) responded to the main effect of phylogenetic similarity (Table 9.2). For both total and aboveground biomass, post hoc contrasts indicated that biomass of plants in monocultures was significantly lower than in among-series mixtures (total biomass $p = 0.008$; aboveground biomass $p = 0.005$); these were the only significant contrast for the four levels of phylogenetic similarity (see Table 9.3A for all pairwise contrasts). In contrast with the niche complementarity hypothesis, where decreasing phylogenetic similarity would increase biomass, these results indicate an inconsistent relationship between phylogenetic similarity and plant biomass (Fig. 9.2). When species interacted

Figure 9.1—Cont'd Productive plants have large contemporary ranges. (A) Total biomass by categories of range size where individual plant values for total biomass are shown. Localized species occupy ranges of <5000 km², regional species occupy ranges of 5000–25,000 km², and widespread species occupy ranges larger than 25,000 km². Letters indicate significant differences in means, adjusting for multiple comparisons by controlling the false discovery rate at 0.05. (B) Range size is positively correlated with plant productivity. Species mean values are shown for 26 species in genus *Eucalyptus*. "Map cell counts" are taken from distribution maps of each species on the island of Tasmania; a grid was overlain on the island and the number of grid cells in which a species was present was tabulated. Each grid cell covers a given amount of area, so widespread species have a greater cell count than regional species, which in turn have a greater cell count than localized species. (C) Range size categories interact with N fertilization to affect plant biomass. Plants occupying widespread ranges show the greatest magnitude of productivity increase in response to N fertilization. Asterisks indicate that N fertilization significantly affected total plant biomass within a given range size category.

Table 9.1 CO_2 concentrations and nitrogen availability interact to affect plant productivity

Fixed effects	(A) No evolutionary history AICc: −369.437			(B) Evolutionary history included AICc: −369.275		
	dfDen	F	p	dfDen	F	p
Range	23.1	5.275	**0.013**	3.5	1.897	0.277
CO_2	727.3	14.258	**<0.001**	727.6	14.302	**<0.001**
Phylogenetic similarity (PS)	739.5	1.785	0.149	739.5	1.778	0.150
N	726.0	18.708	**<0.001**	726.0	18.760	**<0.001**
Range*CO_2	1111.8	1.372	0.254	1112.2	1.353	0.259
Range*PS	1086.2	0.203	0.976	1086.2	0.198	0.977
Range*N	1116.5	3.051	**0.048**	1116.5	3.004	**0.050**
CO_2*PS	726.4	0.455	0.714	726.6	0.446	0.720
CO_2*N	732.4	7.595	**0.006**	732.5	7.660	**0.006**
PS*N	724.1	2.047	0.106	724.3	2.050	0.106
Range*CO_2*PS	1086.5	1.860	0.085	1086.7	1.856	0.085
Range*CO_2*N	1116.9	1.901	0.150	1117.1	1.870	0.155
Range*PS*N	1083.0	2.011	0.062	1083.2	2.019	0.060
CO_2*PS*N	730.8	0.408	0.747	731.0	0.411	0.745
Range*CO_2*PS*N	1088.6	0.439	0.853	1088.8	0.429	0.860
Random effects	Random effects			Random effects		
	Pot, species [range]			Pot, species [subgenus], Subgenus [range]		

When evolutionary history is excluded (A), range size also affects productivity. When evolutionary history is included (B), range size is not significant, suggesting that evolutionary history (subgenus identity) is an important determinant of contemporary range size.
When subgenus identity is added as a random effect (right columns), the main effect of range size is no longer significant. The other fixed effects are mostly unaffected by this change. In these models, range size is a categorical variable with three levels (localized, regional, and widespread) so that other terms can be nested within range size. Bold, italicized p-values are significant at $\alpha = 0.05$.

with other species within their subgenera, phylogenetic dissimilarity results in increasing plant productivity. However, when interactions among species occurred between the two subgenera, productivity declines and suggests that homoplasy (i.e. convergence) may be an important factor that should be considered in phylogenetic community ecology.

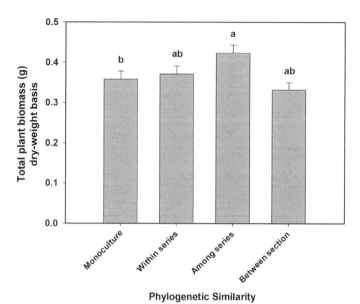

Phylogenetic Similarity

Figure 9.2 Phylogenetic similarity between neighbouring plants affects the total (above- and belowground) biomass produced by the pair of plants. When the interacting plants belong to the same subgenus (all x-axis categories except far right), increasing phylogenetic similarity positively impacts productivity. However, this positive relationship is reduced when between-subgenus mixtures are considered. Different letters above bars show significant differences indicated by post hoc contrasts (all possible pairwise contrasts), controlling the false discovery rate at 0.05.

Evolutionary history determines how plants respond to CO_2 and N. Consistent with much research on progressive N limitation, plant biomass (total, aboveground, and belowground) responded to an interactive effect of the CO_2 and N manipulations (Table 9.2). For aboveground, belowground, and total biomass, post hoc contrasts indicated that plant biomass was significantly higher in the $+CO_2+N$ treatment than in any other treatment, and the $+CO_2$, $+N$, and control treatments did not differ from each other (see Table 9.3B for all pairwise contrasts). These effects suggest that the ability of plants to respond to CO_2 was limited by the availability of N and vice versa. The effects of CO_2, N, and phylogenetic similarity varied by subgenus (*Eucalyptus* vs. *Symphyomyrtus*; Fig. 9.3; Table 9.2). For example, we detected a three-way interaction between subgenus identity, CO_2, and N for plant total and aboveground biomass (Table 9.2). For plants within *Eucalyptus*, all treatments that experienced elevated CO_2 led to increased biomass relative to treatments that experienced ambient CO_2 (Table 9.3C). For plants within *Symphyomyrtus*, the $+CO_2+N$ treatment led to increased biomass relative to all other treatments (Table 9.3C). Additionally, we detected

Table 9.2 Mixed-effect models, using REML (restricted maximum likelihood) to estimate error, show that evolutionary history (subgenus), phylogenetic similarity, CO_2 concentration, and N availability interact in different ways to affect different types of plant biomass and allocation strategies

Fixed effects	df	Aboveground ($N = 1346$) F	p	Belowground ($N = 1346$) F	p	Total biomass ($N = 1346$) F	p	Above-below ratio ($N = 1346$) F	p
Subgenus	1	1.126	0.299	1.768	0.196	1.238	0.277	2.318	0.139
Phylogenetic similarity (PS)	3	2.863	*0.036*	1.264	0.286	2.572	0.053	0.398	0.754
CO_2	1	25.439	*<0.001*	22.264	*<0.001*	26.180	*<0.001*	7.509	*0.006*
N	1	15.717	*<0.001*	3.934	*0.048*	13.146	*<0.001*	2.623	0.106
Subgenus*PS	3	0.743	0.528	0.071	0.975	0.524	0.666	0.751	0.524
Subgenus*CO_2	1	3.589	0.058	2.014	0.156	3.252	0.072	0.328	0.567
Subgenus*N	1	3.742	0.053	3.037	0.082	3.840	0.05	0.019	0.889
PS*CO_2	3	0.130	0.942	0.067	0.978	0.114	0.952	0.211	0.889
PS*N	3	2.175	0.089	0.968	0.407	1.824	0.142	1.590	0.191
CO_2*N	1	14.426	*<0.001*	5.184	*0.023*	12.960	*<0.001*	0.093	0.761
Subgenus*PS*CO_2	3	0.702	0.551	1.408	0.239	0.781	0.505	1.886	0.13
Subgenus*PS*N	3	4.355	*0.005*	2.682	*0.046*	4.070	*0.007*	0.561	0.641
Subgenus*CO_2*N	1	6.019	*0.014*	2.695	0.101	5.759	*0.017*	0.266	0.606
PS*CO_2*N	3	0.359	0.783	0.827	0.479	0.406	0.749	1.917	0.126
Subgenus*PS*CO_2*N	3	1.367	0.252	1.820	0.142	1.541	0.202	0.759	0.517

Continued

Table 9.2 Mixed-effect models, using REML (restricted maximum likelihood) to estimate error, show that evolutionary history (subgenus), phylogenetic similarity, CO_2 concentration, and N availability interact in different ways to affect different types of plant biomass and allocation strategies—cont'd

Random effects	Pct of total	Pct of total	Pct of total	Pct of total
Species identity	32.868	32.578	33.728	7.658
Neighbour species identity	0.365	0.487	0.363	3.011
Pot number	−16.476	2.274	−13.214	19.393

These results suggest that evolutionary history and phylogenetic similarity influence how individual-level plant biomass responds to global change. For fixed effects, significant p-values are bolded and italicized. For random effects, the percent of variation explained by each effect is shown. Species identity, neighbour species identity, and pot number are random effects to control for the species composition of each pot, and for potential differences between pots. The "Pct of Total" column gives the ratio of the variance component for a given effect to the total variance. PS, phylogenetic similarity.

Table 9.3 p-Values for pairwise post hoc contrasts describing the effects of CO_2, N, and phylogenetic similarity on plant biomass traits

		Response variables			
		Aboveground biomass	Belowground biomass	Total biomass	A:B
Group 1	Group 2	p	p	p	p
(A) Phylogenetic similarity					
Monoculture	Within series	0.200	0.754	0.105	0.305
Monoculture	Among series	(+) *0.004*	(+) 0.067	*0.006*	(+) 0.669
Monoculture	Between subgenus	0.139	0.472	0.181	0.849
Within series	Among series	0.109	0.137	0.272	0.553
Within series	Between subgenus	0.897	0.705	0.859	0.387
Among series	Between subgenus	0.125	0.244	0.131	0.803
(B) CO_2*N					
Control	$+CO_2$	0.352	0.066	0.256	0.022
Control	$+N$	0.913	0.824	0.995	0.335
Control	$+CO_2+N$	(+) *<0.001*	(+) *<0.001*	(+) *<0.001*	(+) 0.439
$+CO_2$	$+N$	0.431	0.047	0.278	*0.002* (+)
$+CO_2$	$+CO_2+N$	(+) *<0.001*	(+) *0.004*	(+) *<0.001*	(+) 0.192
$+N$	$+CO_2+N$	(+) *<0.001*	(+) *<0.001*	(+) *<0.001*	(+) 0.107

Continued

Table 9.3 p-Values for pairwise post hoc contrasts describing the effects of CO_2, N, and phylogenetic similarity on plant biomass traits—cont'd

		Response variables						
		Aboveground biomass		Belowground biomass		Total biomass		A:B
Group 1	Group 2	p		p		p		p
(C) Subgenus * CO_2 * N								
Euc. control	Euc. +CO$_2$	0.006	(+)	0.008	(+)	0.005	(+)	(+) 0.128
Euc. control	Euc. +N	0.513		0.854		0.826		0.362
Euc. control	Euc. +CO$_2$+N	<0.001	(+)	0.007	(+)	<0.001	(+)	(+) 0.507
Euc. +CO$_2$	Euc. +N	0.021	(-)	0.007	(-)	0.015	(-)	(-) 0.019
Euc. +CO$_2$	Euc. +CO$_2$+N	0.210		0.704		0.279		0.505
Euc. +N	Euc. +CO$_2$+N	0.001	(+)	0.006	(+)	0.001	(+)	(+) 0.151
Sym. control	Sym. +CO$_2$	0.047		0.708		0.086		0.062
Sym. control	Sym. +N	0.808		0.898		0.785		0.672
Sym. control	Sym. +CO$_2$+N	<0.001	(+)	0.001	(+)	<0.001	(+)	(+) 0.671
Sym. +CO$_2$	Sym. +N	0.088		0.809		0.157		0.025
Sym. +CO$_2$	Sym. +CO$_2$+N	<0.001	(+)	<0.001	(+)	<0.001	(+)	(+) 0.175
Sym. +N	Sym. +CO$_2$+N	<0.001	(+)	<0.001	(+)	<0.001	(+)	(+) 0.414

(D) Subgenus * N * Phylogenetic similarity

Group 1	Group 2							
Euc. Mono	*Euc.* Mono+N		0.212		0.349		0.220	0.804
Euc. WS	*Euc.* WS+N		***0.005***	(+)	0.267		***0.013***	(+) 0.175
Euc. AS	*Euc.* AS+N		0.485		0.693		0.512	0.952
Euc. BS	*Euc.* BS+N		0.203		0.596		0.234	0.187
Sym. Mono	*Sym.* Mono+N	(+)	***<0.001***	(+)	***0.003***	(+)	***<0.001***	(+) 0.629
Sym. WS	*Sym.* WS. +N		***0.002***	(+)	0.014		***0.003***	(+) 0.742
Sym. AS	*Sym.* AS+N		***<0.001***	(+)	0.041		***<0.001***	(+) 0.765
Sym. BS	*Sym.* BS+N		0.986		0.354		0.860	(+) ***0.018***

Groups 1 and 2 are the two categories involved in each contrast. Bold, italicized *p*-values are significant controlling for (within each response variable and group of contrasts) a false discovery rate of 0.05. For significant effects, signs indicate that the mean of Group 2 was greater than (+) or less than (−) Group 1. Abbreviations: *Euc.*, subgenus *Eucalyptus*; *Sym.*, subgenus *Symphyomyrtus*; WS, within series; AS, among series; BS, between subgenus. All contrasts were calculated within the same linear mixed effects model framework (REML model; fixed effects: CO_2, N, phylogenetic similarity, and subgenus identity in a fully factorial design; random effects: pot number, species identity, neighbour species identity).

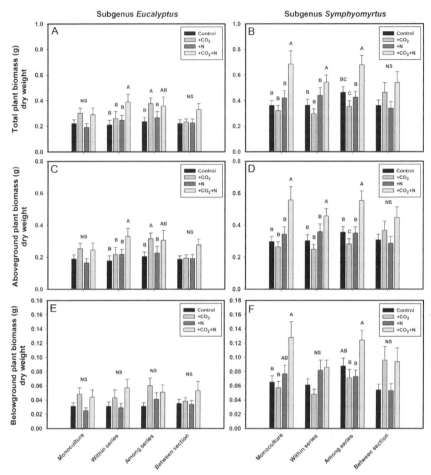

Figure 9.3 The response of total (A, B), aboveground (C, D), and belowground (E, F) biomass of mixtures containing two individual plants from either subgenus *Eucalyptus* (A, C, E) or *Symphyomyrtus* (B, D, F) to the degree of phylogenetic similarity between the plants, elevated CO_2 (ambient vs. 720 ppm), and N fertilization (4 kg/ha vs. 40 kg/ha). Different levels of phylogenetic similarity are listed along the *x*-axis and describe the phylogenetic relationship between neighbouring plants. Significant *p*-values from a three-way analysis of CO_2, N, and phylogenetic similarity are shown in each panel. Different letters indicate post hoc contrasts (controlling false discovery rate at 0.05), within each level of phylogenetic similarity. Non-significant results are denoted by NS. As the figure shows raw data but the model included three random effects (pot number, species identity, and neighbour species identity), the magnitude of difference between means is not always correlated with the results of the contrasts.

a three-way interaction between subgenus identity, N, and phylogenetic similarity for total, aboveground, and belowground biomass (Table 9.2). For plants within *Eucalyptus*, N fertilization increased total and aboveground biomass, but only in within-series mixtures (Table 9.3D). For plants within *Symphyomyrtus*, N fertilization increased total and aboveground biomass in all phylogenetic similarity treatments except between-subgenus mixtures (Table 9.3D), and also increased belowground biomass but only in monocultures. These results show that groups with different evolutionary histories may respond to different combinations of global change factors, and that the response to global change may depend on the composition of affected communities (i.e. phylogenetic similarity).

4.3. Relative effect sizes of phylogenetic similarity, CO_2 manipulation, and N fertilization

The effect sizes (Cohen's *d*) of the main effects of subgenus identity, phylogenetic similarity, CO_2 manipulation, and N fertilization were quantified for mixture-level total, above-, and belowground biomass. For total, aboveground, and belowground biomass, subgenus identity fertilization had the strongest effects and phylogenetic similarity had the weakest effects (Fig. 9.4A). Although the effects of phylogenetic similarity were smaller, they may be considered comparable to the CO_2 and N effects; the effect size of phylogenetic similarity was 42% as large as N fertilization and 48% as large as CO_2 enrichment. Note that phylogenetic similarity incorporates only the genetic distance between two interacting plants and does not incorporate any information about subgenus, clade, or species identity. We further categorized effect size by comparing abiotic factors (CO_2 and N manipulations) with factors related to evolutionary history (subgenus identity and phylogenetic similarity). We examined the strength of the $CO_2 * N$ interaction with the strength of the subgenus identity * phylogenetic similarity interaction and found that factors related to evolutionary history had greater effects on total, aboveground, and belowground biomass than did abiotic factors (Fig. 9.4B), suggesting that the combined effects of lineage splitting and phylogenetic similarity can influence plant biomass at levels at least equal to more traditionally studied abiotic factors.

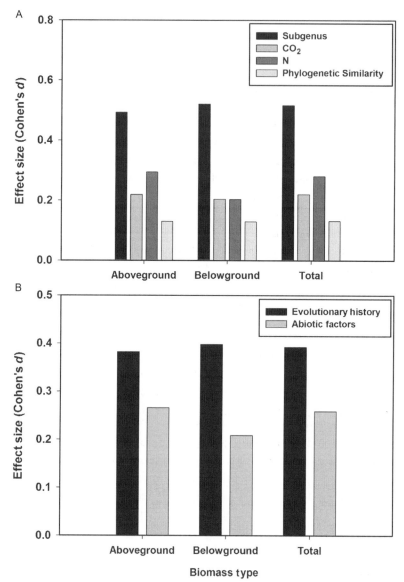

Figure 9.4 (A) The effect size (Cohen's d) of, subgenus, elevated CO_2 (ambient vs. 720 ppm), N fertilization (4 kg/ha vs. 40 kg/ha), and phylogenetic similarity across plant biomass traits (total, above-, and belowground biomass). Subgenus identity consistently had the largest effect size across different measures of plant biomass. Elevated CO_2 and N fertilization have a larger effect than phylogenetic similarity, but the effect sizes are comparable (within a factor of 3). Qualitatively, N fertilization has the largest effect on total and aboveground biomass, while elevated CO_2 has the largest effect on belowground biomass. (B) The combined effect size of evolutionary factors (subgenus identity and phylogenetic similarity) was greater than the combined effect size of abiotic factors (CO_2 and N).

5. DISCUSSION

The last decade has seen large, collaborative attempts aimed at understanding the impacts of human-induced environmental changes on the functioning of ecosystems (e.g. IPCC, 2007; Millennium Ecosystem Assessment, 2005), and research in this area rightly remains active and relatively well funded. In contrast, the role of evolutionary history as a driver of ecosystem functioning is rapidly gaining support (Cadotte et al., 2009; Flynn et al., 2011; Senior et al., 2013; Srivastava et al., 2012) but does not yet have wide recognition in climate change research (but see Davis et al., 2005). Here, we have factorially manipulated evolutionary history, phylogenetic similarity, CO_2 enrichment, and N fertilization using 26 species that also differ in range size and demonstrate that the effects of evolutionary history are at least strong, if not stronger, determinants of plant responses to abiotic factors related to global change. There are four main results that demonstrate the importance of incorporating evolutionary history into global change research: (1) phylogenetically based differences in productivity drive contemporary range size (i.e. range size was correlated with productivity but this relationship disappeared when subgenus identity was added to the model); (2) phylogenetically based biotic interactions influence plant productivity; (3) plant productivity depends upon the interaction of evolutionary and abiotic factors; and (4) the joint effect size (Cohen's d) of evolutionary factors was larger than abiotic factors.

In this study, we constructed the first experiment, to our knowledge, that simultaneously manipulated phylogenetic identity, phylogenetic similarity, and global change factors (CO_2 and N) and show that evolutionary factors deserve to be considered alongside CO_2 and N when predicting how plants will respond to global change. Phylogenetic similarity, CO_2, and N all affected plant productivity, but these factors interacted and the interactions varied across phylogenetic groups. These findings were mostly consistent with previous studies that have shown increases in CO_2 concentrations to increase N use efficiency (Lindroth, 2010), and short-term plant productivity (Cao et al., 2008), with the largest increases in productivity occurring belowground (Griffin et al., 1995; Jach et al., 2000; Nie et al., 2013; Smith et al., 2013; but see also Newingham et al., 2013). While CO_2 sometimes affected plants separately from N fertilization (see "Above–Below Ratio" in Table 9.1), all effects of N fertilization on plant productivity were mediated by CO_2 availability (and subgenus identity, but the following discussion focuses on the interaction of CO_2 and N). This result is consistent with previous studies (e.g. Cao et al., 2008; McMurtrie et al., 2008; Reich and Hobbie, 2012; Zak et al., 2000) that have evaluated plant productivity in response to CO_2 and N enrichment, and illustrates co-limitation of these plants by both CO_2 and N.

These results expand on previous work by demonstrating that all plants are not N limited and that the extent to which N levels limit the response of plants to elevated CO_2 has an evolutionary basis (see also Senior et al., 2013). Plants within the *Eucalyptus* and *Symphyomyrtus* groups respond differently to changes in CO_2 and N (Senior et al., 2013). This is likely because of evolutionary influences on plant strategies for acquiring N, as natural selection may have altered photosynthetic pathways or selected for physiological differences that influence a plant's abilities to acquire N (Hyvönen et al., 2007; Reich et al., 2006). For example, Wallis et al. (2010) recently showed that, compared with leaves from subgenus *Eucalyptus*, leaves from subgenus *Symphyomyrtus* contained fewer tannins and more available (i.e. not bound in larger compounds such as tannins) N. These traits suggest that *Symphyomyrtus* allocates less carbon to defence and produces higher-quality litter than does *Eucalyptus* (Hobbie, 1992), which may lead to faster nutrient cycling and selection for fast-growing plants in the soils beneath *Symphyomyrtus* trees. Adaptation to higher nutrient environments may explain why plants within *Symphyomyrtus* showed a strong, positive response to the $+CO_2+N$ treatment while plants within *Eucalyptus* responded only to elevated CO_2 (Fig. 9.1; Table 9.2). As CO_2 and N levels continue to rise in natural systems, it is important to consider that while species responses are idiosyncratic (e.g. Ainsworth and Long, 2005; Jablonski et al., 2002; Xia and Wan, 2008), larger phylogenetic groups such as subgenera may show similar responses (Senior et al., 2013). Therefore, the selective pressures that result from global change may be acting on entire groups of species that share deep evolutionary history, which may make plant responses to global change more predictable than previously thought.

Speciation and convergent evolution provide mechanisms for phylogenetically based competitive and facilitative interactions that influence plant responses to global change factors. While CO_2 and N availability can alter plant productivity, plant responses may also be modified by diversity and identity of neighbouring plants (Langley and Megonigal, 2010; Reid et al., 2012). For example, Langley and Megonigal (2010) showed that N fertilization promoted the expansion of species that responded less strongly to CO_2, shifting plant communities in a way that limited the impact of elevated CO_2. Recent studies (e.g. Kunstler et al., 2012; Verdú et al., 2012; Violle et al., 2011) have found contrasting answers to the long-standing question (see Darwin, 1859) of how genetic relatedness affects competitive and facilitative interactions among neighbouring plants. Our results indicate that within subgenera, reducing phylogenetic similarity increased total productivity on average; however, when phylogenetic similarity was at its lowest between subgenera, productivity was also at its lowest. Such competitive interactions suggest that phylogenetic niches are not entirely conserved and that radiations within clades result in (1) increased niche diversification leading to complementarity and (2) convergent evolution (homoplasy) among clades resulting in the evolution of similar traits for occupying similar niches. Both represent mechanisms for the relationship between phylogenetic similarity and plant productivity. Although phylogenetic similarity never interacted with CO_2, there were significant main effects of phylogenetic similarity and phylogenetic similarity interacted both with N (total and aboveground biomass; Table 9.1) and N*subgenus identity (total, aboveground, and belowground biomass; Table 9.1). Nitrogen fertilization increased the biomass of plants within *Symphyomyrtus* across most of the phylogenetic similarity treatments, but increased the biomass of plants within *Eucalyptus* in only the within-series treatment. The interaction of phylogenetic similarity and N concentrations suggests that knowledge of how phylogenetic similarity is expected to change in the future could help inform how some species will respond to selective pressures imposed by, and novel interactions created by, global change. The importance of understanding novel species interactions is further underscored by the increased recognition that periods of contemporary (<100 years) evolution are the norm rather than the exception (Carroll et al., 2007; Hendry and Kinnison, 1999; Pelletier et al., 2009), meaning that evolutionary responses to global change may occur quickly.

Models of range shifts are typically based on climate envelope niche-based analyses and generally lack an evolutionary perspective. Elevated CO_2 and N may have different effects on plants that occupy small versus large ranges, if certain suites of traits are associated with range size and if those suites of trait modify responses to CO_2 and N. We found that range size had an evolutionary basis, meaning that historical evolutionary processes (selection, drift, etc.) that determined the traits associated with range size may have also set the stage for how related groups of plants will respond to global change. Species with "widespread" geographic ranges showed a greater positive response to elevated N than did species that occupy regional or local ranges (Table 9.1). This suggests that smaller-ranged species, which are already more vulnerable to extinction (Gaston, 2003), may not benefit (from a productivity standpoint) from greater rates of N deposition to the same degree that species with larger range sizes do. Studies should build upon this work by examining how range sizes vary across phylogenies, and by considering evolutionary processes as drivers of variation in range size.

We recognize that it may be difficult to parameterize phylogenetic similarity in the sense that there are not firm predictions for how phylogenetic similarity will change over the next 100 years. However, with species loss and range shifts, the contemporary and predicted ranges of species will collide resulting in unique interactions among more and less distantly related species. We show that decreasing similarity between neighbouring plants can result in significant changes in the combined productivity of the interacting plants. Although phylogenetic similarity (a property of pairs or communities of individuals) and global change factors (properties of ecosystems or the environment) may be difficult to compare, both factors can be viewed in the same framework by asking: how have phylogenetic similarity, CO_2, and N changed over the last 100 years, and how are they predicted to change over the next 100? On one hand, CO_2 and N concentrations have risen and will almost certainly continue to do so (IPCC, 2007). Predicting the future of phylogenetic similarity is more speculative, but it is known that human activities have a negative effect on most species, while positively impacting a small minority (biotic homogenization, see review by McKinney and Lockwood, 1999), and that population-level genetic diversity is quickly declining (Luck et al., 2003). Given these changes, it seems likely that phylogenetic similarity of communities will decrease as certain species and populations become more dominant. For species within the subgenus *Eucalyptus*, changes in phylogenetic similarity, CO_2, and N could therefore present multiple issues—not only do plants with different evolutionary histories (e.g. plants in subgenus *Symphyomyrtus*) benefit more from changes to CO_2 and N (at least in terms of production) but also increases in phylogenetic similarity may reduce biomass production. Many phylogenetic groups other than the subgenus *Eucalyptus* likely face a similar situation, reinforcing the need to understand evolutionary history as an important factor in global change studies.

ACKNOWLEDGEMENTS

We thank the University of Tasmania for use of greenhouse facilities and Tracy Winterbottom for expert assistance in the greenhouse. Thanks to Camilla Bloomfield and Justin Bloomfield for assistance in the greenhouse and lab and to Mark Hovenden for technical advice and equipment for elevating CO_2. The Australian Research Council Future Fellowship (to J. K. B.) and the University of Tasmania provided funding for the experiment.

APPENDIX A

This list shows the replicate numbers and species combinations at each level of phylogenetic similarity. "Between section" is analogous to "between subgenus".

Focal species	Neighbour species	Phylogenetic similarity	Replicates
E. amygdalina	E. amygdalina	a—mono	11
E. barberi	E. barberi	a—mono	12
E. brookeriana	E. brookeriana	a—mono	12
E. cordata	E. cordata	a—mono	12
E. dalrympleana	E. dalrympleana	a—mono	12
E. delegatensis	E. delegatensis	a—mono	12
E. globulus	E. globulus	a—mono	12
E. gunnii	E. gunnii	a—mono	12
E. johnstonii	E. johnstonii	a—mono	12
E. morrisbyi	E. morrisbyi	a—mono	12
E. nitida	E. nitida	a—mono	12
E. obliqua	E. obliqua	a—mono	12
E. ovata	E. ovata	a—mono	12
E. perriniana	E. perriniana	a—mono	12
E. pulchella	E. pulchella	a—mono	12
E. radiata	E. radiata	a—mono	11
E. regnans	E. regnans	a—mono	12
E. risdonii	E. risdonii	a—mono	12
E. rodwayi	E. rodwayi	a—mono	12
E. rubida	E. rubida	a—mono	12
E. sieberi	E. sieberi	a—mono	12
E. subcrenulata	E. subcrenulata	a—mono	11
E. tenuiramis	E. tenuiramis	a—mono	12

E. urnigera	*E. urnigera*	a—mono	12
E. vernicosa	*E. vernicosa*	a—mono	10
E. viminalis	*E. viminalis*	a—mono	12
E. amygdalina	*E. coccifera*	b—within series	2
E. amygdalina	*E. nitida*	b—within series	1
E. amygdalina	*E. pulchella*	b—within series	2
E. amygdalina	*E. radiata*	b—within series	2
E. amygdalina	*E. risdonii*	b—within series	3
E. amygdalina	*E. tenuiramis*	b—within series	2
E. barberi	*E. brookeriana*	b—within series	4
E. barberi	*E. ovata*	b—within series	4
E. barberi	*E. rodwayi*	b—within series	4
E. brookeriana	*E. barberi*	b—within series	4
E. brookeriana	*E. ovata*	b—within series	4
E. brookeriana	*E. rodwayi*	b—within series	4
E. cordata	*E. dalrympleana*	b—within series	1
E. cordata	*E. globulus*	b—within series	2
E. cordata	*E. gunnii*	b—within series	2
E. cordata	*E. johnstonii*	b—within series	1
E. cordata	*E. morrisbyi*	b—within series	2
E. cordata	*E. rubida*	b—within series	1
E. cordata	*E. subcrenulata*	b—within series	1
E. cordata	*E. urnigera*	b—within series	2
E. dalrympleana	*E. dalrympleana*	b—within series	2
E. dalrympleana	*E. gunnii*	b—within series	1
E. dalrympleana	*E. morrisbyi*	b—within series	3
E. dalrympleana	*E. subcrenulata*	b—within series	1
E. dalrympleana	*E. urnigera*	b—within series	1
E. dalrympleana	*E. vernicosa*	b—within series	1
E. dalrympleana	*E. viminalis*	b—within series	2

E. delegatensis	E. obliqua	b—within series	4
E. delegatensis	E. regnans	b—within series	4
E. delegatensis	E. sieberi	b—within series	4
E. globulus	E. dalrympleana	b—within series	1
E. globulus	E. gunnii	b—within series	1
E. globulus	E. johnstonii	b—within series	2
E. globulus	E. perriniana	b—within series	2
E. globulus	E. rubida	b—within series	1
E. globulus	E. subcrenulata	b—within series	1
E. globulus	E. urnigera	b—within series	1
E. globulus	E. viminalis	b—within series	3
E. gunnii	E. cordata	b—within series	2
E. gunnii	E. globulus	b—within series	2
E. gunnii	E. johnstonii	b—within series	1
E. gunnii	E. perriniana	b—within series	2
E. gunnii	E. subcrenulata	b—within series	1
E. gunnii	E. urnigera	b—within series	2
E. gunnii	E. vernicosa	b—within series	1
E. johnstonii	E. cordata	b—within series	1
E. johnstonii	E. dalrympleana	b—within series	2
E. johnstonii	E. globulus	b—within series	2
E. johnstonii	E. gunnii	b—within series	1
E. johnstonii	E. morrisbyi	b—within series	1
E. johnstonii	E. perriniana	b—within series	1
E. johnstonii	E. rubida	b—within series	1
E. johnstonii	E. subcrenulata	b—within series	1
E. johnstonii	E. viminalis	b—within series	2
E. morrisbyi	E. cordata	b—within series	2
E. morrisbyi	E. globulus	b—within series	4
E. morrisbyi	E. gunnii	b—within series	1

E. morrisbyi	*E. rubida*	b—within series	2
E. morrisbyi	*E. subcrenulata*	b—within series	1
E. morrisbyi	*E. urnigera*	b—within series	1
E. morrisbyi	*E. viminalis*	b—within series	1
E. nitida	*E. amygdalina*	b—within series	3
E. nitida	*E. pulchella*	b—within series	2
E. nitida	*E. radiata*	b—within series	4
E. nitida	*E. risdonii*	b—within series	1
E. nitida	*E. tenuiramis*	b—within series	2
E. obliqua	*E. delegatensis*	b—within series	4
E. obliqua	*E. regnans*	b—within series	4
E. obliqua	*E. sieberi*	b—within series	4
E. ovata	*E. barberi*	b—within series	4
E. ovata	*E. brookeriana*	b—within series	4
E. ovata	*E. rodwayi*	b—within series	4
E. perriniana	*E. cordata*	b—within series	1
E. perriniana	*E. dalrympleana*	b—within series	2
E. perriniana	*E. gunnii*	b—within series	1
E. perriniana	*E. johnstonii*	b—within series	1
E. perriniana	*E. morrisbyi*	b—within series	3
E. perriniana	*E. subcrenulata*	b—within series	1
E. perriniana	*E. urnigera*	b—within series	1
E. perriniana	*E. vernicosa*	b—within series	2
E. pulchella	*E. amygdalina*	b—within series	2
E. pulchella	*E. coccifera*	b—within series	2
E. pulchella	*E. nitida*	b—within series	3
E. pulchella	*E. radiata*	b—within series	1
E. pulchella	*E. risdonii*	b—within series	3
E. pulchella	*E. tenuiramis*	b—within series	1
E. radiata	*E. amygdalina*	b—within series	3

E. radiata	*E. coccifera*	b—within series	2
E. radiata	*E. nitida*	b—within series	1
E. radiata	*E. pulchella*	b—within series	2
E. radiata	*E. risdonii*	b—within series	2
E. radiata	*E. tenuiramis*	b—within series	2
E. regnans	*E. delegatensis*	b—within series	4
E. regnans	*E. obliqua*	b—within series	4
E. regnans	*E. sieberi*	b—within series	4
E. risdonii	*E. amygdalina*	b—within series	1
E. risdonii	*E. coccifera*	b—within series	1
E. risdonii	*E. nitida*	b—within series	3
E. risdonii	*E. pulchella*	b—within series	1
E. risdonii	*E. radiata*	b—within series	4
E. risdonii	*E. tenuiramis*	b—within series	2
E. rodwayi	*E. barberi*	b—within series	4
E. rodwayi	*E. brookeriana*	b—within series	4
E. rodwayi	*E. ovata*	b—within series	4
E. rubida	*E. cordata*	b—within series	1
E. rubida	*E. globulus*	b—within series	1
E. rubida	*E. morrisbyi*	b—within series	3
E. rubida	*E. perriniana*	b—within series	1
E. rubida	*E. subcrenulata*	b—within series	2
E. rubida	*E. urnigera*	b—within series	3
E. rubida	*E. viminalis*	b—within series	1
E. sieberi	*E. delegatensis*	b—within series	4
E. sieberi	*E. obliqua*	b—within series	4
E. sieberi	*E. regnans*	b—within series	4
E. subcrenulata	*E. dalrympleana*	b—within series	2
E. subcrenulata	*E. gunnii*	b—within series	2
E. subcrenulata	*E. morrisbyi*	b—within series	2

E. subcrenulata	*E. perriniana*	b—within series	1
E. subcrenulata	*E. rubida*	b—within series	1
E. subcrenulata	*E. urnigera*	b—within series	3
E. subcrenulata	*E. vernicosa*	b—within series	1
E. tenuiramis	*E. amygdalina*	b—within series	2
E. tenuiramis	*E. nitida*	b—within series	1
E. tenuiramis	*E. pulchella*	b—within series	4
E. tenuiramis	*E. radiata*	b—within series	2
E. tenuiramis	*E. risdonii*	b—within series	3
E. urnigera	*E. cordata*	b—within series	1
E. urnigera	*E. globulus*	b—within series	2
E. urnigera	*E. gunnii*	b—within series	3
E. urnigera	*E. johnstonii*	b—within series	1
E. urnigera	*E. morrisbyi*	b—within series	1
E. urnigera	*E. vernicosa*	b—within series	3
E. urnigera	*E. viminalis*	b—within series	1
E. vernicosa	*E. cordata*	b—within series	1
E. vernicosa	*E. globulus*	b—within series	2
E. vernicosa	*E. gunnii*	b—within series	1
E. vernicosa	*E. johnstonii*	b—within series	2
E. vernicosa	*E. morrisbyi*	b—within series	1
E. vernicosa	*E. perriniana*	b—within series	1
E. vernicosa	*E. rubida*	b—within series	1
E. vernicosa	*E. urnigera*	b—within series	1
E. vernicosa	*E. viminalis*	b—within series	2
E. viminalis	*E. cordata*	b—within series	1
E. viminalis	*E. globulus*	b—within series	2
E. viminalis	*E. gunnii*	b—within series	1
E. viminalis	*E. johnstonii*	b—within series	2
E. viminalis	*E. perriniana*	b—within series	1

E. viminalis	*E. subcrenulata*	b—within series	1
E. viminalis	*E. urnigera*	b—within series	4
E. amygdalina	*E. delegatensis*	c—among series	3
E. amygdalina	*E. obliqua*	c—among series	4
E. amygdalina	*E. regnans*	c—among series	3
E. amygdalina	*E. sieberi*	c—among series	2
E. barberi	*E. cordata*	c—among series	1
E. barberi	*E. dalrympleana*	c—among series	1
E. barberi	*E. globulus*	c—among series	2
E. barberi	*E. morrisbyi*	c—among series	1
E. barberi	*E. perriniana*	c—among series	1
E. barberi	*E. rubida*	c—among series	2
E. barberi	*E. subcrenulata*	c—among series	2
E. barberi	*E. vernicosa*	c—among series	1
E. barberi	*E. viminalis*	c—among series	1
E. brookeriana	*E. cordata*	c—among series	2
E. brookeriana	*E. globulus*	c—among series	3
E. brookeriana	*E. johnstonii*	c—among series	1
E. brookeriana	*E. perriniana*	c—among series	1
E. brookeriana	*E. rubida*	c—among series	1
E. brookeriana	*E. subcrenulata*	c—among series	2
E. brookeriana	*E. vernicosa*	c—among series	2
E. cordata	*E. barberi*	c—among series	4
E. cordata	*E. brookeriana*	c—among series	1
E. cordata	*E. ovata*	c—among series	3
E. cordata	*E. rodwayi*	c—among series	4
E. dalrympleana	*E. barberi*	c—among series	2
E. dalrympleana	*E. brookeriana*	c—among series	4
E. dalrympleana	*E. ovata*	c—among series	2
E. dalrympleana	*E. rodwayi*	c—among series	4

E. delegatensis	E. amygdalina	c—among series	1
E. delegatensis	E. coccifera	c—among series	2
E. delegatensis	E. nitida	c—among series	1
E. delegatensis	E. pulchella	c—among series	1
E. delegatensis	E. radiata	c—among series	2
E. delegatensis	E. risdonii	c—among series	2
E. delegatensis	E. tenuiramis	c—among series	3
E. globulus	E. barberi	c—among series	4
E. globulus	E. brookeriana	c—among series	4
E. globulus	E. ovata	c—among series	4
E. gunnii	E. barberi	c—among series	3
E. gunnii	E. brookeriana	c—among series	4
E. gunnii	E. ovata	c—among series	1
E. gunnii	E. rodwayi	c—among series	4
E. johnstonii	E. barberi	c—among series	4
E. johnstonii	E. brookeriana	c—among series	1
E. johnstonii	E. ovata	c—among series	4
E. johnstonii	E. rodwayi	c—among series	3
E. morrisbyi	E. barberi	c—among series	4
E. morrisbyi	E. brookeriana	c—among series	2
E. morrisbyi	E. ovata	c—among series	3
E. morrisbyi	E. rodwayi	c—among series	3
E. nitida	E. delegatensis	c—among series	4
E. nitida	E. obliqua	c—among series	3
E. nitida	E. regnans	c—among series	2
E. nitida	E. sieberi	c—among series	3
E. obliqua	E. amygdalina	c—among series	2
E. obliqua	E. coccifera	c—among series	2
E. obliqua	E. pulchella	c—among series	1
E. obliqua	E. radiata	c—among series	1

E. obliqua	*E. risdonii*	c—among series	2
E. obliqua	*E. tenuiramis*	c—among series	4
E. ovata	*E. dalrympleana*	c—among series	2
E. ovata	*E. globulus*	c—among series	1
E. ovata	*E. gunnii*	c—among series	2
E. ovata	*E. morrisbyi*	c—among series	1
E. ovata	*E. perriniana*	c—among series	1
E. ovata	*E. subcrenulata*	c—among series	3
E. ovata	*E. vernicosa*	c—among series	1
E. ovata	*E. viminalis*	c—among series	1
E. perriniana	*E. barberi*	c—among series	3
E. perriniana	*E. brookeriana*	c—among series	3
E. perriniana	*E. ovata*	c—among series	3
E. perriniana	*E. rodwayi*	c—among series	3
E. pulchella	*E. delegatensis*	c—among series	2
E. pulchella	*E. obliqua*	c—among series	3
E. pulchella	*E. regnans*	c—among series	3
E. pulchella	*E. sieberi*	c—among series	4
E. radiata	*E. delegatensis*	c—among series	3
E. radiata	*E. obliqua*	c—among series	2
E. radiata	*E. regnans*	c—among series	3
E. radiata	*E. sieberi*	c—among series	4
E. regnans	*E. amygdalina*	c—among series	1
E. regnans	*E. coccifera*	c—among series	2
E. regnans	*E. nitida*	c—among series	1
E. regnans	*E. pulchella*	c—among series	2
E. regnans	*E. radiata*	c—among series	3
E. regnans	*E. risdonii*	c—among series	1
E. regnans	*E. tenuiramis*	c—among series	2
E. risdonii	*E. delegatensis*	c—among series	2

E. risdonii	E. obliqua	c—among series	3
E. risdonii	E. regnans	c—among series	3
E. risdonii	E. sieberi	c—among series	4
E. rodwayi	E. cordata	c—among series	1
E. rodwayi	E. globulus	c—among series	1
E. rodwayi	E. gunnii	c—among series	2
E. rodwayi	E. johnstonii	c—among series	1
E. rodwayi	E. morrisbyi	c—among series	1
E. rodwayi	E. rubida	c—among series	1
E. rodwayi	E. subcrenulata	c—among series	1
E. rodwayi	E. urnigera	c—among series	1
E. rodwayi	E. vernicosa	c—among series	2
E. rodwayi	E. viminalis	c—among series	1
E. rubida	E. barberi	c—among series	4
E. rubida	E. brookeriana	c—among series	1
E. rubida	E. ovata	c—among series	4
E. rubida	E. rodwayi	c—among series	3
E. sieberi	E. amygdalina	c—among series	2
E. sieberi	E. coccifera	c—among series	1
E. sieberi	E. pulchella	c—among series	2
E. sieberi	E. radiata	c—among series	3
E. sieberi	E. risdonii	c—among series	3
E. sieberi	E. tenuiramis	c—among series	1
E. subcrenulata	E. barberi	c—among series	2
E. subcrenulata	E. brookeriana	c—among series	4
E. subcrenulata	E. ovata	c—among series	2
E. subcrenulata	E. rodwayi	c—among series	4
E. tenuiramis	E. delegatensis	c—among series	4
E. tenuiramis	E. obliqua	c—among series	1
E. tenuiramis	E. regnans	c—among series	4

E. tenuiramis	*E. sieberi*	c—among series	3
E. urnigera	*E. barberi*	c—among series	3
E. urnigera	*E. brookeriana*	c—among series	2
E. urnigera	*E. ovata*	c—among series	4
E. urnigera	*E. rodwayi*	c—among series	2
E. vernicosa	*E. barberi*	c—among series	4
E. vernicosa	*E. brookeriana*	c—among series	4
E. vernicosa	*E. ovata*	c—among series	1
E. vernicosa	*E. rodwayi*	c—among series	3
E. viminalis	*E. barberi*	c—among series	3
E. viminalis	*E. brookeriana*	c—among series	2
E. viminalis	*E. ovata*	c—among series	3
E. viminalis	*E. rodwayi*	c—among series	4
E. amygdalina	*E. barberi*	d—between section	1
E. amygdalina	*E. cordata*	d—between section	2
E. amygdalina	*E. globulus*	d—between section	1
E. amygdalina	*E. gunnii*	d—between section	2
E. amygdalina	*E. johnstonii*	d—between section	1
E. amygdalina	*E. morrisbyi*	d—between section	1
E. amygdalina	*E. rubida*	d—between section	2
E. amygdalina	*E. subcrenulata*	d—between section	1
E. amygdalina	*E. viminalis*	d—between section	1
E. barberi	*E. amygdalina*	d—between section	2
E. barberi	*E. coccifera*	d—between section	1
E. barberi	*E. delegatensis*	d—between section	2
E. barberi	*E. obliqua*	d—between section	1
E. barberi	*E. pulchella*	d—between section	1
E. barberi	*E. radiata*	d—between section	2
E. barberi	*E. regnans*	d—between section	1
E. barberi	*E. risdonii*	d—between section	1

E. barberi	E. sieberi	d—between section	1
E. brookeriana	E. amygdalina	d—between section	3
E. brookeriana	E. coccifera	d—between section	1
E. brookeriana	E. delegatensis	d—between section	1
E. brookeriana	E. nitida	d—between section	1
E. brookeriana	E. pulchella	d—between section	1
E. brookeriana	E. radiata	d—between section	1
E. brookeriana	E. regnans	d—between section	1
E. brookeriana	E. risdonii	d—between section	1
E. brookeriana	E. sieberi	d—between section	1
E. brookeriana	E. tenuiramis	d—between section	1
E. cordata	E. amygdalina	d—between section	2
E. cordata	E. coccifera	d—between section	1
E. cordata	E. delegatensis	d—between section	1
E. cordata	E. nitida	d—between section	2
E. cordata	E. pulchella	d—between section	3
E. cordata	E. radiata	d—between section	3
E. dalrympleana	E. amygdalina	d—between section	1
E. dalrympleana	E. coccifera	d—between section	1
E. dalrympleana	E. delegatensis	d—between section	1
E. dalrympleana	E. nitida	d—between section	2
E. dalrympleana	E. obliqua	d—between section	3
E. dalrympleana	E. pulchella	d—between section	1
E. dalrympleana	E. radiata	d—between section	1
E. dalrympleana	E. regnans	d—between section	1
E. dalrympleana	E. risdonii	d—between section	1
E. delegatensis	E. barberi	d—between section	1
E. delegatensis	E. brookeriana	d—between section	1
E. delegatensis	E. dalrympleana	d—between section	1
E. delegatensis	E. globulus	d—between section	2

E. delegatensis	*E. gunnii*	d—between section	1
E. delegatensis	*E. ovata*	d—between section	2
E. delegatensis	*E. subcrenulata*	d—between section	3
E. delegatensis	*E. urnigera*	d—between section	1
E. globulus	*E. amygdalina*	d—between section	1
E. globulus	*E. coccifera*	d—between section	1
E. globulus	*E. nitida*	d—between section	2
E. globulus	*E. obliqua*	d—between section	1
E. globulus	*E. pulchella*	d—between section	1
E. globulus	*E. radiata*	d—between section	3
E. globulus	*E. regnans*	d—between section	1
E. globulus	*E. tenuiramis*	d—between section	2
E. gunnii	*E. amygdalina*	d—between section	2
E. gunnii	*E. delegatensis*	d—between section	2
E. gunnii	*E. nitida*	d—between section	1
E. gunnii	*E. obliqua*	d—between section	1
E. gunnii	*E. radiata*	d—between section	2
E. gunnii	*E. risdonii*	d—between section	2
E. gunnii	*E. tenuiramis*	d—between section	2
E. johnstonii	*E. amygdalina*	d—between section	2
E. johnstonii	*E. coccifera*	d—between section	1
E. johnstonii	*E. delegatensis*	d—between section	2
E. johnstonii	*E. pulchella*	d—between section	1
E. johnstonii	*E. radiata*	d—between section	1
E. johnstonii	*E. risdonii*	d—between section	2
E. johnstonii	*E. sieberi*	d—between section	2
E. johnstonii	*E. tenuiramis*	d—between section	1
E. morrisbyi	*E. amygdalina*	d—between section	2
E. morrisbyi	*E. coccifera*	d—between section	1
E. morrisbyi	*E. delegatensis*	d—between section	3

E. morrisbyi	*E. pulchella*	d—between section	1
E. morrisbyi	*E. regnans*	d—between section	3
E. morrisbyi	*E. risdonii*	d—between section	2
E. nitida	*E. brookeriana*	d—between section	4
E. nitida	*E. dalrympleana*	d—between section	1
E. nitida	*E. morrisbyi*	d—between section	2
E. nitida	*E. ovata*	d—between section	2
E. nitida	*E. perriniana*	d—between section	2
E. nitida	*E. urnigera*	d—between section	1
E. obliqua	*E. dalrympleana*	d—between section	3
E. obliqua	*E. globulus*	d—between section	1
E. obliqua	*E. gunnii*	d—between section	2
E. obliqua	*E. ovata*	d—between section	1
E. obliqua	*E. rubida*	d—between section	1
E. obliqua	*E. subcrenulata*	d—between section	2
E. obliqua	*E. urnigera*	d—between section	1
E. obliqua	*E. viminalis*	d—between section	1
E. ovata	*E. amygdalina*	d—between section	2
E. ovata	*E. nitida*	d—between section	2
E. ovata	*E. obliqua*	d—between section	2
E. ovata	*E. pulchella*	d—between section	2
E. ovata	*E. regnans*	d—between section	1
E. ovata	*E. risdonii*	d—between section	2
E. ovata	*E. sieberi*	d—between section	1
E. perriniana	*E. amygdalina*	d—between section	3
E. perriniana	*E. nitida*	d—between section	1
E. perriniana	*E. obliqua*	d—between section	1
E. perriniana	*E. pulchella*	d—between section	2
E. perriniana	*E. radiata*	d—between section	1
E. perriniana	*E. tenuiramis*	d—between section	4
E. pulchella	*E. brookeriana*	d—between section	2

E. pulchella	E. globulus	d—between section	1
E. pulchella	E. ovata	d—between section	2
E. pulchella	E. perriniana	d—between section	2
E. pulchella	E. subcrenulata	d—between section	1
E. pulchella	E. vernicosa	d—between section	1
E. pulchella	E. viminalis	d—between section	3
E. radiata	E. brookeriana	d—between section	1
E. radiata	E. gunnii	d—between section	2
E. radiata	E. ovata	d—between section	1
E. radiata	E. rubida	d—between section	2
E. radiata	E. subcrenulata	d—between section	2
E. radiata	E. urnigera	d—between section	1
E. radiata	E. vernicosa	d—between section	1
E. radiata	E. viminalis	d—between section	2
E. regnans	E. cordata	d—between section	2
E. regnans	E. globulus	d—between section	1
E. regnans	E. gunnii	d—between section	1
E. regnans	E. morrisbyi	d—between section	1
E. regnans	E. ovata	d—between section	1
E. regnans	E. perriniana	d—between section	1
E. regnans	E. rodwayi	d—between section	1
E. regnans	E. rubida	d—between section	3
E. regnans	E. urnigera	d—between section	1
E. risdonii	E. cordata	d—between section	1
E. risdonii	E. gunnii	d—between section	1
E. risdonii	E. morrisbyi	d—between section	3
E. risdonii	E. ovata	d—between section	3
E. risdonii	E. rodwayi	d—between section	1
E. risdonii	E. rubida	d—between section	2
E. risdonii	E. urnigera	d—between section	1
E. rodwayi	E. amygdalina	d—between section	1

E. rodwayi	*E. coccifera*	d—between section	1
E. rodwayi	*E. delegatensis*	d—between section	1
E. rodwayi	*E. nitida*	d—between section	4
E. rodwayi	*E. regnans*	d—between section	1
E. rodwayi	*E. risdonii*	d—between section	1
E. rodwayi	*E. sieberi*	d—between section	1
E. rodwayi	*E. tenuiramis*	d—between section	2
E. rubida	*E. delegatensis*	d—between section	1
E. rubida	*E. nitida*	d—between section	1
E. rubida	*E. obliqua*	d—between section	2
E. rubida	*E. pulchella*	d—between section	3
E. rubida	*E. regnans*	d—between section	1
E. rubida	*E. risdonii*	d—between section	2
E. rubida	*E. sieberi*	d—between section	1
E. rubida	*E. tenuiramis*	d—between section	1
E. sieberi	*E. brookeriana*	d—between section	1
E. sieberi	*E. globulus*	d—between section	1
E. sieberi	*E. gunnii*	d—between section	1
E. sieberi	*E. johnstonii*	d—between section	1
E. sieberi	*E. morrisbyi*	d—between section	1
E. sieberi	*E. perriniana*	d—between section	1
E. sieberi	*E. rubida*	d—between section	1
E. sieberi	*E. urnigera*	d—between section	1
E. sieberi	*E. vernicosa*	d—between section	3
E. sieberi	*E. viminalis*	d—between section	1
E. subcrenulata	*E. amygdalina*	d—between section	2
E. subcrenulata	*E. delegatensis*	d—between section	1
E. subcrenulata	*E. nitida*	d—between section	2
E. subcrenulata	*E. obliqua*	d—between section	2
E. subcrenulata	*E. pulchella*	d—between section	1
E. subcrenulata	*E. radiata*	d—between section	1

E. subcrenulata	E. sieberi	d—between section	2
E. subcrenulata	E. tenuiramis	d—between section	1
E. tenuiramis	E. brookeriana	d—between section	2
E. tenuiramis	E. cordata	d—between section	2
E. tenuiramis	E. globulus	d—between section	1
E. tenuiramis	E. morrisbyi	d—between section	1
E. tenuiramis	E. ovata	d—between section	2
E. tenuiramis	E. perriniana	d—between section	1
E. tenuiramis	E. rubida	d—between section	2
E. tenuiramis	E. urnigera	d—between section	1
E. urnigera	E. amygdalina	d—between section	2
E. urnigera	E. delegatensis	d—between section	1
E. urnigera	E. nitida	d—between section	1
E. urnigera	E. obliqua	d—between section	2
E. urnigera	E. pulchella	d—between section	1
E. urnigera	E. radiata	d—between section	1
E. urnigera	E. risdonii	d—between section	2
E. urnigera	E. tenuiramis	d—between section	2
E. vernicosa	E. coccifera	d—between section	1
E. vernicosa	E. delegatensis	d—between section	1
E. vernicosa	E. nitida	d—between section	1
E. vernicosa	E. obliqua	d—between section	1
E. vernicosa	E. pulchella	d—between section	1
E. vernicosa	E. radiata	d—between section	2
E. vernicosa	E. risdonii	d—between section	1
E. vernicosa	E. sieberi	d—between section	2
E. vernicosa	E. tenuiramis	d—between section	2
E. viminalis	E. amygdalina	d—between section	1
E. viminalis	E. coccifera	d—between section	1
E. viminalis	E. delegatensis	d—between section	2
E. viminalis	E. nitida	d—between section	1

E. viminalis	E. pulchella	d—between section	1
E. viminalis	E. radiata	d—between section	2
E. viminalis	E. regnans	d—between section	1
E. viminalis	E. sieberi	d—between section	3

APPENDIX B

Here, we test whether our use of residuals has been an effective method of removing the signature of phylogenetic identity such that the analysis can be focused on phylogenetic similarity, CO_2, and N. If we include focal subgenus, neighbour subgenus, focal species within focal subgenus, and neighbour species within neighbour subgenus alongside our main predictors (phylogenetic similarity, CO_2, and N in a fully factorial design) as predictors of *residual* biomass data, none of the phylogenetic terms should be significant predictors and the significance of our main predictors should be essentially unchanged from a model with only the main predictors and no phylogenetic terms. We also observe this pattern, as shown in the table below.

This table shows the F score and p value for two different models—one model (left F and p columns) includes only our main predictors; this is the model we use in the chapter. The second model (right F and p columns) also includes the phylogenetic information that we attempted to control for by using the residual approach. Bold and dark grey cells are significant at $\alpha = 0.05$, and italic and light grey cells are significant at $\alpha = 0.10$. The results of the two models are nearly identical, indicating that we have been almost entirely successful at pulling out species- and subgenus-level variation using our residual approach. All biomasses were originally measured at the mixture level (i.e. the focal and neighbour plant were summed) before residuals were calculated.

	F	p	F	p
Residual aboveground biomass				
Species [subgenera]	NA	NA	0.0922	1
Interacting species [interaction subgenus]	NA	NA	0.2055	1
Subgenera	NA	NA	0.0056	0.9405

Interaction subgenus	NA	NA	0.1602	0.6891
Interaction	**2.6906**	**0.0454**	**3.1258**	**0.0254**
CO_2 treatment	**32.2711**	**<0.0001**	**31.976**	**<0.0001**
Interaction*CO_2 treatment	0.2296	0.8758	0.171	0.916
Fertilizer	**50.3959**	**<0.0001**	**50.4439**	**<0.0001**
Interaction*fertilizer	1.7248	0.1605	2.0071	0.1117
CO_2 treatment*fertilizer	**27.7514**	**<0.0001**	**27.454**	**<0.0001**
Interaction*CO_2 treatment*fertilizer	0.7825	0.5039	0.8424	0.471
Residual belowground biomass				
Species [subgenera]	NA	NA	0.0355	1
Interacting species [interaction subgenus]	NA	NA	0.1033	1
Subgenera	NA	NA	0.0017	0.9674
Interaction subgenus	NA	NA	0.0224	0.8811
Interaction	1.0509	0.3694	1.3062	0.2714
CO_2 treatment	**23.5428**	**<0.0001**	**22.9792**	**<0.0001**
Interaction*CO_2 treatment	0.7355	0.531	0.6644	0.5741
Fertilizer	**13.9998**	**0.0002**	**14.0591**	**0.0002**
Interaction*fertilizer	0.9821	0.4006	1.1006	0.3482
CO_2 treatment*fertilizer	**8.3516**	**0.004**	**8.3057**	**0.0041**
Interaction*CO_2 treatment*fertilizer	1.2141	0.3036	1.3151	0.2684
Residual total biomass				
Species [subgenera]	NA	NA	0.0798	1
Interacting species [interaction subgenus]	NA	NA	0.1872	1
Subgenera	NA	NA	0.0047	0.9453
Interaction subgenus	NA	NA	0.1232	0.7257
Interaction	***2.3205***	***0.0741***	***2.7316***	***0.043***
CO_2 treatment	**31.9263**	**<0.0001**	**31.5381**	**<0.0001**

Interaction * CO_2 treatment	0.2922	0.831	0.2282	0.8768
Fertilizer	**42.1511**	**<0.0001**	**42.1984**	**<0.0001**
Interaction * fertilizer	1.5877	0.191	1.8386	0.1388
CO_2 treatment * fertilizer	**23.4795**	**<0.0001**	**23.2397**	**<0.0001**
Interaction * CO_2 treatment * fertilizer	0.8751	0.4536	0.9576	0.4123

APPENDIX C

Treatment means and standard errors for all global change * phylogenetic similarity combinations. Means are presented for both subgenera, and for all three biomass response variables. "Between section" is analogous to "between subgenus".

Global change treatment	Phylogenetic similarity	Subgenus	Response variable (pot level)	Mean	Std error
H-CO_2, H-N	a—mono	E	Aboveground biomass	0.4893	0.0772
H-CO_2, L-N	a—mono	E	Aboveground biomass	0.5016	0.0622
L-CO_2, H-N	a—mono	E	Aboveground biomass	0.3243	0.0505
L-CO_2, L-N	a—mono	E	Aboveground biomass	0.3773	0.0490
H-CO_2, H-N	b—within series	E	Aboveground biomass	0.6911	0.0807
H-CO_2, L-N	b—within series	E	Aboveground biomass	0.4377	0.0827
L-CO_2, H-N	b—within series	E	Aboveground biomass	0.5100	0.0688
L-CO_2, L-N	b—within series	E	Aboveground biomass	0.3547	0.0560
H-CO_2, H-N	c—among series	E	Aboveground biomass	0.5902	0.0879

H-CO$_2$, L-N	c—among series	E	Aboveground biomass	0.6319	0.0506
L-CO$_2$, H-N	c—among series	E	Aboveground biomass	0.4497	0.0808
L-CO$_2$, L-N	c—among series	E	Aboveground biomass	0.3930	0.0411
H-CO$_2$, H-N	d—between section	E	Aboveground biomass	0.5057	0.0804
H-CO$_2$, L-N	d—between section	E	Aboveground biomass	0.5792	0.0943
L-CO$_2$, H-N	d—between section	E	Aboveground biomass	0.4977	0.0789
L-CO$_2$, L-N	d—between section	E	Aboveground biomass	0.5409	0.0493
H-CO$_2$, H-N	a—mono	S	Aboveground biomass	1.0596	0.1553
H-CO$_2$, L-N	a—mono	S	Aboveground biomass	0.4854	0.0605
L-CO$_2$, H-N	a—mono	S	Aboveground biomass	0.6177	0.0950
L-CO$_2$, L-N	a—mono	S	Aboveground biomass	0.5630	0.0622
H-CO$_2$, H-N	b—within series	S	Aboveground biomass	0.8928	0.0829
H-CO$_2$, L-N	b—within series	S	Aboveground biomass	0.4985	0.0609
L-CO$_2$, H-N	b—within series	S	Aboveground biomass	0.7008	0.0854
L-CO$_2$, L-N	b—within series	S	Aboveground biomass	0.5856	0.0625
H-CO$_2$, H-N	c—among series	S	Aboveground biomass	1.1065	0.0789
H-CO$_2$, L-N	c—among series	S	Aboveground biomass	0.5520	0.0542
L-CO$_2$, H-N	c—among series	S	Aboveground biomass	0.6989	0.0625

L-CO$_2$, L-N	c—among series	S	Aboveground biomass	0.7248	0.0560
H-CO$_2$, H-N	d—between section	S	Aboveground biomass	0.8306	0.0857
H-CO$_2$, L-N	d—between section	S	Aboveground biomass	0.5079	0.0607
L-CO$_2$, H-N	d—between section	S	Belowground biomass	0.4586	0.0480
L-CO$_2$, L-N	d—between section	S	Belowground biomass	0.4627	0.0397
H-CO$_2$, H-N	a—mono	E	Belowground biomass	0.0885	0.0203
H-CO$_2$, L-N	a—mono	E	Belowground biomass	0.0926	0.0182
L-CO$_2$, H-N	a—mono	E	Belowground biomass	0.0453	0.0086
L-CO$_2$, L-N	a—mono	E	Belowground biomass	0.0629	0.0108
H-CO$_2$, H-N	b—within series	E	Belowground biomass	0.1271	0.0236
H-CO$_2$, L-N	b—within series	E	Belowground biomass	0.0881	0.0195
L-CO$_2$, H-N	b—within series	E	Belowground biomass	0.0755	0.0165
L-CO$_2$, L-N	b—within series	E	Belowground biomass	0.0624	0.0118
H-CO$_2$, H-N	c—among series	E	Belowground biomass	0.0955	0.0144
H-CO$_2$, L-N	c—among series	E	Belowground biomass	0.1203	0.0198
L-CO$_2$, H-N	c—among series	E	Belowground biomass	0.0819	0.0214
L-CO$_2$, L-N	c—among series	E	Belowground biomass	0.0587	0.0102
H-CO$_2$, H-N	d—between section	E	Belowground biomass	0.0717	0.0241

H-CO$_2$, L-N	d—between section	E	Belowground biomass	0.1346	0.0287
L-CO$_2$, H-N	d—between section	E	Belowground biomass	0.0904	0.0208
L-CO$_2$, L-N	d—between section	E	Belowground biomass	0.0942	0.0138
H-CO$_2$, H-N	a—mono	S	Belowground biomass	0.2432	0.0438
H-CO$_2$, L-N	a—mono	S	Belowground biomass	0.1129	0.0173
L-CO$_2$, H-N	a—mono	S	Belowground biomass	0.1346	0.0254
L-CO$_2$, L-N	a—mono	S	Belowground biomass	0.1207	0.0190
H-CO$_2$, H-N	b—within series	S	Belowground biomass	0.1672	0.0208
H-CO$_2$, L-N	b—within series	S	Belowground biomass	0.0964	0.0142
L-CO$_2$, H-N	b—within series	S	Belowground biomass	0.1603	0.0276
L-CO$_2$, L-N	b—within series	S	Belowground biomass	0.1171	0.0191
H-CO$_2$, H-N	c—among series	S	Belowground biomass	0.2488	0.0257
H-CO$_2$, L-N	c—among series	S	Belowground biomass	0.1389	0.0220
L-CO$_2$, H-N	c—among series	S	Belowground biomass	0.1446	0.0171
L-CO$_2$, L-N	c—among series	S	Belowground biomass	0.1696	0.0196
H-CO$_2$, H-N	d—between section	S	Belowground biomass	0.1856	0.0316
H-CO$_2$, L-N	d—between section	S	Belowground biomass	0.1190	0.0228
L-CO$_2$, H-N	d—between section	S	Belowground biomass	0.0816	0.0120

L-CO$_2$, L-N	d—between section	S	Belowground biomass	0.0915	0.0138
H-CO$_2$, H-N	a—mono	E	Total biomass	0.5778	0.0955
H-CO$_2$, L-N	a—mono	E	Total biomass	0.5943	0.0790
L-CO$_2$, H-N	a—mono	E	Total biomass	0.3695	0.0581
L-CO$_2$, L-N	a—mono	E	Total biomass	0.4402	0.0579
H-CO$_2$, H-N	b—within series	E	Total biomass	0.8182	0.1033
H-CO$_2$, L-N	b—within series	E	Total biomass	0.5258	0.0961
L-CO$_2$, H-N	b—within series	E	Total biomass	0.5855	0.0844
L-CO$_2$, L-N	b—within series	E	Total biomass	0.4171	0.0666
H-CO$_2$, H-N	c—among series	E	Total biomass	0.6857	0.1013
H-CO$_2$, L-N	c—among series	E	Total biomass	0.7522	0.0682
L-CO$_2$, H-N	c—among series	E	Total biomass	0.5316	0.1005
L-CO$_2$, L-N	c—among series	E	Total biomass	0.4516	0.0506
H-CO$_2$, H-N	d—between section	E	Total biomass	0.5773	0.0998
H-CO$_2$, L-N	d—between section	E	Total biomass	0.7138	0.1213
L-CO$_2$, H-N	d—between section	E	Total biomass	0.5881	0.0991
L-CO$_2$, L-N	d—between section	E	Total biomass	0.6351	0.0615
H-CO$_2$, H-N	a—mono	S	Total biomass	1.3028	0.1963
H-CO$_2$, L-N	a—mono	S	Total biomass	0.5982	0.0750
L-CO$_2$, H-N	a—mono	S	Total biomass	0.7522	0.1196
L-CO$_2$, L-N	a—mono	S	Total biomass	0.6836	0.0782

H-CO$_2$, H-N	b—within series	S	Total biomass	1.0600	0.1017
H-CO$_2$, L-N	b—within series	S	Total biomass	0.5950	0.0744
L-CO$_2$, H-N	b—within series	S	Total biomass	0.8611	0.1113
L-CO$_2$, L-N	b—within series	S	Total biomass	0.7028	0.0802
H-CO$_2$, H-N	c—among series	S	Total biomass	1.3553	0.1025
H-CO$_2$, L-N	c—among series	S	Total biomass	0.6909	0.0744
L-CO$_2$, H-N	c—among series	S	Total biomass	0.8435	0.0781
L-CO$_2$, L-N	c—among series	S	Total biomass	0.8944	0.0742
H-CO$_2$, H-N	d—between section	S	Total biomass	1.0162	0.1147
H-CO$_2$, L-N	d—between section	S	Total biomass	0.6270	0.0808
L-CO$_2$, H-N	d—between section	S	Total biomass	0.5402	0.0588
L-CO$_2$, L-N	d—between section	S	Total biomass	0.5542	0.0519

REFERENCES

Ainsworth, E.A., Long, S.P., 2005. What have we learned from 15 years of free-air CO$_2$ enrichment (FACE)? A meta-analytic review of the responses of photosynthesis, canopy properties and plant production to rising CO$_2$. New Phytol. 165, 351–372.

Bailey, J.K., Genung, M.A., Ware, I., Gorman, C., Van Nuland, M.E., Long, H., Schweitzer, J.A., 2014. Indirect genetic effects: an evolutionary mechanism linking feedbacks, genotypic diversity and coadaptation in a climate change context. Func. Ecol. 28, 87–95.

Brown, J.H., Stevens, G.C., Kaufman, D.M., 1996. The geographic range: Size, shape, boundaries, and internal structure. Ann. Rev. Ecol. Systemat. 27, 597–623.

Brooker, M.J.H., 2000. A new classification of the genus Eucalyptus L'Her. (Myrtaceae). Aust. Syst. Biol. 13, 79–148.

Cadotte, M.W., Cavender-Bares, J., Tilman, D., Oakley, T.H., 2009. Using phylogenetic, functional and trait diversity to understand patterns of plant community productivity. PLoS One 4, e5695.

Cao, B., Dang, Q.L., Yü, X., Zhang, S., 2008. Effects of CO_2 and nitrogen on morphological and biomass traits of white birch Betula papyrifera seedlings. For. Ecol. Manage. 254, 217–224.

Carroll, S.P., Hendry, A.P., Reznick, D.N., Fox, C.W., 2007. Evolution on ecological time-scales. Funct. Ecol. 21, 387–393.

Cooper, N., Freckleton, R.P., Jetz, W., 2011. Phylogenetic conservatism of environmental niches in mammals. Proc. Biol. Sci. 278, 2384–2391.

Darwin, C., 1859. On the Origin of Species. Murray, London, UK.

Davis, M.B., Shaw, R.G., Etterson, J.R., 2005. Evolutionary responses to changing climate. Ecology 86, 1704–1714.

Davis, C.C., Willis, C.G., Primack, R.B., Miller-Rushing, A.J., 2010. The importance of phylogeny to the study of phenological response to global climate change. Philos. Trans. R. Soc. Lond. B Biol. Sci. 365, 3201–3213.

Díaz, S., Cabido, M., 2001. Vive la difference: plant functional diversity matters to ecosystem processes. Trends Ecol. Evol. 16, 646–655.

Edwards, E.J., Still, C.J., Donoghue, M.J., 2007. The relevance of phylogeny to studies of global change. Trends Ecol. Evol. 22, 243–249.

Flynn, D.F., Mirotchnick, N., Jain, M., Palmer, M.I., Naeem, S., 2011. Functional and phylogenetic diversity as predictors of biodiversity-ecosystem-function relationships. Ecology 92, 1573–1581.

Galloway, J.N., Dentener, F.J., Capone, D.G., et al., 2004. Nitrogen cycles: past, present, and future. Biogeochemistry 70, 153–226.

Gaston, K.J., 2003. The Structure and Dynamics of Geographic Ranges. Oxford University Press, Oxford, UK.

Gause, G.F., 1934. The Struggle for Existence. Williams and Wilkins, Baltimore.

Griffin, K.L., Winner, W.E., Strain, B.R., 1995. Growth and dry matter partitioning in loblolly and ponderosa pine seedlings in response to carbon and nitrogen availability. New Phytol. 129, 547–556.

Hardin, G., 1960. Competitive exclusion principle. Science 131, 1292–1297.

Hendry, A.P., Kinnison, M.T., 1999. The pace of modern life: measuring rates of contemporary microevolution. Evolution 53, 1637–1653.

Hobbie, S.E., 1992. Effects of plant species on nutrient cycling. Trends Ecol. Evol. 7, 336–339.

Hooper, D.U., Adair, E.C., Cardinale, B.J., et al., 2012. A global synthesis reveals biodiversity loss as a major driver of ecosystem change. Nature 486, 59–67.

Hyvönen, R., Ågren, G.I., Linder, S., et al., 2007. The likely impact of elevated [CO2], nitrogen deposition, increased temperature and management on carbon sequestration in temperate and boreal forest ecosystems: a literature review. New Phytol. 173, 463–480.

IPCC, 2007. Climate change 2007: the scientific basis. In: Solomon, S., Qin, D., Manning, M., Chen, Z., Marquis, M., Averyt, K.B., Tignor, M., Miller, H.L. (Eds.), Contributions of Working Group I to the Fourth Assessment Report of the Intergovernmental Panel on Climate Change: "The Physical Science Basis". Cambridge University Press, Cambridge, UK, pp. 1–18.

Jablonski, L.M., Wang, X., Curtis, P.S., 2002. Plant reproduction under elevated CO_2 conditions: a meta-analysis of reports on 79 crop and wild species. New Phytol. 156, 9–26.

Jaccoud, D., Peng, K., Feinstein, D., Killian, A., 2001. Diversity arrays: a solid-state technology for sequence information independent genotyping. Nucleic Acids Res. 29, e25.

Jach, M.E., Laureysens, I., Ceulemans, R., 2000. Above- and below-ground production of young Scots pine (Pinus sylvestris L.) trees after three years of growth in the field under elevated CO_2. Ann. Bot. 85, 789–798.

Kenward, M.G., Roger, J.H., 1997. Small sample inference for fixed effects from restricted maximum likelihood. Biometrics 53, 983–997.

Kunstler, G., Lavergne, S., Courbaud, B., et al., 2012. Competitive interactions between forest trees are driven by species' trait hierarchy, not phylogenetic or functional similarity: implications for forest community assembly. Ecol. Lett. 15, 831–840.

Langley, J.A., Megonigal, J.P., 2010. Ecosystem response to elevated CO_2 levels limited by nitrogen-induced plant species shift. Nature 466, 96–99.

Lindroth, R.L., 2010. Impacts of elevated atmospheric CO_2 and O_3 on forests: phytochemistry, trophic interactions, and ecosystem dynamics. J. Chem. Ecol. 36, 2–21.

Luck, G.W., Daily, G.C., Ehrlich, P.R., 2003. Population diversity and ecosystem services. Trends Ecol. Evol. 18, 331–336.

McKinney, M.L., Lockwood, J.L., 1999. Biotic homogenization: a few winners replacing many losers in the next mass extinction. Trends Ecol. Evol. 14, 450–453.

McKinnon, G.E., Vaillancourt, R.E., Steane, D.A., Potts, B.M., 2008. An AFLP marker approach to lower-level systematics in Eucalyptus (Myrtaceae). Am. J. Bot. 95, 368–380.

McLauchlan, K.K., Williams, J.J., Craine, J.M., Jeffers, E.S., 2013. Changes in global nitrogen cycling during the Holocene epoch. Nature 495, 352–355.

McMurtrie, R.E., Norby, R.J., Medlyn, B.E., Dewar, R.C., Pepper, D.A., Reich, P.B., Barton, C.V., 2008. Why is plant-growth response to elevated CO_2 amplified when water is limiting, but reduced when nitrogen is limiting? A growth-optimisation hypothesis. Funct. Plant Biol. 35, 521–534.

Millennium Ecosystem Assessment, 2005. Ecosystems and Human Well-Being: Synthesis. Island Press, Washington, DC.

Morueta-Holme, N., Enquist, B.J., McGill, B.J., Boyle, B., Jørgenson, P.M., Ott, J.E., Peet, R.K., Šímová, I., Sloat, L.L., Thiers, B., Violle, C., Wiser, S.K., Dolins, S., Donoghue II, J.C., Kraft, N.J.B., Regetz, J., Schildhauer, M., Spencer, N., Svenning, J.-C., 2013. Habitat area and climate stability determine geographical variation in plant species range sizes. Ecol. Lett. 16, 1446–1454.

Newingham, B.A., Vanier, C.H., Charlet, T.N., Ogle, K., Smith, S.D., Nowak, R.S., 2013. No cumulative effect of 10 years of elevated [CO_2] on perennial plant biomass components in the Mojave Desert. Glob. Chang. Biol. 19, 2168–2181.

Nie, M., Lu, M., Bell, J., Raut, S., Pendall, E., 2013. Altered root traits due to elevated CO2: a meta-analysis. Glob. Ecol. Biogeogr 22, 1095–1105. http://dx.doi.org/10.1111/geb.12062.

Norby, R.J., Warren, J.M., Iversen, C.M., Medlyn, B.E., McMurtrie, R.E., 2010. CO_2 enhancement of forest productivity constrained by limited nitrogen availability. Proc. Natl. Acad. Sci. U.S.A. 107, 19368–19373.

Ohlemüller, R., Anderson, B.J., Araújo, M.B., Butchart, S.H.M., Kudrna, O., Ridgely, R.S., Thomas, C.D., 2008. The coincidence of climate and species rarity: high risk to small-range species from climate change. Biol. Lett. 4, 568–572.

Pelletier, F., Garant, D., Hendry, A.P., 2009. Eco-evolutionary dynamics. Philos. Trans. R. Soc. Lond. B Biol. Sci. 364, 1483–1489.

Pike, N., 2010. Using false discovery rates for multiple comparisons in ecology and evolution. Methods Ecol. Evol. 2, 278–282.

Reich, P.B., Hobbie, S.E., 2012. Decade-long soil nitrogen constraint on the CO_2 fertilization of plant biomass. Nat. Clim. Chang. 3, 278–282.

Reich, P.B., Hobbie, S.E., Lee, T., et al., 2006. Nitrogen limitation constrains sustainability of ecosystem response to CO_2. Nature 440, 922–925.

Reid, J.P., Adair, E.C., Hobbie, S.E., Reich, P.B., 2012. Biodiversity, nitrogen deposition, and CO_2 affect grassland soil carbon cycling but not storage. Ecosystems 15, 580–590.

Senior, J.K., Schweitzer, J.A., O'Reilly-Wapstra, J., Chapman, S.K., Steane, D., Langley, A., Bailey, J.K., 2013. Phylogenetic responses of forest trees to global change. PLoS One 8, e60088.

Smith, A.R., Lukac, M., Bambrick, M., Miglietta, F., Godbold, D.L., 2013. Tree species diversity interacts with elevated CO_2 to induce a greater root system response. Glob. Chang. Biol. 19, 217–228.

Srivastava, D.S., Cadotte, M.W., MacDonald, A.A.M., Marushia, R.G., Mirotchnick, N., 2012. Phylogenetic diversity and the functioning of ecosystems. Ecol. Lett. 15, 637–648.

Steane, D.A., Nicolle, D., Sansoloni, C.P., et al., 2011. Population genetic analysis and phylogeny reconstruction in *Eucalyptus* (Myrtaceae) using high-throughput, genome-wide sequencing. Mol. Phylogenet. Evol. 59, 206–224.

Stone, C., Simpson, J.A., Gittins, R., 1998. Differential impact of insect herbivores and fungal pathogens on the Eucalyptus subgenera Symphyomyrtus and Monocalyptus and genus Corymbia. Aust. J. Bot. 46, 723–724.

Tilman, D., Reich, P.B., Isbell, F., 2012. Biodiversity influences ecosystems as much as resources, disturbance, or herbivory. Proc. Natl. Acad. Sci. U.S.A. 109, 10394–10397.

Verdú, M., Gómez-Aparicio, L., Valiente-Banuet, A., 2012. Phylogenetic relatedness as a tool in restoration ecology: a meta-analysis. Proc. R. Soc. Lond. B Biol. Sci. 279, 1761–1767.

Verhoeven, K.J.F., Simonsen, K.L., McIntyre, L.M., 2005. Implementing false discovery rate control: increasing your power. Oikos 108, 643–647.

Violle, C., Nemergut, D.R., Pu, Z., Jiang, L., 2011. Phylogenetic limiting similarity and competitive exclusion. Ecol. Lett. 14, 782–787.

Vitousek, P.M., Mooney, H.A., Lubchenco, J., Melillo, J.M., 1997. Human domination of Earth's ecosystems. Science 277, 494–499.

Wallis, I.R., Nicolle, D., Foley, W.J., 2010. Available and not total nitrogen in leaves explains key chemical differences between eucalypt subgenera. For. Ecol. Manage. 260, 814–821.

Webb, C.O., Ackerly, D.D., McPeek, M.A., Donoghue, M.J., 2002. Phylogenies and community ecology. Annu. Rev. Ecol. Syst. 33, 475–505.

Wiens, J.J., Graham, C.H., 2005. Niche conservatism: integrating ecology, evolution, and conservation biology. Annu. Rev. Ecol. Syst. 36, 519–539.

Williams, K., Potts, B., 1996. The natural distribution of *Eucalyptus* species in Tasmania. Tasforests 8, 39–165.

Willis, K.J., Bailey, R.M., Bhagwat, S.A., Birks, H.J.B., 2010. Biodiversity baselines, thresholds and resilience: testing predictions and assumptions using paleoecological data. Trends Ecol. Evol. 25, 583–591.

Xia, J., Wan, S., 2008. Global response patterns of terrestrial plant species to nitrogen addition. New Phytol. 179, 428–439.

Zak, D.R., Pregitzer, K.S., Curtis, P.S., Vogel, C.S., Holmes, W.E., Lussenhop, J., 2000. Atmospheric CO_2, soil-N availability, and allocation of biomass and nitrogen by Populus tremuloides. Ecol. Appl. 10, 34–46.

INDEX

Note: Page numbers followed by "*f*" indicate figures and "*t*" indicate tables.

ADVANCES IN ECOLOGICAL RESEARCH VOLUME 1–50

CUMULATIVE LIST OF TITLES

Aerial heavy metal pollution and terrestrial ecosystems, **11**, 218

Age determination and growth of Baikal seals (*Phoca sibirica*), **31**, 449

Age-related decline in forest productivity: pattern and process, **27**, 213

Allometry of body size and abundance in 166 food webs, **41**, 1

Analysis and interpretation of long-term studies investigating responses to climate change, **35**, 111

Analysis of processes involved in the natural control of insects, **2**, 1

Ancient Lake Pennon and its endemic molluscan faun (Central Europe; Mio-Pliocene), **31**, 463

Ant-plant-homopteran interactions, **16**, 53

Anthropogenic impacts on litter decomposition and soil organic matter, **38**, 263

Arctic climate and climate change with a focus on Greenland, **40**, 13

Arrival and departure dates, **35**, 1

Assessing the contribution of micro-organisms and macrofauna to biodiversity-ecosystem functioning relationships in freshwater microcosms, **43**, 151

A belowground perspective on Dutch agroecosystems: how soil organisms interact to support ecosystem services, **44**, 277

The benthic invertebrates of Lake Khubsugul, Mongolia, **31**, 97

Biodiversity, species interactions and ecological networks in a fragmented world **46**, 89

Biogeography and species diversity of diatoms in the northern basin of Lake Tanganyika, **31**, 115

Biological strategies of nutrient cycling in soil systems, **13**, 1

Biomanipulation as a restoration tool to combat eutrophication: recent advances and future challenges, **47**, 411

Biomonitoring of human impacts in freshwater ecosystems: the good, the bad and the ugly, **44**, 1

Bray-Curtis ordination: an effective strategy for analysis of multivariate ecological data, **14**, 1

Body size, life history and the structure of host-parasitoid networks, **45**, 135

357

CPI Antony Rowe
Eastbourne, UK
August 19, 2014